INTERNATIONAL SERIES OF MONOGRAPHS IN
ANALYTICAL CHEMISTRY

GENERAL EDITORS: R. BELCHER AND H. FREISER

VOLUME 38

ION EXCHANGE
IN ANALYTICAL CHEMISTRY

ION EXCHANGE
IN
ANALYTICAL CHEMISTRY

by

WILLIAM RIEMAN III, Ph.D.

Professor Emeritus of Chemistry, Rutgers, The State University
Professor Visitante, Universidad Nacional de Trujillo, Perú

and

HAROLD F. WALTON

Professor of Chemistry, University of Colorado

PERGAMON PRESS

Oxford · New York · Toronto
Sydney · Braunschweig

Pergamon Press Ltd., Headington Hill Hall, Oxford

Pergamon Press (Scotland) Ltd., 2 & 3 Teviot Place, Edinburgh 1

Pergamon Press Inc., Maxwell House, Fairview Park, Elmsford, New York 10523

Pergamon of Canada Ltd., 207 Queen's Quay West, Toronto 1

Pergamon Press (Aust.) Pty. Ltd., 19a Boundary Street,
Rushcutters Bay, N.S.W. 2011, Australia

Vieweg & Sohn GmbH, Burgplatz 1, Braunschweig

First edition 1970

Library of Congress Catalog Card No. 74–105870

*Printed in Great Britain by **A.** Wheaton & Co., Exeter*

08 015511 1

CONTENTS

Chapter 6   Theory of Ion-exchange Chromatography   88

Chapter 9   Salting-out Chromatography and Related Methods

Chapter 10   Less Common Ion Exchangers

PREFACE

THE purpose of this book is to provide analytical chemists with a broad survey of the role that ion exchange can and should play in chemical analysis. In order that the readers of the book may properly understand the recommended methods, considerable space is devoted to the preparation, structure, and properties of ion-exchange materials. These include not only the well-known resins but also the inorganic, cellulosic, and liquid ion exchangers.

The plate-equilibrium theory of chromatography is emphasized. Although this is less rigorous theoretically than the mass-transfer theory, it is much more helpful to the analyst who wishes to calculate from the data of a few elutions the concentration and pH of eluent that will give the most efficient separation of a given mixture.

WILLIAM RIEMAN III
HAROLD F. WALTON

CHAPTER 1

INTRODUCTION

A. History of Ion Exchange

The earliest recorded application of ion exchange may have been the "miracle" by which Moses rendered the saline water of the spring of Marah potable,[1] about 3000 years B.C. If the tree that Yahweh pointed out to Moses was dead and if part of its cellulose had been oxidized so as to form carboxyl groups, it is possible that the "bitter" Epsom salt was removed by an ion-exchange reaction

$$Mg^{2+} + SO_4^{2-} + 2RCOOH \rightarrow (RCOO)_2Mg + H^+ + HSO_4^-$$

followed by a neutralization of the resulting sulphuric acid by deposits of limestone,

$$H^+ + HSO_4^- + CaCO_3 \rightarrow CaSO_4 + H_2O + CO_2$$

It is interesting to note that in the Spanish Bible the object that Moses threw into the spring is called *un madero* (a log), not *un árbol* (a tree).

Even if this explanation of the "miracle" be true, Moses was too busy with his manifold duties as priest, guide, legislator, judge, and commander in chief of the army to conduct research on the phenomenon.

The first report (1850) of research on ion exchange was the work of an English agricultural chemist, H. S. Thompson.[2] He found that on passage of a solution of ammonium sulphate through soil some or all of the ammonium ion was removed and replaced by calcium ion. He called this phenomenon base exchange. It must be remembered that Arrhenius' concept of ions was not published until 1887.

In the same year (1850) another Englishman, T. J. Way,[3] published a more extensive report entitled "On the power of soils to absorb manure". In one of his 96 experiments he passed a solution of ammonium sulphate through a column of soil 18 in. in length. He wrote: "Three ounces of the liquid percolated without a trace of ammoniacal salt, but sulphuric acid was present in the very first portions, combined with lime. So strongly was the liquid impregnated with sulphate of lime that the salt separated from it in flakes after standing a few minutes." After describing similar experiments with other salts he wrote: "Thus, then, it has been proved that for bases of the salts of ammonia, potash, magnesia, and soda, the absorptive power of the soil is available."

Way's research also provided a clue regarding the constituent of the soil that was responsible for the ion exchange. He stated that soil consists chiefly of sand, clay, and vegetable matter. Experiments with pure sand convinced him that it possessed no ion-exchange property. Then he experimented with soil previously "burnt in a covered Hessian crucible in the laboratory furnace". He found that the ion-exchange capacity of this soil was greatly diminished but could not decide whether the exchange capacity of the clay had

1

been "greatly diminished by burning it" or whether the remaining exchange capacity was due to charcoal. He next destroyed the organic matter in soil by several evaporations with concentrated nitric acid and tested this soil for its exchange ability. Although less than that of the untreated soil, it was considerably greater than that of the "burnt" soil. He concluded, therefore, that some compound in the clay was responsible for the exchange.

Two years later, Way[4] reported his researches on the nature of the ion-exchange compound of soil. He tested the exchange ability of finely ground albite, of finely ground feldspar and of a calcium silicate that he prepared by mixing solutions of sodium silicate and a calcium salt. All of these results were negative. Then he prepared a gelatinous sodium aluminum silicate by mixing solutions of sodium aluminate and sodium silicate. After partial drying, the composition of the material corresponded approximately to $Na_2O \cdot Al_2O_3 \cdot 3SiO_2 \cdot 2H_2O$. From this he prepared the analogous compound of potassium and calcium by ion exchange. Each of these double silicates absorbed ammonium ion from solution. He concluded that the ion-exchange compound of soils is a double silicate of aluminum with sodium, potassium, or calcium.

Way believed, therefore, that the "bases" will replace each other in the double silicate in the order Na, K, Ca, Mg, NH_4 (the last being most tightly bound) regardless of the relative concentrations. He can hardly be criticized for this because Guldberg and Waage[5] did not publish their classic paper on reversibility until 1867. This posed another problem. How could the ammonium ion, after it was once bound by the exchanger, be available to supply nitrogen to plants? To answer this question, Way postulated that the solubility of the double silicate was sufficient to meet the needs of plants, especially since its solubility is increased by the presence of soluble salts in the water of soil. (Some present-day chemists incur appreciable errors by neglecting the effect of ionic strength on the solubility products of precipitates and on the ionization constants of weak acids.)

Way's work was bitterly attacked by the great Justus von Liebig.[6] The latter believed that the solubility of the ammonium zeolite was insufficient to supply the ammonium ion needed by growing crops. Since the principle of the reversibility of chemical reactions[5] had not been stated yet, he came to the logical conclusion that Way's work must be fallacious. After the paper of Guldberg and Waage, another German investigator, Wiegner,[7] wrote in support of Way.

The investigations of geochemists also gave evidence of ion exchange. Lemberg,[8] for example, analyzed two samples of oligoclase—one slightly weathered the other extensively weathered. Since he took the samples from the same geological formation and from very nearly the same point, he believed that they had had the same composition before weathering. Table 1.1 shows the molar and equivalent ratios calculated from his analysis on the basis of 22.00 mmol of silicon dioxide. It is seen that the chemical effect of the weathering is the replacement of sodium by calcium, magnesium, and potassium and to a lesser extent the replacement of aluminum by iron.

In a later paper[9] he reported experimental evidence that ion exchange can occur in minerals. He analyzed a sample of analcite and a sample of the same material that he had kept in contact with aqueous potassium chloride for 3 months with frequent changes of the solution. Table 1.2 indicates the number of moles of the various oxides associated with 2.000 moles of silica in both the original and the altered mineral. Within the limits of small experimental errors, a complete and stoichiometric exchange of potassium for sodium occurred, accompanied by the loss of almost a mole of water. He demonstrated that this

TABLE 1.1. EFFECT OF WEATHERING ON THE COMPOSITION OF OLIGOCLASE

Oxide	Slightly weathered sample			Badly weathered sample		
	%	mmol	meq	%	mmol	meq
H_2O	1.50	1.75	3.50	1.71	2.00	4.00
Na_2O	4.71	1.595	3.19	1.69	0.572	1.14
K_2O	6.99	1.560	3.12	8.48	1.891	3.78
CaO	0.77	0.288	0.58	2.23	0.835	1.67
MgO	2.14	1.115	2.23	2.34	1.220	2.44
Al_2O_3	18.04	3.719	22.31	16.78	3.464	20.78
Fe_2O_3	2.88	0.380	2.28	3.90	0.513	3.08
SiO_2	62.97	22.000		62.87	22.000	

(Data of J. Lemberg, used by permission from *Zeitschrift der deutschen geologischen Gesellschaft*.)

and similar exchanges were reversible and that they occur when the mineral is brought in contact with an aqueous salt solution or a fused salt.

In the early history of ion exchange, the name of R. Gans stands out as an exponent of the application of zeolites, both natural and synthetic, to industrial problems.

TABLE 1.2. EFFECT OF AQUEOUS POTASSIUM CHLORIDE ON ANALCITE

Oxide	Moles of oxide per 2.000 moles of SiO_2	
	Original	Altered
H_2O	1.042	0.233
Al_2O_3	0.460	0.474
CaO	0.019	0.017
K_2O	0.000	0.448
Na_2O	0.450	0.000
SiO_2	2.000	2.000

(Data of J. Lemberg, used by permission from *Zeitschrift der deutschen geologischen Gesellschaft*.)

In the first decade of this century, many countries granted Gans numerous patents on the manufacture of synthetic zeolites, the softening of water, and the replacement of potassium in molasses by calcium to aid in the crystallization of the sugar, and similar subjects. These two fields are the only ones in which the silicate exchangers have been extensively used.

An important advance was made in 1935 when Adams and Holmes[10] published the first paper on the synthesis of ion-exchange resins. They polymerized various polyhydroxyl-benzenes with formaldehyde. By virtue of the phenolic hydroxyl groups, the products were cation exchangers of the very weakly acidic type. Hence their capacity to react with cations in neutral solution was very limited. These investigators also polymerized m-phenylene-diamine with formaldehyde. The resin contained aromatic amino groups and was therefore an anion exchanger of the very weakly basic type.

Although the resins of Adams and Holmes are of no practical value today, their paper is very important because it pointed the way to the synthesis of other and much better resinous ion exchangers. In less than a decade, patents were granted for the synthesis of phenolformaldehyde resins with sulphonic acid groups attached to the rings through methylene groups[11] or directly, i.e. bifunctional resins with one strongly acidic group able to exchange cations at low pH. D'Alelio's[12] synthesis of the sulphonated copolymer of styrene and divinylbenzene yielded the first monofunctional strong-acid resin. The synthesis of anion exchangers of both the weak-base and strong-base types also followed shortly after the pioneering work of Adams and Holmes.

The advantages of ion-exchange resins over the silicate type of exchangers are the following. (1) The resinous exchangers have greater physical strength so that they suffer much less physical degradation under agitation. (2) They can be prepared as spheres of fairly uniform size presenting less resistance to the flow of solutions through them. (3) Their exchange capacities, measured either in milliequivalents per gram or in milliequivalents per millilitre of column (bed) volume, are much greater. (4) Their rate of exchange is much faster. (5) Unlike the silicate exchangers, they are chemically stable at very low pH values and hence can be used to exchange hydrogen ions. (6) By the art of the synthetic organic chemist, they can be prepared with a wide range of properties such as acidic or basic strength or degree of crosslinking in order to meet various needs.

These advantages led to a great increase in research on the theoretical and practical aspects of ion exchange, so that 1935 may be said to mark the renaissance of ion exchange.

B. Synthesis of Ion-exchange Resins

An ion-exchange resin is a crosslinked polymer containing ionized or ionizable groups such as $-SO_3H$, $-SO_3Na$, $-COOH$, $-NH_2$, $-NH_3Cl$, or $-NMe_3Cl$. Resins with a very large degree of crosslinking such as 24% have a dense impenetrable structure. Ions from the external solution can not diffuse through such resins, and exchange can occur only on the surface of resin particles. Hence the specific exchange capacity of these resins is very small. Ion-exchange resins with smaller degrees of crosslinking imbibe water and swell when placed in contact with water or an aqueous solution. Then exchanging ions can diffuse through the internal water. Thus all ionogenic groups of the resin can undergo exchange, and the specific exchange capacity may be large. Ion-exchange resins with very small degrees of crosslinking imbibe so much water that they behave like particles of soft jelly. The physical properties of such resins render them ill suited for either laboratory or industrial ion-exchange reactions. The swelling of resins is discussed more fully on p. 24.

B.I. SYNTHESIS OF CATION-EXCHANGE RESINS
B.I.a. *Condensation Polymers*

In a condensation polymerization, a small molecule, usually water, is eliminated for every pair of monomers that react. An example is the reaction between a *p*-substituted phenol and formaldehyde to yield a linear polymer.

where X represents an alkyl group. No crosslinking can occur because each molecule of the *p*-substituted phenol can react with formaldehyde only at the two positions ortho to the hydroxyl group. Unsubstituted phenol, on the other hand, can react with formaldehyde in three places, the two ortho and one para. Thus the reaction between phenol and formaldehyde in a mole ratio of 2:3 yields a very highly crosslinked resin.

When a mixture of *p*-substituted phenol and an unsubstituted phenol is treated with the proper amount of formaldehyde, the molecules of the substituted phenol form polymeric chains, and those of the unsubstituted phenol form crosslinks between these chains. Thus the degree of crosslinking can be controlled by the ratio of *p*-substituted to unsubstituted phenol.

This is a cation-exchange resin. The hydrogen atoms of the hydroxyl groups may be replaced by metals. However, the ionization constant of phenol is so small ($pK \simeq 10$) that the exchange can occur only at high pH. In other words, this is a cation-exchange resin of extremely weak acid properties.

If X in the foregoing resin is the sulphonic acid group, the resin has the properties of both a strong and a weak acid. The sulphonic hydrogen can be replaced by metals at any pH, but the phenolic hydrogen undergoes ion exchange only in alkaline solution. A resin of this type is said to be bifunctional.

Another bifunctional resin would result if the substituent in the foregoing formula were the carboxyl group. This would be a combination of a moderately weak and a very weak acid.

In the foregoing examples, the ionogenic group was present in at least one of the monomers. It is possible also to introduce an ionogenic group during the polymerization. For example, the reaction of sodium phenolate, sodium sulphite, and formaldehyde produces a resin with —ONa and —CH_2—SO_3Na groups. Still another possibility is to introduce the ionogenic group into the preformed resin. Thus a resin with both phenolic and aryl sulphonic acid groups can be made by sulphonating a phenolic resin.

Condensation polymers played an important role in the early development of ion exchange, and the syntheses of the resins outlined above represent only a small fraction of such syntheses described in the literature. Nevertheless, almost all of the analytically important ion-exchange resins of the present day are prepared by addition polymerization. In comparison with the latter method, the condensation procedure has four major disadvantages. (1) It yields products less resistant toward oxidation. (2) It normally yields products in bulk, which are then ground to an appropriate size; in contrast, the products of the addition method usually consist of little spheres of nearly uniform size. (3) Most of the condensation resins are bifunctional, whereas the addition process yields monofunctional exchangers. (4) It is generally more difficult to control the degree of crosslinking and the specific capacity of the product in the condensation method than in the addition method. However, resins of the condensation type are more resistant to γ-radiation.

B.I.b. *Addition Polymers*

B.I.b.1. *Preparation of pearls of crosslinked polystyrene.* In addition polymerization, the monomers combine by a free-radical mechanism without the elimination of any small molecules. For example, styrene polymerizes under the influence of a catalytic agent such as benzoyl peroxide and mild heat. The product is linear polystyrene

$$nC_6H_5\!-\!CH\!=\!CH_2 \longrightarrow \left(\begin{array}{c} -CH-CH_2- \\ | \\ C_6H_5 \end{array}\right)_n$$

If a mixture of styrene and divinylbenzene is warmed with a catalyst, the molecules of styrene form linear polymeric chains while the molecules of divinylbenzene (DVB) become crosslinks between the chains.[12]

Thus the degree of crosslinking is controlled by the ratio of DVB to styrene in the original mixture. Pure DVB is too expensive to use for this purpose, and a mixture of about equal weights of DVB and ethylstyrene is used industrially. Both the DVB and the ethylstyrene, in turn, consist of a mixture of their three isomers. The conventional numerical measure of the degree of crosslinking of the hydrocarbon copolymer and of the ion-exchange resins made from it is the *nominal DVB content*, which means the mole percent of pure DVB in the original mixture. Of course, the presence of ethylstyrene in the mixture of monomers results in the presence of ethyl substituents on some of the benzene rings in the final product. Variations in the composition of commercial divinylbenzene account in part for the differences observed among different batches of supposedly identical resins.

Ion-exchange resins in the form of small spheres are more convenient for most purposes than the same resin in the form of irregular particles resulting from the grinding of a massive piece of resin. Small spheres[12, 13] of the polystyrene resin are produced by emulsifying a mixture of styrene, DVB, and catalyst in an aqueous solution of an emulsifying agent and heating the mixture to about 90°. The average size of the spheres (called pearls or beads) depends on the concentration and nature of the emulsifying agent and the vigor of the stirring.

B.I.b.2. *Sulphonation of pearls of polystyrene resin.* The next step in the manufacture of a strong-acid cation-exchange resin is the sulphonation of the beads of crosslinked polystyrene. This can be accomplished by treatment of the beads with concentrated sulphuric acid for several hours at about 100°. Thus each benzene ring of the polymer is sulphonated regardless of whether it has one or two —CH—CH$_2$— substituents or whether it has an ethyl group (derived from ethylstyrene).

Although the reaction appears very simple, difficulties are encountered if care is not taken. The sulphonation starts at the surface of the beads and proceeds inward. It is accompanied by the evolution of heat and swelling. The tendency of the outer layers to expand before the inner layers produces strains, which may crack the beads. The difficulty is avoided by treating the beads prior to sulphonation with a solvent that causes them to swell. Toluene, tetrachloroethylene, trichloroethylene, or dichloromethane may be used for this purpose. The relative volume increase induced by these solvents in 20 hr is 15, 60, 68, and 72 % respectively.[14]

After sulphonation, the beads are filtered on a glass filter. If the sulphuric acid is washed from the surface and internal pores of the resin too quickly, the sudden expansion of the beads causes cracking. The beads may be washed on the filter with a very fine spray of water so that about 24 hr is required for the complete removal of the acid, or they may be washed first with 26 % aqueous sodium chloride and then gradually with more dilute solutions.[14]

Chlorosulphonic acid, ClSO$_3$H is a more modern sulphonating agent. Seifert[15] treated the pearls of the hydrocarbon copolymer with this acid at about 60° to produce polystyrene sulphonylchloride. This is readily hydrolyzed to the sulphonic acid by water at 50°.

Most of the sulphonated polystyrene resins have one sulphonic acid group on almost every benzene ring. For example, it can be readily calculated that in a sulphonated resin with a specific capacity of 5.20 meq/g prepared from a mixture of 84.00 mole % of styrene and 8.00 mole % each of DVB and ethylstyrene, the molar ratio of sulphonic acid groups to benzene rings is 0.97.

B.I.b.3. *Sulphonated polystyrene resins of low capacity.* Less completely sulphonated resins are sometimes desired for industrial[16] or analytical[17, 18] purposes or for testing certain theories of resin behavior.[19, 20] The properties of the partly sulphonated resins are profoundly influenced by the distribution of the sulphonate groups within the resin bead. For Small's process of gel-liquid extraction,[21] each resin bead should be completely or nearly completely sulphonated throughout a very thin layer at the surface, the rest of the bead being completely unsulphonated. For the other purposes, the sulphonate groups should be distributed randomly or uniformly throughout the bead. There is no demand for resins with a distribution of sulphonate groups between these two extremes.

Staudinger and Hutchinson[22] devised a method for the sulphonation of the surface of plastics by exposure to sulphur trioxide, sulphuric acid, or sulphuryl chloride in either the gaseous or liquid state. The process depends on the very slow diffusion of the sulphonating agent inside the resin. Small[21] applied this method to the superficial sulphonation of copolymer beads as follows. To a known weight of concentrated sulphuric acid maintained at 100°, he added an equal weight of the dry copolymer beads. He agitated the mixture at 100° until the exchange capacity of the product was about 0.005 meq/g. The time required for this degree of sulphonation varies with the percentage of DVB in the resin and with the concentration of the acid. Trial runs are necessary to find the proper time for the acid and resin in question. As a rough guide, Small reported that 20 min sufficed in the case of a copolymer with 2% of DVB. At the end of the sulphonation period, he filtered the mixture and washed the beads with water to remove the sulphuric acid. He determined the exchange capacity of the product by potentiometric titration (p. 22).

The preparation of randomly sulphonated resins of low capacity presents a more difficult problem. There are three different approaches.

The first preparation of such a resin was accomplished by Boyd[19] and his coworkers in 1954. Bauman[23] had previously observed that Dowex 50 loses capacity when heated in water at high temperature. This reaction is appreciable at 150° for the resin in the hydrogen form and at 175° for the resin in the sodium form. Boyd[19] reasoned that the loss of sulphonated groups was very likely a random process and that the partial desulphonation of ordinary Dowex 50 at high temperature should yield a uniformly sulphonated resin of low capacity.

Several difficulties are encountered in this preparation. In the first place, the reaction of either the hydrogen form or the sodium form is difficult to control. The investigators avoided this trouble by heating a resin in the mixed sodium-hydrogen form with water in sealed tubes. The desulphonation is not simply the reverse of the sulphonation reaction

$$-C_6H_5 + H_2SO_4 \rightleftarrows -C_6H_4SO_3H + H_2O$$

because the odor of mercaptans was generally noted when the sealed tubes were opened after the desulphonation. Probably the most serious disadvantage of the desulphonation method is that the degree of crosslinking of the resin is diminished.[19]

Reichenberg[24] has modified the procedure by heating the resin with concentrated hydrochloric acid at 165°. This eliminates the formation of mercaptans.

The second approach to the problem of preparing a randomly sulphonated low-capacity resin is exemplified in the work of Graydon and his students.[25] They emulsified a mixture of styrene, commercial DVB and the butyl ester of *p*-sulphostyrene and added benzoyl peroxide to catalyze the copolymerization. Then they treated the resin beads with 1.2 M sodium hydroxide to convert the ester group into the sodium salt, thus obtaining the partly sulphonated ion-exchange resin. The dry hydrogen-form resin had a capacity of 2.91 meq/g. Obviously the capacity of the resin depends on the percentage of the ester of styrene sulphonic acid in the original mixture. They used the sulphonic ester because free styrene sulphonic acid is not miscible with DVB. The partly sulphonated resin could be further sulphonated in the usual way to yield a resin with a specific capacity of 4.87 meq/g rather similar in its properties to the ordinary sulphonic cation exchangers.

Graydon[26] also copolymerized mixtures of N,N-dimethyl-p-sulphonamidostyrene, DVB, and styrene. He treated the products with boiling 25% hydrochloric acid for 120 hr to substitute the hydroxyl group for the dimethylamido group, thus obtaining cation exchange resins with capacities between 0.94 and 3.12 meq/g.

The third approach to the problem of preparing a randomly sulphonated low-capacity resin is illustrated by the method of Reichenberg.[24] He treated beads of the copolymer of styrene and DVB with nitrobenzene to make them swell. Then he added concentrated sulphuric acid to the mixture at a temperature too low for appreciable sulphonation. After the sulphuric acid had diffused into the resin, he raised the temperature of the mixture, causing random sulphonation. The extent of the sulphonation depends on the concentration of the sulphuric acid in the nitrobenzene and on the time and temperature of the reaction. This method is not applicable to resins containing more than 15% DVB.

The Dow Chemical Company markets a randomly sulphonated cation-exchange resin with a capacity of about 1.2 meq/g. It is known as ET-561. Although the details of the manufacture of this resin have not been published, the method is probably similar to that of Reichenberg.[24]

B.I.b.4. *Polystyrene resins with phosphonate or phosphinate groups.* If crosslinked polystyrene is treated with phosphorus trichloride and aluminum chloride as a catalyst, the —PCl_2 group is substituted in the benzene rings. Subsequent hydrolysis yields phosphinated polystyrene.

$$—CH—CH_2$$
$$|$$
$$C_6H_4—PHO(OH)$$

Oxidation of this product yields phosphonated polystyrene,[27]

$$—CH—CH_2$$
$$|$$
$$C_6H_4—PO(OH)_2$$

B.I.b.5. *Chelating resins.* The best-known example of this type has the group —$CH_2N(CH_2COOH)_2$ attached to crosslinked polystyrene. It can be made by the following series of reactions:[28]

$$—CH—CH_2— \quad \xrightarrow[+ ZnCl_2]{ClCH_2OMe} \quad —CH—CH_2 \quad \xrightarrow{NH_3} \quad —CH—CH_2-$$
$$| \qquad\qquad\qquad\qquad\qquad\qquad | \qquad\qquad\qquad\qquad |$$
$$C_6H_5 \qquad\qquad\qquad\qquad\qquad C_6H_4—CH_2Cl \qquad\qquad\quad C_6H_4—CH_2NH_2$$

$$\xrightarrow{ClCH_2COOH} \quad —CH—CH_2—$$
$$|$$
$$C_6H_4—CH_2—N(CH_2COOH)_2$$

B.I.b.6. *Carboxylic resins.* Although carboxyl groups can be incorporated in polystyrene resins, the most popular carboxylic resins are made by the copolymerization of methacrylic acid (or its esters) with DVB.[29, 30] As in the case of the polystyrene resins, the DVB serves to provide the crosslinks. Because of the solubility of methacrylic acid in water, this copolymerization can not be used to produce beads of the resin. However, the use of esters of methacrylic acid permits copolymerization in emulsion and yields beads of the resin with esterified carboxylic groups. Then hydrolysis yields the resin with free carboxylic acid groups.

B.II. SYNTHESIS OF ANION-EXCHANGE RESINS

B.II.a. *Condensation Polymers*

The reaction of formaldehyde with aromatic amines is analogous to its reactions with phenols, yielding resins with aromatic amino groups. These are very weak bases; they can react with aqueous solutions of acids and then perform anion-exchange reactions.

$$ArNH_2 + H^+ + Cl^- \rightleftarrows ArNH_3Cl$$

$$ArNH_3Cl + NO_3^- \rightleftarrows ArNH_3NO_3 + Cl^-$$

These resins are of little importance, however, because the anion exchangers made from addition polymers have the advantages of greater stability and greater basic strength.

B.II.b. *Addition Polymers*

Beads of crosslinked polystyrene may be considered the starting point for the manufacture of anion-exchange resins. These beads are first swollen in chloromethyl ether (more properly chloromethyl methyl ether) and then caused to react with the ether by the addition of a suitable catalyst such as anhydrous stannic chloride or zinc chloride.[31] The reaction

$$
\begin{array}{ccc}
-CH-CH_2 & & -CH-CH_2- \\
| & + \ MeOCH_2Cl \rightarrow & | \\
C_6H_5 & & C_6H_4-CH_2Cl
\end{array} + \ MeOH \qquad (1.a)
$$

causes the substitution of the chloromethyl group in about 56% of the benzene rings. The beads of chloromethylated resin are then swollen in a suitable solvent and treated with excess ammonia or (more frequently) an aliphatic amine. All of the chloromethyl groups react according to the equations

In each case, the product is an anion-exchange resin in the chloride form. If these resins are treated with a sufficient quantity of sodium hydroxide, the respective products are:

I $-CH-CH_2-$
    $\quad\quad$ |
    $C_6H_4-CH_2-NH_3^+OH^-$

$\quad\quad$ or $-CH-CH_2-$
$\quad\quad\quad\quad\quad$ |
$\quad\quad\quad\quad\quad$ $C_6H_4-CH_2-NH_2$

II $-CH-CH_2-$
    $\quad\quad$ |
    $C_6H_4-CH_2-NH_2Me^+OH^-$

$\quad\quad$ or $-CH-CH_2-$
$\quad\quad\quad\quad\quad$ |
$\quad\quad\quad\quad\quad$ $C_6H_4-CH_2-NHMe$

III $-CH-CH_2-$
    $\quad\quad$ |
    $C_6H_4-CH_2-NHMe_2^+OH^-$

$\quad\quad$ or $-CH-CH_2-$
$\quad\quad\quad\quad\quad$ |
$\quad\quad\quad\quad\quad$ $C_6H_4-CH_2-NMe_2$

IV $-CH-CH_2-$
    $\quad\quad$ |
    $C_6H_4-CH_2-NMe_3^+OH^-$

V $-CH-CH_2-$
    $\quad\quad$ |
    $C_6H_4-CH_2-NMe_2(CH_2CH_2OH)^+OH^-$

The first three of these with primary, secondary, or tertiary nitrogen atoms are weak-base resins. The fourth and fifth with quaternary nitrogens are strong-base resins. They are sometimes designated as strong-base resins of types 1 and 2 respectively.

More problems are encountered in the synthesis of polystyrene-based anion exchangers than cation exchangers. In the first place, chloromethyl ether may react with crosslinked polystyrene not only according to reaction (1.a) but also according to (1.b)

$-CH-CH_2-$
$\quad$ |
$C_6H_4-CH_2Cl$

$\quad\quad$ +

C_6H_5
$\quad$ |
$-CH-CH_2-$

$\longrightarrow$

$-CH-CH_2-$
$\quad$ |
C_6H_4
$\quad$ |
CH_2
$\quad$ |
C_6H_4
$\quad$ |
$CH-CH_2-$

$\quad + HCl$ $\quad\quad\quad$ (1.b)

The latter introduces additional crosslinking, the extent of which can be calculated from the amount of chloromethyl ether consumed in both reactions and the specific exchange capacity of the final product. This calculation is not applied to commercial resins; rather, the conditions of the chloromethylation reaction are adjusted so as to favor reaction (1.a) relative to (1.b).

As the average distance between two polymer chains in a polystyrene bead is increased, the probability of the occurrence of reaction (1.b) is decreased; therefore the beads should be swollen as much as possible when the reaction is started.[32] Analogous reasoning indicates that increases in the DVB content of the original polystyrene bead cause increases in the extent of reaction (1.b). It is interesting to note in this connection that the Dow Chemical Co. markets sulphonated polystyrene with DVB contents from 2 to 16% but that their most highly crosslinked strong-base anion-exchange resin contains only 10% DVB. The rate of reaction (1.b) obviously depends on the concentration of the $-C_6H_4-CH_2Cl$

groups, and this concentration increases as reaction (1.a) progresses. Therefore the rate of (1.b) relative to (1.a) increases as the percentage of chloromethylated rings increases, and the reaction should be stopped before the chloromethylation reaction approaches completion. This has been confirmed experimentally by Pepper et al.[32] As previously mentioned, the commercial anion-exchange resins are manufactured from crosslinked polystyrene in which only about 56% of the benzene rings have been chloromethylated. Other conditions that influence the relative rates are the temperature, the nature, and amount of the catalyst, the amount and purity of the chloromethyl ether, and the presence or absence of solvents.[33]

In the second place, undesired side reactions may occur in the reaction between the chloromethylated polystyrene and the amine. For example, if methylamine is used, the desired reaction

$$
\begin{array}{c}
-CH-CH_2 \\
| \\
C_6H_4-CH_2Cl
\end{array}
\;+\; NH_2Me \;\longrightarrow\;
\begin{array}{c}
-CH-CH_2- \\
| \\
C_6H_4-CH_2NHMe
\end{array}
\;+\; HCl
\tag{1.c}
$$

may be followed by an undesired reaction

$$
\begin{array}{c}
-CH-CH_2- \\
| \\
C_6H_4 \\
| \\
CH_2Cl \\
+ \\
NMeH \\
| \\
CH_2 \\
| \\
C_6H_4 \\
| \\
-CH-CH_2-
\end{array}
\;\longrightarrow\;
\begin{array}{c}
-CH-CH_2- \\
| \\
C_6H_4 \\
| \\
CH_2 \\
| \\
NMe \\
| \\
CH_2 \\
| \\
C_6H_4 \\
| \\
-CH-CH_2
\end{array}
\;+\; HCl
\tag{1.d}
$$

Reaction (1.d) not only causes an undetermined amount of additional crosslinking but also introduces a second ionogenic group ($>$NMe), thus making the resin bifunctional. However, the titration curve (p. 23) shows only one jump, indicating that the basic strengths of the groups —NHMe and $>$NMe are not very different. Reactions analogous to (1.c) and (1.d) can occur when ammonia or other primary amine is used instead of methylamine. It is also possible that a secondary amine can act analogously yielding a quaternary nitrogen as the crosslink. However, Pepper et al.[32] did not find any quaternary groups in the anion exchanger made from chloromethylated crosslinked polystyrene and dimethylamine.

A molecule of tertiary amine can react with only one chloromethyl group. Therefore no additional crosslinks are produced in the synthesis of strong-base anion-exchange resins by the reaction between the chloromethylated resin and the tertiary amine. Furthermore, we should expect the strong-base resins to be strictly monofunctional except for the minor disturbance to be discussed in the next section. On the contrary, a relatively small amount of weak-base groups is found in many strong-base resins. For example, Kraus and Moore[34] found evidence for the presence in Dowex 1 of weak-base groups to the extent of $<0.3\%$ of the total anion-exchange groups. These may be the result of a slight decomposition according to reaction (2.b).

B.III. POLYFUNCTIONALITY OF POLYSTYRENE RESINS

Although the strong-acid resins prepared by the sulphonation of crosslinked polystyrene are generally classified as monofunctional, it is widely recognized that these resins are not truly *monofunctional* in the strict sense of this word. There are two causes of the polyfunctionality.

B.III.a. *Position Isomerism*

The sulphonation of a benzene ring derived from styrene can occur in the ortho, meta, or para position relative to the bond between the ring and the hydrocarbon chain. Therefore three different kinds of sulphonate groups exist on the styrene rings alone. Furthermore the rings derived from the DVB are also sulphonated and give rise to six additional isomerically different sulphonate groups. If the ethylstyrene in the impure DVB is considered, ten more isomerically distinguishable sulphonate groups must be taken into account.

Each one of the nineteen different kinds of sulphonate groups probably exerts slightly different attractions for any given pair of exchanging cations, sodium and hydrogen for example. In other words, the selectivity coefficient, i.e. the classical equilibrium constant of the reaction

$$HR + Na^+ \rightleftarrows NaR + H^+$$

$$E = \frac{[NaR][H^+]}{[HR][Na^+]}$$

is the composite of nineteen individual selectivity coefficients, one for each type of sulphonate group. In view of the complexity of this situation, it is not surprising that the selectivity coefficient is not constant but varies with the molar ratio of sodium to hydrogen in the resin. See Chapter 3.

The situation is entirely analogous in the case of anion-exchange resins because the chloromethyl group and hence the ionogenic group can occupy nineteen different positions.

To demonstrate the importance of isomerism even in unsulphonated crosslinked polystyrene, Wiley et al.[35] prepared three samples of resin by copolymerizing styrene with (1) commercial DVB, (2) pure p-, and (3) pure m-DVB. They used 8% of DVB in each case. The increases in volume of a bead caused by putting the dry bead in benzene were respectively 1.84, 1.53, and 1.77%.

B.III.b. *Nonrandom Crosslinking*

During the copolymerization of styrene and DVB, the chains of copolymer grow by the addition of the monomeric units at the ends. If the kinetic constants of this reaction were the same for DVB as for styrene, the DVB (and therefore the crosslinks) would be randomly distributed along each chain of the resin and hence in the resin bead as a whole. The fact is, however, that the kinetic constant is greater for DVB.[36] Therefore the molar ratio of unreacted DVB to styrene decreases as the polymerization proceeds. This means that those portions of a resin bead where the polymerization occurred early in the process are more highly crosslinked than those portions where the polymerization occurred later. The bead is not uniform or random with respect to crosslinking.

It is well known that the selectivity coefficient of any given pair of ions increases as the degree of crosslinking is increased (Chapter 3). It follows therefore that within any one resin bead the selectivity coefficient varies from region to region depending on the crosslinking. This is another cause of polyfunctionality, probably more serious than that due to isomerism.

C. List of Ion-exchange Resins

Table 1.3 presents a list of most of the ion-exchange resins that are mentioned in this text. Although other brands are on the market, this table includes the simple resins that are most frequently used in analytical chemistry in the English-speaking world. Some less common resins and non-resinous exchangers are discussed at appropriate places throughout the text.

TABLE 1.3. ION-EXCHANGE RESINS

Type	Structure	Functional group[a]	pH limits[b]	Approximate specific capacity		Trade names
				meq/g[c]	meq/ml[d]	
Strong acid	Crosslinked polystyrene	—SO₃H	0–14	5.2	1.8	Amberlite IR-120 ⟋ Amberlite 200 Dowex 50[e] AG 50[f] Dowex 50W[e] AG 50W[f] ZeoKarb 225
Moderately strong acid	Crosslinked polystyrene	—PO(OH)₂	4–14	8		Duolite C-60 Duolite C-61
Weak acid	Polymerized acrylic acid	—COOH	6–14	10	3.5	Amberlite IRC-50 ZeoKarb 226
Chelating resin	Crosslinked polystyrene	CH₂COONa | —CH₂N | CH₂COONa	6–14	3[i]	i	Dowex A-1
Strong base[g]	Crosslinked polystyrene	—CH₂NMe₃Cl (type 1)	0–14	3.3 3.3 4.2	1.2 1.2 1.2	Amberlite Dowex 1[e] AG 1 Dowex 21K
		CH₂CH₂OH | —CH₂NMe₂Cl (type 2)	0–14	3.3	1.2	Dowex 2[e] AG 2
Weak base	Crosslinked polystyrene	—CH₂NHMe₂OH and —CH₂NH₂MeOH	0–7	5.0	2.0	Amberlite IR-45 Dowex 3
Bi-functional	Phenol-formaldehyde polymer	phenolic groups and —SO₂Na or —CH₂S₃ONa	h	h		ZeoKarb 215
Ion-retardation	Linear polyacrylic acid trapped in Dowex 1	—COONa and —CH₂NMe₃Cl	5–14			Retardion 11A 8 AG 11A 8

[a] The exchangeable ion in the functional group indicates the form in which the resin is generally sold.
[b] The pH limits within which the resin undergoes ion exchange.
[c] Capacity in milliequivalents per gram of dried resin.
[d] Capacity in milliequivalents per millilitre of resin bed, if resin is about 8% crosslinked.
[e] Most of the Dow resins are named to indicate the percentage of nominal divinylbenzene in the hydrocarbon mixture that was co-polymerized. Thus Dowex 50-X8 signifies 8% divinylbenzene and Dowex 50-X4 signifies 4%.
[f] The AG resins (analytical grade) are Dowex resins purified and marketed by Bio-Rad Laboratories.
[g] The functional groups indicated in column 3 are, of course, neutral salts. The strongly basic properties become apparent when the chloride ion is replaced by hydroxide ion.
[h] The sulphonate group functions as an exchanger at any pH, whereas the phenolate group can exchange cations only at very high pH. Hence the capacity varies greatly with the pH.
[i] The capacity varies with the exchanging ions and the pH. The manufacturer gives the value 0.33 mmol of $Cu(NH_3)_4^{2+}$ per ml of bed of the sodium form.

References

1. *Exodus*, **15**, 24 and 25.
2. H. S. THOMPSON, *J. Roy. Agr. Soc. Eng.*, **11,** 68 (1850).
3. T. J. WAY, *J. Roy. Agr. Soc. Eng.*, **11,** 313 (1850).
4. T. J. WAY, *J. Roy. Soc. Eng.*, **13,** 123 (1852).
5. C. M. GULDBERG and P. WAAGE, *Études sur les affinités chimiques*, Christiania, 1867.
6. J. VON LIEBIG, *Liebigs Ann. Chem. Pharm.*, **94,** 373 (1855) and **105,** 109 (1858).
7. G. WIEGNER, *J. Landwirtschaft*, **60,** 197 (1912).
8. J. LEMBERG, *Z. deut. geol. Ges.*, **22,** 335 (1870).
9. J. LEMBERG, *Z. deut. geol. Ges.*, **28,** 519 (1876).
10. B. A. ADAMS and E. L. HOLMES, *J. Soc. Chem. Ind. (London)*, **54T,** 1 (1935).
11. I. G. FARBENINDUSTRIE, German Patent 734,279 (1943).
12. G. F. D'ALELIO, U.S. Patent 2,366,007 (1944); *Chem. Abs.*, **39,** 4418³ (1945).
13. K. W. PEPPER, *J. Appl. Chem. (London)*, **1,** 124 (1954).
14. R. M. WHEATON and D. E. HARRINGTON, *Ind. Eng. Chem.*, **44,** 1796 (1952).
15. H. SEIFERT, German Patent 1,016,445 (1957); *Chem. Abs.*, **54,** 8155a (1960).
16. THE DOW CHEMICAL COMPANY, Brochure "ET-561", Midland, Mich., 1961.
17. F. JAKOB, K. C. PARK, J. CIRIC and W. RIEMAN, *Talanta*, **8,** 431 (1961).
18. A. VARON, F. JAKOB, K. C. PARK, J. CIRIC and W. RIEMAN, *Talanta*, **9,** 573 (1962).
19. G. E. BOYD, B. A. SOLDANO and O. D. BONNER, *J. Phys. Chem.*, **58,** 456 (1954).
20. S. LINDENBAUM, C. F. JUMPER and G. E. BOYD, *J. Phys. Chem.*, **63,** 1924 (1959).
21. H. SMALL, *J. Inorg. Nucl. Chem.*, **18,** 232 (1961).
22. H. P. STAUDINGER and H. M. HUTCHINSON, U.S. Patent 2,400,720 (1946); *Chem., Abs.*, **40,** 5295² (1946).
23. W. C. BAUMAN, J. R. SKIDMORE and R. H. OSMUN, *Ind. Eng. Chem.*, **40,** 1350 (1948).
24. D. REICHENBERG, private communication (1962).
25. I. H. SPINNER, J. CIRIC and W. F. GRAYDON, *Can. J. Chem.*, **32,** 143 (1954).
26. W. F. GRAYDON, U.S. Patent 2,877,191 (1959); *Chem. Abs.*, **53,** 16417i (1959).
27. T. R. E. KRESSMAN and F. L. TYE, British Patent 726,918 (1955); *Chem. Abs.*, **49,** 11209b.
28. K. W. PEPPER and D. K. HALE, in Society of Chemical Industry, London, *Ion Exchange and Its Applications*, 1955, p. 20.
29. G. F. D'ALELIO, U.S. Patent 2,340,111 (1944); *Chem. Abs.*, **38,** 4078¹ (1944).
30. D. K. HALE and D. REICHENBERG, *Disc. Faraday Soc.*, no. 7, 79 (1949).
31. W. C. BAUMAN and R. MCKELLAR, U.S. Patent 2,614,099 (1952); *Chem. Abs.*, **47,** 2401c (1953).
32. K. W. PEPPER, H. M. PAISLEY, and M. A. YOUNG, *J. Chem. Soc.*, 4097 (1953).
33. R. E. ANDERSON, *Ind. Eng. Chem. Prod. Res. Develop.*, **3,** 85 (1964).
34. K. A. KRAUS and G. E. MOORE, *J. Am. Chem. Soc.*, **75,** 1457 (1953).
35. R. H. WILEY, J. K. ALLEN, S. P. CHANG, K. E. MUSSELMAN, and T. K. VENKATACHALAM, *J. Phys. Chem.*, **68,** 1776 (1964).
36. R. H. WILEY and E. E. SALE, *J. Polymer Sci.*, **42,** 491 (1960).

CHAPTER 2

GENERAL PROPERTIES
OF ION-EXCHANGE RESINS

A. Stability

A.I. THERMAL STABILITY

High temperatures affect ion-exchange resins in two distinct manners: (1) loss of cross-links, and (2) loss of ionogenic groups. The extent of these reactions depends not only on the temperature and the duration of heating but also on the ionic form of the resin and on the environment.

A.I.a. *Loss of Crosslinks*

When the hydrogen form of a sulphonated polystyrene with 8% or more of DVB is heated in a sealed tube, very little reaction[1] occurs below 150°. On the other hand, irreversible change,[2] presumably loss of crosslinks, has been observed on heating a sulphonated polystyrene with 0.5% DVB at 50° in a dry atmosphere. Although the resin with the greater degree of crosslinkage would probably be slightly more stable under any condition, the salient point in this striking comparison is the diminished stability of the resin in the dry state.

A.I.b. *Loss of Ionogenic Groups*

The most extensive study of the thermal loss of ionogenic groups has been performed by Marinsky and Potter.[3] The hydrogen form of Amberlite IR-120 (a sulphonated polystyrene) when heated under water loses sulphonate groups essentially according to a first-ordor reaction. However, about 15% of the sulphonate groups react more rapidly and are lost almost completely in 12 days at 150° or in one day at 180°. The lithium form of the same resin is much more stable. The decomposition follows first-order kinetics with half-lives over 10 times those of the hydrogen form. The rapid loss of a fraction of the sulphonate groups does not occur, as in the case of the hydrogen form. The stability of other salt forms of the resin is similar to that of the lithium form.

The thermal decomposition under water of the hydroxide form of Amberlite IRA-400, a polystyrene with trimethylbenzylammonium groups, is more complicated. As in the case of the hydrogen form of the corresponding cation exchanger, there is a rapid loss of about 15% of the quaternary-ammonium groups followed by a much slower first-order decomposition. There are, however, two distinct reactions occurring in both the rapid and the slow decomposition:

$$-CH_2NMe_3OH \rightarrow -CH_2OH + NMe_3 \qquad (2.a)$$

$$-CH_2NMe_3OH \rightarrow -CH_2NMe_2 + MeOH \qquad (2.b)$$

The relative rates of these reactions remain about the same, 60% decomposing by the first reaction and 40% by the latter, regardless of the temperature and regardless of whether the rapid initial or the slower second decomposition predominates.[3, 4] It should be noted that the second of these reactions does not involve a loss of exchange capacity but rather the conversion of strong-base groups to weak-base groups.

Figure 2.1 summarizes the data of Marinsky and Potter. Note that the phosphate form is much more stable than the hydroxide form. It is very likely that less easily hydrolyzed

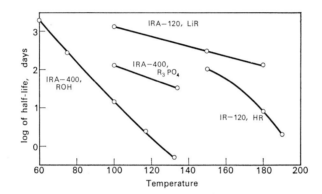

FIG. 2.1. Thermal stability of ion-exchange resins. (Data used by permission of Ionics Inc.)

forms such as RCl and RNO_3 are even more stable because the decomposition of the resin in the phosphate form is aggravated by the partial conversion to the hydroxide form.

$$R_3PO_4 + HOH \rightleftarrows R_2HPO_4 + ROH$$

Hydrolysis of the carbonate and bicarbonate forms also occurs, especially if the resin is heated in an open vessel so as to permit the escape of carbon dioxide.[4]

$$2RHCO_3 + H_2O \rightleftarrows R_2CO_3 + H_2O + CO_2$$

$$R_2CO_3 + H_2O \rightleftarrows RHCO_3 + ROH$$

A sample of the carbonate form of Amberlite IRA-400 lost 12% of its strong-base groups when heated at 90° for 30 days in an open vessel. A similar experiment in a closed vessel caused the loss of only 3%.

The data of Fig. 2.1 are in satisfactory agreement with the very small amount of quantitative data of other investigators on thermal stability. W. C. Bauman and his collaborators[5] found no appreciable decomposition upon heating the sodium form of Dowex 50 under water for 16 hr at 180°. The data of Fig. 2.1 indicate that 0.82% of the lithium form was destroyed under these conditions. Under the same conditions, the hydrogen form loses 10% of its capacity according to Bauman and 13% according to Fig. 2.1. Extrapolation of the graph of the hydrogen form to 140° indicates that 0.50% of its capacity would be lost in 16 hr; Bauman found no loss under these conditions. Figure 2.1 indicates that the loss of strong-base capacity of Amberlite IRA-400 should be 67% when it is heated under water at 90° for 30 days: according to Miss Baumann,[4] the loss was 47%.

Resins containing the ionogenic group —$CH_2NMe_2(CH_2CH_2OH)OH$ are slightly less stable than those with the group —CH_2NMe_3OH. Weak-base resins in the free-base form are more stable than strong-base resins in the hydroxide form.

From the foregoing discussion it is obvious that it is difficult to obtain a resin in the perfectly dry state, especially a strong-base resin in the hydroxide form and a strong-acid resin in the hydrogen form. Fortunately, it is seldom necessary thoroughly to dry a resin before using it in some experiment such as a titration (p. 22) or a determination of selectivity coefficient. In such cases it suffices to know what weight of dry resin is represented by the weight of moist resin taken in the experiment. This is readily determined as follows:

A quantity of resin is partly dried by exposure to air at room temperature or by heating in vacuum at such a temperature and for such a time interval that appreciable change can not occur. This resin is then thoroughly mixed, and a suitable weighed sample is dried in vacuum to constant weight. The dried sample is then thrown away because it may have suffered some decomposition. Thus the percentage of moisture in the partly dried resin can be calculated. Pollio[5a] recommends the Karl Fischer titration for the determination of water in resins.

A.II. RESISTANCE TO REAGENTS

The resins of Table 1.3 are not decomposed by strong bases nor by strong nonoxidizing acids except in so far as conversion to the hydrogen or hydroxide form hastens thermal decomposition. They also resist the action of ordinary oxidizing agents but are attacked by strong oxidizing agents such as dichromate, permanganate, and hot concentrated nitric acid above 2.5 M. Phenolformaldehyde resins are less stable in this respect than the polystyrene resins.

Although hydrogen peroxide does not have a very large redox potential, it attacks resins slowly, breaking the crosslinks and reacting with the ionogenic group of strong-base resins in the hydroxide form. Even oxygen of the air has a very slow action on resins. This phenomenon has been explained[6] as an attack of oxygen on a "weak link" of the resin to form a peroxide and free radicals. The reaction then continues by a free-radical chain,

$$ROOH \rightarrow RO\cdot + HO\cdot$$

$$RO\cdot + R'H \rightarrow ROH + R'\cdot$$

$$HO\cdot + R'H \rightarrow H_2O + R'\cdot$$

where R and R′ represent aliphatic radicals. The oxygen is necessary only to initiate the chain. This reaction is catalyzed very strongly by copper, iron, and manganese[6] and by some aminoacids.[7] As in the case of thermal decomposition, the products are both —CH_2NMe_2 and —CH_2OH. A sample of Amberlite IRA-400 in the hydroxide form lost 10% of its strong-base groups when treated with 5% hydrogen peroxide for 16 hr at 30° in the dark.

A.III. RESISTANCE TO RADIATION

It is well known that the sulphonated polystyrene resins in the hydrogen form and the strong-base resins in the hydroxide form undergo rapid deterioration when they are used to deionize the circulating water of a nuclear reactor. Marinsky and Potter[3a] subjected these resins to 1.0×10^7 roentgens without observing any decomposition, but Hall and Streat[3c] found that the hydroxide form of a strong-base resin of type 1 is damaged by dosages of γ-radiation of 1.0×10^8 roentgen or more.

They put samples of about 1.5 g of this resin in stoppered test-tubes with water and placed the test-tubes 3 cm from a kilocurie source of cobalt-60. At various time intervals they determined both the strong-base and weak-base capacity of the resin and also determined various nitrogen compounds in the solution. The results are summarized in Table 2.1.

TABLE 2.1. EFFECT OF GAMMA RAYS ON A STRONG-BASE ANION-EXCHANGE RESIN

Dose (megaroentgen)	Specific capacity (meq/g)			Quantity in solution (mmol)			
	Strong base	Weak base	Total	NMe_3	$NHMe_2$	NH_2Me	NH_3
0	4.0	0.2	4.2	0.0	0.0	0.0	0.0
100	2.7	0.6	3.3	6.9	2.3	0.7	0.3
200	2.1	0.8	2.9	8.7	3.6	1.9	1.4
300	1.7	0.9	2.6	9.6	4.2	3.1	2.3
400	1.3	1.0	2.3	10.4	4.3	3.6	4.0
500	0.8	1.2	2.0	10.9	4.3	3.8	5.4

Data of Hall and Streat, used by permission from the *Journal of the Chemical Society*.

The losses in total and strong-base capacities, together with the gain in weak-base capacity, can be explained by the same reactions (2.a) and (2.b) that occur in the thermal decomposition of this resin. The presence in the solution of ammonia and primary and secondary amines, however, proves that other reactions were also occurring, either in the resin or in the solution. The authors suggest the following reactions for the formation of dimethylamine:

$$-CH_2NMe_2 + H_2O \rightarrow -CH_2OH + NHMe_2$$
$$-CH_2NMe_3OH \rightarrow -CHO + CH_4 + NHMe_2$$

They account for the presence of monomethylamine by the reactions

$$NMe_3 \rightarrow CH_4 + MeNCH_2$$
$$MeNCH_2 + H_2O \rightarrow NH_2Me + HCHO$$

and

$$NHMe_2 \rightarrow CH_2NMe + 2H$$
$$CH_2NMe + H_2O \rightarrow NH_2Me + HCHO$$

Analogous reactions explain the formation of ammonia

$$NHMe_2 \rightarrow HNCH_2 + CH_4$$
$$HNCH_2 + H_2O \rightarrow NH_3 + HCHO$$

and

$$NH_2Me \rightarrow HNCH_2$$
$$HNCH_2 + H_2O \rightarrow NH_3 + HCHO$$

A.IV. MECHANICAL STABILITY

In industrial work where large beds of resin are subjected to frequent backwashing, the mechanical disintegration of the beads is a consideration of some importance in the life of the resin. In work on the laboratory scale, on the other hand, mechanical injury of the resin presents no problem provided that rapid swelling is avoided (see p. 24).

B. Equivalence of Exchange Reactions

The principle of electroneutrality indicates that the number of milliequivalents of an ion taken up by an exchanger should be exactly equal to the number of milliequivalents of similarly charged ion released by the resin.

$$HR + K^+ + Cl^- \rightarrow KR + H^+ + Cl^-$$

$$2HR + Ca^{2+} + 2Cl^- \rightarrow CaR_2 + 2H^+ + 2Cl^-$$

There are ion-exchange reactions, however, in which the stoichiometric relationship between the reagents and the products seems to be violated. For example, exactly 1 g (dry weight) of a hydrogen form polystyrene resin was treated with a solution of magnesium chloride until all the hydrogen ion was displaced from the resin. It was found[8] that the resin had taken up 5.23 meq of magnesium and had released only 4.90 meq of hydrogen ion. The explanation is probably that some of the exchange sites of the resin were occupied by $Mg^{2+} + Cl^-$. Calcium behaves similarly to magnesium, but the univalent cations, Li^+, Na^+, K^+, and NH_4^+, exchange for hydrogen ion (in the absence of anions of weak acids) on a strictly stoichiometric basis.[8]

Another apparent exception to stoichiometric exchange is encountered when a hydrogen resin is treated with the acetate of sodium, potassium, or ammonium. The quantity of hydrogen ion (both free and as nonionized acetic acid) gained by the solution is less than the loss of sodium ion. The following reactions occur:

$$HR + Na^+ \rightleftharpoons NaR + H^+$$

$$H^+ + C_2H_3O_2^- \rightleftharpoons HC_2H_3O_2$$

and some of the nonionized acetic acid is absorbed by the resin (see p. 30).

C. Reversibility of Ion-exchange Reactions

Ion-exchange reactions are generally reversible. For example, if some hydrogen-form resin is treated with an equivalent quantity of sodium chloride in aqueous solution, the composition of both phases after equilibrium is reached is the same as it would be if the sodium-form resin had been treated with hydrochloric acid.

D. Conversion of the Resin from One Form to Another

The conversion of a resin to any desired form is generally accomplished quite readily by treatment of the resin in a chromatographic tube with a solution of the appropriate electrolyte. The completion of the conversion is indicated by a negative qualitative test in the effluent for the ion originally in the resin or by quantitative determinations that show the same composition for both influent and effluent. For example, conversion of a cation exchanger from the hydrogen to the potassium form may be accomplished by passage of a solution of potassium chloride through the resin until the effluent has no appreciable acidity. In converting an anion exchanger from the nitrate to the chloride form, sodium chloride may be used until the effluent has the same chloride-ion concentration as the influent.

There are some cases, however, where this simple treatment is impracticable either because of a very unfavorable equilibrium or because of a very slow reaction. An example of an unfavorable equilibrium is seen in the attempt to convert the hydrogen form of a weak-acid resin to the sodium form.

$$HR + Na^+ + Cl^- \rightleftharpoons NaR + H^+ + Cl^-$$

Treatment with sodium chloride is utterly impracticable. However, by using sodium hydroxide and thus taking advantage of the small value of K_w, the conversion is readily accomplished.

$$HR + Na^+ + OH^- \rightleftharpoons NaR + H_2O$$

Analogously, the free-base forms of weak-base resins are converted to other forms by the use of acids (not salts) of the desired anions.

$$ROH + H^+ + NO_3^- \rightleftharpoons RNO_3 + H_2O$$

Hydrolytic reactions may introduce another difficulty if it is desired to wash the resin free of the excess salt that was used for the conversion. Four examples of such reactions follow:

Weak-acid resin: $NaR + H_2O \rightleftharpoons HR + Na^+ + OH^-$

Weak-base resin: $RCl + H_2O \rightleftharpoons ROH + H^+ + Cl^-$

Strong-acid resin: $NH_4R \rightleftharpoons HR + NH_3$

Strong-base resin:† $RBO_2 + H_2O \rightleftharpoons ROH + HBO_2$

In these cases, the water used to wash the excess electrolyte from the resin causes partial conversion as indicated. The extent of these reactions depends on the quantity of water used in the washing and on the ionization constant of the weak acid or weak base (either in the aqueous phase or in the resin) that is formed in the hydrolysis. Of course, the disturbance is aggravated if both the acid and base are weak as in the case of the acetate form of a weak-base resin. It follows that extensive washing should be avoided when hydrolysis is likely to occur.

Difficulties in converting one form to another due to slow reactions are found chiefly with chelating resins such as Dowex A-1. The equilibrium

$$CaR_2 + Mg^{2+} \rightleftharpoons MgR_2 + Ca^{2+}$$

is favorable for the conversion of the calcium form to the magnesium form. Nevertheless, the reaction is so slow that much time is saved by doing this conversion by the circuitous route indicated by the following reactions:

$$CaR_2 + 2H^+ + 2Cl^- \rightarrow 2HR + Ca^{2+} + 2Cl^-$$

$$2HR + 2Na^+ + 2OH^- \rightarrow 2NaR + 2H_2O$$

$$2NaR + Mg^{2+} + 2Cl^- \rightarrow MgR_2 + 2Na^+ + 2Cl^-$$

---

† In this text, HBO_2 and BO_2^- rather than H_3BO_3 and $H_2BO_3^-$ will be used to represent orthoboric acid and orthoborate anion because the former formulae indicate the monoprotic nature of boric acid.

E. Titration Curves and Capacity

Let us consider a suspension of a hydrogen-form cation exchanger in pure water to which successive increments of pure sodium hydroxide are added. After each addition of the base, the mixture is stirred until equilibrium is established; then the resin is allowed to settle, and the pH of the supernatant solution is measured potentiometrically. Theoretically, the pH of the solution should remain constant at 7.0 until the amount of base is equivalent to the amount of resin because all of the base is used in the reaction

$$HR + Na^+ + OH^- \rightarrow NaR + H_2O$$

After the equivalence point, the graph rises as though sodium hydroxide were being added to pure water. Because of traces of salts in both the water and the sodium hydroxide, the experimental curve is never exactly like Fig. 2.2 but shows some of the characteristics of curve *A*, *B*, or *C* of Fig. 2.3.

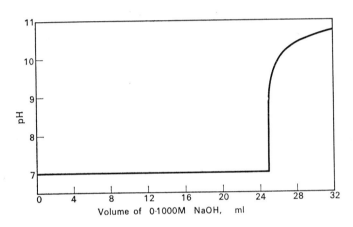

FIG. 2.2. Ideal titration graph of 2.500 meq of strong-acid resin in absence of salt.

If some hydrogen-form sulphonated polystyrene is suspended in 1 M potassium chloride, the solution becomes acidic because of the exchange reaction. If the suspension is titrated as described above, the graph resembles curve *A* of Fig. 2.3. Note the similarity of the graph to that of the titration of a strong acid. The capacity of the resin corresponds to the location of the inflection point on the horizontal axis.

Curve *C* is the titration graph of a phosphonic resin suspended in 1 M sodium chloride. Since there are two ionizable hydrogen atoms in the phosphonate group with different ionization constants, the curve shows two jumps and is very similar to the titration of a soluble phosphonic acid such as methanephosphonic acid. At a pH of 9 or above, both hydrogen atoms are ionized, and the resin has its maximum capacity of about 8 meq/g. At a pH of 5, only one hydrogen is ionized, and the resin has about one-half of its maximum capacity.

Curve *B* represents the titration of a carboxylic resin suspended in 1 M potassium chloride. The maximum capacity of the resin, 10 meq/g, is available only at the pH of the jump, 9, or at a higher pH. This curve is very similar to the titration graph of acetic acid. Thus it is seen that the titration graph of any cation-exchange resin resembles that of the monomer, provided that sufficient neutral salt is present.

Curve *D* is the titration graph of Amberlite IRA-400, hydroxide form, in 1 M potassium chloride. It resembles the titration of any strong base with hydrochloric acid. The capacity of this sample of resin was only 2.0 meq/g, unusually small for a resin of this type.

Curve *E* shows the titration of a weak-base resin in 1 M potassium chloride. There is no distinct inflection point. The reason is that the resin contained both tertiary and secondary amine groups with different basic strengths. The graph resembles that of the titration of a mixture of monomeric amines of different ionization constants.

If 1.00 g of dry hydrogen-form resin with a capacity of 5.20 meq/g is converted to the dry sodium form, there is a gain in weight of 5.20 (23.0 − 1.0) or 114 mg. Thus the weight of the sodium-form resin is 1.114 g, and its capacity is 5.20/1.114 or 4.67 meq/g. In order to avoid the confusion of having a change in capacity with every ion-exchange reaction, it has been agreed to refer the capacity of cation exchangers to the dry hydrogen form. Analogously, it is conventional to express the capacity of anion exchangers as the dry chloride form.

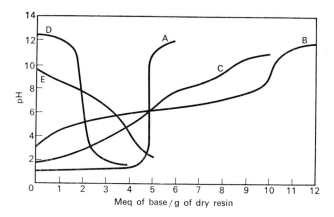

FIG. 2.3. Actual titration graphs of ion-exchange resins: *A*, Amberlite IR-120 in 1 M potassium chloride[9] (reproduced by permission from *Analytical Chemistry*). *B*, Amberlite IR-50 in 1 M potassium chloride[10] (reproduced by permission from *Industrial and Engineering Chemistry*). *C*, Phosphonic resin in 1 M sodium chloride[11] (reproduced by permission of the American Chemical Society). *D*, Amberlite IRA-400 in 1 M potassium chloride[12] (reproduced by permission from *Industrial and Engineering Chemistry*). *E*, Weak-base resin in 1 M potassium chloride[13] (reproduced by permission of John Wiley & Sons Inc.).

E.I. PROCEDURE FOR THE TITRATION OF A RESIN

The procedure outlined at the beginning of this section for the titration of a resin, either in the absence or presence of neutral salt, is inefficient because several hours are required for the equilibration of the system after each addition of titrant, especially near the end point. Much time can be saved by weighing about a dozen equal samples of dry resin (or resin of known moisture content) into separate flasks, adding to each the same quantity of water or salt solution, adding a different amount of titrant to each, agitating all the samples at one time, and measuring the pH of each solution after the resin has settled. Near the equivalence point, the mixtures should be agitated, allowed to settle and measured several times until successive readings of pH agree.

E.II. PROCEDURE FOR THE DETERMINATION OF CAPACITY

Although a potentiometric titration of a resin may serve as a determination of capacity, there are more efficient methods of accomplishing this purpose. A known quantity of hydrogen-form cation exchanger in a column may be treated by the passage of a moderate excess of standard sodium hydroxide. After rinsing the column with water, the excess base in the effluent is titrated. An anion exchanger in the chloride form is conveniently treated in the column with unstandardized 1 M sodium nitrate until all the chloride is removed from the resin. The eluate is then titrated with silver nitrate.

F. Crosslinking and Swelling

Consider a bead of dry sulphonated polystyrene resin in the hydrogen form that is put into water. The water penetrates into the resin and hydrates both the hydrogen and sulphonate ions. These ions may then be considered to be dissolved in their water of hydration, thus forming a very concentrated aqueous solution inside the resin. Osmotic pressure then drives more water into the resin. Thus the bead swells, and the valence bonds between the carbon atoms of the hydrocarbon chains and crosslinks are stretched and bent to accommodate the invading water. This distortion of the valence bonds exerts a pressure on the imbibed water, tending to drive it out of the resin. The amount of water imbibed at equilibrium depends on the degree of crosslinking of the resin and on the exchangeable ion of the resin. These facts are illustrated in Table 2.2. As is to be expected, the swelling decreases with increasing crosslinkage and decreasing size of the hydrated radius of the exchangeable ion.

TABLE 2.2. SWELLING OF STRONG-ACID RESINS IN WATER[14]

% of DVB	Swelling (mg of H_2O per meq of resin)				
	HR	LiR	NaR	KR	CsR
2	943	625	513	500	345
4	417	357	303	294	233
8	219	196	172	167	144
12	145	130	115	112	100
16	128	119	99	95	86
24	96	80	71	69	59
a	9.0	6.0	4.2	3.0	2.5

a=Radius (Å) of the hydrated cation.[15]
(Data used by permission of the American Chemical Society.)

The approximate molarities of the sulphonates in the internal solutions are readily calculated from the data of Table 2.2. They are 1000 times the reciprocals of the values given in the table. Thus these molarities vary from 1.06 for the hydrogen form of the least crosslinked resin to 17 for the cesium form of the most highly crosslinked.

Calmon[16] has studied the changes of bed volume that accompany the exchange reactions of a sulphonated polystyrene with 1% of crosslinking. In each case, he measured the bed volume after backwashing the resin with water and allowing it to settle. With twelve different bivalent cations in the resin, the volumes ranged from 18 to 63% of the volume when the resin was in the hydrogen form. With tervalent cations, the figures were 19–26%. The bed volume of the resin in the thorium form was only 11% of the bed volume in the hydrogen form. These data indicate the important effect of the valence of the exchangeable ion on the swelling of the resin, especially if the crosslinkage is small. Ions of large valence attract the sulphonate groups of the resin more strongly and thus permit less swelling. A study of the swelling of resins containing univalent and divalent ions in varying proportions has recently appeared.[17]

Table 2.3 shows the swelling data for strong-base resins. It will be observed that these resins swell less than the sulphonated polystyrene resins but they show the same trend of decreasing swelling with increasing crosslinkage. Although there is little difference in the radii of the hydrated halide ions, the table indicates that the anions with the smaller selectivity coefficients (see p. 43) cause the most swelling.

Weak-acid resins in any salt form behave rather similarly to the strong-acid resins in regard to swelling. However, the hydrogen form of these resins swells very little because the nonionized carboxyl group has very little ability to attract water. Analogously, weak-base resins swell appreciably in all forms except the free-base form.

TABLE 2.3. SWELLING OF RESINS[18] WITH THE BENZYLDIMETHYLETHANOL-AMMONIUM ION

% of DVB	Capacity of dry RCl	mg H_2O per meq			
		RF	RCl	RBr	RI
1	3.71	424	333	215	86
2	3.79	345	264	181	85
4	3.47	294	233	159	83
8	3.10	185	139	100	67
16	1.91	128	114	87	65
a		3.5	3.0	3.0	3.0

a = Radius (Å) of the hydrated anion.[15] (Data used by permission from *J. Phys. Chem.*)

If a resin is immersed in a solution containing an electrolyte whose cation (or anion) is the same as the exchangeable ion of the resin (so that no ion exchange can occur), the difference between the osmotic pressures inside and outside of the resin is less than if the resin were immersed in pure water. Therefore the resin swells less. This is illustrated in Table 2.4.

When water penetrates into a dry resin bead, the swelling occurs first at the surface layers. Thus mechanical strains are generated, which are often great enough to crack the bead, especially if it is large. Because of this risk of mechanical injury to the beads, it is often advisable to let a dry resin swell gradually by immersing it successively in solutions of decreasing concentration or by exposing it to moist air. A resin will eventually absorb as much water from air of 100% relative humidity as from pure water, but the rate of absorption is less in the former case. Because of risk that large beads will crack on swelling, commercial ion-exchange resins are seldom prepared in sizes larger than 20-mesh ($D = 0.84$ mm). The

TABLE 2.4. ABSORPTION OF WATER AND HYDROCHLORIC ACID BY SULPHONATED POLYSTYRENE
RESINS

Molarity of external HCl	2% DVB		5% DVB		10% DVB		15% DVB		25% DVB	
	w	L	w	L	w	L	w	L	w	L
0.00	3.26	0.00	1.57	0.00	0.88	0.00	0.60	0.00	0.36	0.00
0.11	3.14	0.04	1.52	0.02	0.88	0.00	0.60	0.00	0.36	0.00
0.52	2.80	0.39	1.48	0.06	0.86	0.01	0.60	0.01	0.35	0.00
1.05	2.41	1.03	1.38	0.19	0.84	0.04	0.59	0.02	0.35	0.00
2.30	1.81	2.31	1.24	0.70	0.78	0.18	0.57	0.04	0.35	0.00

w is the amount of water (g) absorbed by 1 g of dry resin.
L is the amount of hydrochloric acid (mmol) absorbed by 1 g of dry resin.
(Data used by permission from the *Journal of the Chemical Society.*)

size specifications of the Amberlite resins refer to the dry ion-exchange resin, but those of the Dowex resins refer to the hydrocarbon copolymer before the introduction of the ionogenic group. The sizes of the wet resins are greater, of course. Table A.1 (p. 277) of the Appendix shows the mesh openings of various standard sieves.

Ion-exchange resins also swell when immersed in some nonaqueous liquids. Such swelling is usually less than in water. In general, hydrophylic liquids cause more swelling than hydrophobic liquids, and hydrocarbons cause only insignificant swelling. On the other hand, the styrene-divinylbenzene copolymer without ionogenic groups is swollen by hydrocarbons but not by water.

F.I. MEASUREMENT OF SWELLING

Swelling is usually expressed as the weight of water (or other liquid) absorbed per gram or per milliequivalent of dry resin rather than as increase in resin volume.

In principle, the determination is very simple; it is necessary only to weigh a quantity of resin that has come to equilibrium with the liquid in question and to weigh the same resin again after vacuum drying.

Of course, the liquid that clings to the surface of the resin must be removed before the first weighing. This is usually accomplished[19, 20] by centrifuging the resin in a tube provided with a filter disk. A correction must be applied for the water that adheres to the surface of the resin. The correction may be evaluated by centrifuging a similar quantity of glass beads or bead of surface-sulphonated polystyrene resin. Its magnitude depends on the centrifugal force and the time of spinning. According to Kraus,[20] this correction is 33 mg for each milliliter of resin bed. Pepper's[19] correction was 50 mg of water per gram of wet resin. If the density of the resin is taken as 1.0 and if the interstitial volume is taken as 38% of the bed volume (see p. 134), this correction is equivalent to 29 mg of water per milliliter of resin bed. The accuracy[21] of the determination is about ±3%.

F.II. FLOTATION TEST FOR UNIFORMITY OF CROSSLINKING

It sometimes happens that a given batch of resin contains beads that are crosslinked to different degrees. Nonuniformity such as this can be easily detected by the following test. This test depends on the fact that the density of a swollen cation-exchange resin is governed chiefly by the degree of crosslinking, the more highly crosslinked particles having the greater density.

A glass vessel[22] is used that consists of two bulbs, each about 100 ml in capacity, connected to each other by a glass tube about 18 cm long and 2.5 cm in diameter. The vessel is held vertically, and the lower bulb along with about half of the tube is filled with a solution of sodium tungstate denser than the swollen resin which is to be tested. The rest of the vessel is gently filled with a less concentrated solution of sodium tungstate, less dense than the swollen resin. The tube is held vertically in a thermostat at $25° \pm 0.01°$ for 2 hr. A stable gradient of concentration and density is thus established in the tube.

Now about 50 mg of the air-dried sodium-form resin is dropped into the tube. Each bead of resin settles slowly until it reaches the level where the density of the solution is equal to that of the bead. This takes an hour or less. A capillary tube is inserted into the mixture to any desired level and used to withdraw a small volume of the solution. The density of these samples of solution can then be determined; or, preferably, their refractive index is determined, and a standard curve of density or concentration vs. refractive index is consulted.

Sodium tungstate was selected as the solute because it yields clear, colorless solutions of high density and because the anion is large, by virtue of polymerization, and hence prevents Donnan invasion (see next section). Disodium monolead ethylenediaminetetraacetate can also be used.

The authors[22] found that Dowex 50W-X4 and Dowex 50W-X8 gave two sharp, distinct bands in the column, the former being the higher band. This indicates that each of these resins was nearly uniform. On the other hand, a test with three experimental batches of crosslinked sulphonated polystyrenes containing 7, 10, and 17% DVB gave three diffuse bands.

G. Donnan Equilibrium

G.I. SULPHONATED POLYSTYRENE RESINS

Let us consider a system consisting of an ion-exchange resin immersed in a solution of a strong electrolyte having one ion in common with the resin. To be specific, let us consider a sulphonated polystyrene resin in hydrochloric acid. In addition to the swelling, discussed above, another important phenomenon occurs: an appreciable amount of the hydrochloric acid diffuses into the internal solution of the resin.

Since there was originally no chloride ion inside the resin, it is obvious that there will be a tendency for this species to diffuse from the external to the internal solution. On the other hand, as indicated on p. 24, the internal solution of sulphonic acid is concentrated; therefore the hydrogen ions *per se* have no tendency to diffuse from the external dilute solution into the internal concentrated solution. The necessity for electroneutrality requires that hydrogen and chloride ions migrate into the resin in equal amounts. The final result is that a comparatively small amount of chloride ions does diffuse into the internal solution dragging with it an equal amount of hydrogen ions in spite of the unfavorable concentration gradient.

The well-known principle of the Donnan equilibrium is applicable here, and we may write

$$\overline{(H^+)}\,\overline{(Cl^-)} = (H^+)\,(Cl^-) \tag{2.1}$$

where the parentheses denote the activities of the enclosed species and the bars indicate the resin phase. Since this text will not be concerned with the difficulties involved in the evaluation of activity coefficients in the concentrated internal solution, we shall simplify the foregoing equation at the sacrifice of accuracy by substituting concentrations for activities.

$$[\overline{H^+}]\,[\overline{Cl^-}] = [H^+]\,[Cl^-]$$

Let M_r and M_d denote respectively the molarity of sulphonate ions and of the chloride ions in the internal solution. Let M_s denote the concentration of hydrochloric acid in the external solution at equilibrium. Then

$$[\overline{H^+}] = M_r + M_d$$

$$[H^+] = [Cl^-] = M_s$$

By combination of the last three equations

$$(M_r + M_d)M_d = M_s^2$$

$$M_d = \sqrt{(M_s^2 + \tfrac{1}{4}M_r^2)} - \tfrac{1}{2}M_r \qquad (2.2)$$

Table 2.4 shows the data[19] on the absorption of both water and hydrochloric acid by a series of sulphonated polystyrene resins from various solutions of hydrochloric acid.

From these data, one may calculate $M_d = \frac{L}{w}$. The assumption involved in this equation that 1 ml of internal solution results from each gram of imbibed water does not introduce a serious error. In Table 2.5 the values of the Donnan invasion, M_d, calculated from the experimental data of Table 2.4, are compared with the values calculated by eqn. (2.2). To find the values of M_r it was assumed that the capacity of all the resins was 5.2 meq per dry g.

TABLE 2.5. DONNAN INVASION OF SULPHONATED POLYSTYRENE RESINS BY HYDROCHLORIC ACID

Molarity of external HCl	Molarity of internal hydrochloric acid									
	2% DVB		5% DVB		10% DVB		15% DVB		25% DVB	
	Obsd.	Calcd.	Obsd.	Calcd.	Obsd.	Calcd.	Obsd.	Calcd.	Obsd.	Calcd.
0.00	0.000	0.000	0.000	0.000	0.000	0.000	0.000	0.000	0.000	0.000
0.01	0.013	0.007	0.013	0.004	0.000	0.002	0.000	0.000	0.000	0.001
0.52	0.139	0.135	0.041	0.078	0.012	0.045	0.017	0.031	0.000	0.018
1.05	0.427	0.428	0.138	0.272	0.048	0.172	0.034	0.124	0.000	0.074
2.30	1.28	1.28	0.565	1.014	0.23	0.71	0.070	0.549	0.000	0.348

The discrepancies between the calculated and observed values of M_d are within the experimental error for the resin with 2% DVB. The discrepancies, in general, increase with increasing crosslinkage and with increasing concentration of external hydrochloric acid. Nevertheless, eqn. (2.2) is qualitatively correct in indicating that the Donnan invasion increases with decreasing crosslinkage and with increasing external concentration. The greatest cause of the discrepancies is the error involved in substituting concentrations for activities in the derivation of eqn. (2.2).

However, Glueckauf and Watts[23] have demonstrated that even eqn. (2.1) is not in good agreement with experimental data. According to these authors, the discrepancy is due to nonuniformity (p. 13) within the resin. They believe that different parts of a single resin bead have different degrees of crosslinking and hence different values of w and M_r, and therefore different concentrations of the invading electrolyte. In the derivation of eqn. (2.2), it was tacitly assumed that the resin was uniformly crosslinked.

The foregoing equation applies to the invasion by strong electrolytes of the 1–1 valence type. Equations for invasion by strong electrolytes of other valence types can readily be derived by the reader, or he may refer to an equation given elsewhere.[24]

G.II. MEASUREMENT OF DONNAN INVASION

The determination of the extent of the Donnan invasion is very similar to the determination of swelling (p. 26). After equilibrating the resin with the electrolytic solution, the mixture is separated by centrifugation. The resin is weighed and then washed with water to remove the electrolyte from the internal solution. Finally, the electrolyte that has been washed from the resin is determined by an appropriate method. The weight of dry resin and the weight of invading electrolyte are subtracted from the weight of resin after centrifugation to find the weight of imbibed water. Of course, a correction must be applied for the incomplete separation of the liquid from the resin by the centrifuge (p. 26).

Since the invading electrolyte is always more concentrated in the external solution than in the internal solution, small errors in estimating the correction for the liquid not removed by centrifugation cause larger errors in observed quantity of invasion. An ingenious method of avoiding this difficulty has been devised by Glueckauf and Watts.[23] After equilibrating the resin with the electrolytic solution, they removed most of the solution from the resin by blotting with filter paper and then immersed the resin in a bath of stirred water. At frequent time intervals they determined the quantity of the electrolyte in this solution. They plotted these quantities against the time elapsed since immersion. Applying the equations of diffusion, they extrapolated these data to zero time. The quantity of electrolyte in the solution at zero time represents the quantity that was on the surface of the blotted resin. It was subtracted from the amount eventually found in solution to yield the amount of electrolyte that had entered the resin. This method is more successful with resin membranes than with resin beads.

G.III. STRONG-BASE ANION-EXCHANGE RESINS

Data are given in Table 2.6 on the uptake of both water and hydrochloric acid from aqueous solutions by a series of strong-base anion-exchange resins.[25] These data were obtained by the method outlined above.

The data indicate that the imbibition of water follows the normal pattern; within a small experimental error, the swelling decreases regularly as the crosslinking is increased and is

TABLE 2.6. ABSORPTION OF WATER and HYDROCHLORIC ACID

Resin	0.01 M hydrochloric acid		0.1 M hydrochloric acid	
	H_2O (g per g of dry resin)	HCl (meq per g of dry resin)	H_2O (g per g of dry resin)	HCl (meq per g of dry resin)
Dowex 1-X2	3.45	0.0135	3.41	0.083
Dowex 1-X4	1.55	.0018	1.47	.016
Dowex 1-X8	0.54	.0025	0.64	.0158
Dowex 1-X10	0.59	.0069	0.59	.0176

(Data used by permission from *Analytica Chimica Acta*.)

slightly larger in 0.01 M acid than in 0.1 M. The uptake of hydrochloric acid is consistently greater from 0.1 M solution than from 0.01 M, in qualitative agreement with the principles of the Donnan equilibrium. On the other hand, the degree of crosslinking of the resin seems to have only an erratic influence on the absorption of hydrochloric acid; contrary to expectations, Dowex 1-X10 takes up much more of the acid than Dowex 1-X4.

This peculiar behavior is probably due to the presence of a small amount of tertiary nitrogen atoms in the resins,[25] the quantity varying from one batch of resin to another. These weakly basic groups combine with hydrochloric acid when the resin is immersed in a solution of this acid and slowly release the acid when the resin is washed with water.

$$-CH_2NMe_2 + H^+ + Cl^- \rightleftharpoons -CH_2NMe_2H^+Cl^-$$

$$\quad\text{resin}\qquad\qquad\text{aqueous}\qquad\qquad\text{resin}$$

Thus the uptake of hydrochloric acid in Table 2.6 represents the Donnan invasion *plus* the amount of weakly basic groups in the resin. Other investigators[20, 26] have also found evidence for the existence of tertiary amino groups in Dowex 1.

H. Absorption of Nonelectrolytes

We shall now consider some data of Reichenberg and Wall[27] on the absorption of both water and acetic acid by sulphonated polystyrene resins in the hydrogen form. These data were obtained by a method essentially similar to that used for studying the Donnan invasion. In these experiments, acetic acid can be considered as a nonelectrolyte. In 0.01 M external solution, it is only 4.1% ionized. The ionization is even less in the internal solution because of the large concentration of hydrogen ion from the sulphonic acid groups. Furthermore, no ion exchange can occur between the ions of acetic acid and the hydrogen-form resin.

These investigators found that resins with 15, 10, and 5.5% DVB absorbed respectively 0.034, 0.059, and 0.13 mmol of acetic acid per gram of dry resin when equilibrated with 0.101 M acetic acid at 25°. The corresponding quantities of water absorbed were 0.60, 0.84, and 1.49 g. It is not surprising that the absorption of acetic acid increases with decreasing crosslinkage; the less highly crosslinked resins absorb more water, and the sorbed acid can be considered to be dissolved in the internal water.

Indeed, one is tempted to postulate that the concentration of acetic acid should be the same in the internal solution as in the external solution. If this be true, a plot of the internal concentration of acetic acid *vs.* the external concentration should be a straight line for all degrees of crosslinkage. This line should pass through the origin and have a slope of exactly 1.00. The three lowest graphs of Fig. 2.4 indicate that this is not the case. Each resin gives a slightly curved graph, quite distinct from the other two.

Reichenberg and Wall explain this discrepancy by assuming that the fixed and exchangeable ions of the internal solution exert a salting-out effect on the acetic acid, driving a part of it from the internal to the external solution. They further assume that the salting-out effect is due to the hydration of hydrogen and sulphonate ions and that the water of hydration has no solvent power for other solutes. The data indicate that each sulphonate group holds four molecules of water. If this quantity of water is subtracted from the internal water to find the corrected values of the molarity of the internal acetic acid, the points for all three resins lie fairly well on one almost linear graph, whose slope is very close to 1.00 (Fig. 2.4).

Other nonelectrolytes, however, show wide differences from 1.00 in their values of

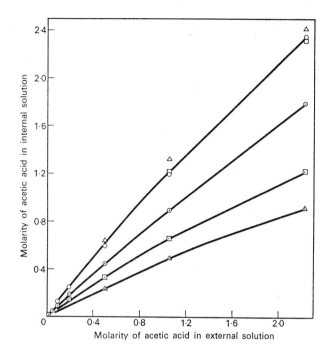

FIG. 2.4. Absorption of acetic acid by sulphonated polystyrene resins. ⊙ 5.5% DVB. ▫ 10% DVB. △ 15% DVB. ○ 5.5% DVB corrected for hydration of ions. □ 10% DVB corrected for hydration of ions. △ 15% DVB corrected for hydration of ions. (Used by permission of the *Journal of the Chemical Society*.)

K_a, i.e. the ratio of the internal to the external concentration. This is illustrated in Table 2.7. These values[28] pertain to an external concentration of about 50 g/l. Although they have not been corrected for hydration of the ions of the resin, it is obvious that such a correction will not cause all the values of K_a to take a value of 1.00.

Clearly there are other factors beside the salting-out effect that influence the absorption of nonelectrolytes by ion-exchange resins. The London (or van der Waals) forces of attraction[27] between the hydrocarbon part of the resin and the hydrocarbon part of the solute is

TABLE 2.7. VALUES OF K_a

Solute	Resin		
	Dowex 50-X8		Dowex 1-X7.5
	HR	NaR	RCl
Glycerol	0.49	0.56	1.12
Ethylene glycol	0.67	0.33	
Methanol	0.61		0.61
Acetone	1.20		1.08
Phenol	3.08		17.7

(Data used by permission of the New York Academy of Sciences.)

undoubtedly one of these. The fact that the values of K_a generally increase with the hydrophobic nature of the nonelectrolyte indicate that this is an important consideration. The dependence of K_a on the nature of the resin and of the exchangeable ion can be explained as due to an attraction[27] between the dipoles of the solutes and the ions of the resin.

In summary, the sorption of nonelectrolytes by ion-exchange resins depends on the following phenomena. (1) The resins absorb water, and the nonelectrolytes dissolve in the internal water. (2) London or van der Waals forces of attraction cause large absorption of the more hydrophobic solutes. (3) Attractions between the dipoles of the nonelectrolytes and the ions of the resin also increase the sorption. (4) The salting-out effect of these ions works in the opposite direction. There is no satisfactory equation for taking all of these factors into consideration. Nevertheless, the ratio of internal to external concentration of nonelectrolyte is approximately constant, at least over small concentration ranges, for any given solute and resin. It should also be mentioned that large molecules are excluded from highly crosslinked resins—or at least absorbed very slowly—because they can not penetrate through the narrow pores.

The sorption of nonelectrolytes by ion-exchange resins is markedly influenced by the exchange capacity of the resin. Table 2.8 gives data on the sorption of water[28 a] and phenol[29] by hydrogen-form resins of various capacities. All these resins contained 8% DVB and were uniformly sulphonated. The external solutions were 0.00032 M at equilibrium. The "weight-swelling", i.e. the grams of water absorbed by 1 g of dry resin is denoted by w. The partition ratio K_b denotes the quantity of solute absorbed per gram of dry resin divided by the quantity in one ml of solution.

$$K_a w = K_b \qquad\qquad (2.3)$$

TABLE 2.8. EFFECT OF EXCHANGE CAPACITY ON THE SORPTION OF
PHENOL

Capacity	w	K_b	K_a
5.2	1.09	5.7	5.2
3.02	0.85	32	38
2.03	0.56	39	70
0.76	0.32	28	87

(Data used by permission from *Zeitschrift für physikalische Chemie* (*Frankfurt*) and *Journal of Physical Chemistry*.)

Within the range of capacities listed, the ratio K_a of internal to external molarity rises steadily with decreasing capacity, probably because the phenol is much more strongly attracted by van der Waals forces to an unsulphonated benzene ring than to a sulphonated ring. As the capacity of the resin is decreased from the normal value of 5.2, the partition ratio K_a rises at first because of the increasing concentration of the phenol in the internal solution. A maximum is reached, and further decreases in capacity cause decreases in K_a because of the diminishing water content of the resin.

H.I. ABSORPTION OR ADSORPTION

The removal of molecules or ions from solution by ion-exchange resins is sometimes called *adsorption*. However, it can very easily be proved that this removal is not a surface phenomenon because grinding a resin does not increase either its ion-exchange capacity or its sorptive capacity (except toward very large molecules or ions). Therefore the term *adsorption* should be replaced by *absorption* or at least by the noncommittal term *sorption*.

I. Catalysis by Ion-exchange Resins

Many reactions that are catalyzed in aqueous solution by hydrogen or hydroxide ion are also catalyzed when the aqueous solution of the reagents is brought in contact with a strong-acid or a strong-base resin. This is not surprising since the reagents can diffuse into the resins and thus come in contact with the catalytically active hydrogen or hydroxide ion. Among the many examples, may be mentioned the hydrolysis of sucrose,[30] esters,[31] and amides.[32]

The catalytic effect of ion-exchange resins is not limited to reagents in aqueous solution. When derivatives of benzaldehyde and of benzyl cyanide are boiled in ethanolic solution with Amberlite IRA-410, the corresponding derivatives of phenylcinnamic nitrile, $C_6H_5CH=C(C_6H_5)CN$, are formed.[33] Octaacetylsucrose and the tetra-acetyl-derivatives of glucose and fructose can be prepared by treating a solution of the sugar in acetic anhydride with a strong-acid resin.[34]

Resins in forms other than hydrogen and hydroxide are occasionally used catalytically. The mercuric form of Dowex 50 causes the rapid hydration of the triple bond.[35]

The chief advantage of ion-exchange resins over solutions of catalytically active electrolytes is that they may be removed very easily from the reaction products. This may be accomplished by filtration or decantation, or the solution of the reagents may be passed through an ion-exchange column. Often the resins give less undesired side products, and they are useful in studying the mechanism of catalytic reactions. Their major limitations are the higher cost and their lack of stability toward high temperature and strong oxidizing agent.

Metals that catalyze reactions between gases, such as hydrogenation, are often supported in a finely divided state on ion-exchange resins. These catalysts are prepared by treating the appropriate metallic form of the cation-exchange resin with a suitable reducing agent.

J. Ion-exchange Resins as Desiccants

Boyd and Soldano[36] found that the sulphonated polystyrene resins have very low aqueous tensions at room temperature when their water content is small. This fact led Wymore[37] to investigate the use of these resins in drying liquids. He recommended that the liquids be passed through a column of potassium-form, Dowex 50-X8, 20-50 mesh. After the resin has absorbed too much water to function as a drying agent, it may be regenerated by heating between 115 and 140° in an oven or by passing air through the column at this temperature. In contrast, a molecular sieve (Chapter 10, p. 223) requires a temperature over 200°. For any given activity of water in the external phase, the lithium form absorbs more water than either the sodium or potassium form. Nevertheless, he recommended the use of the potassium form because this form absorbs the water most rapidly. This resin readily dries triethylene glycol, ethanol, and propanol-2 to less than 50 parts per million of water. It dries hydrophobic liquids such as hexane, carbon tetrachloride, and dichloromethane to 5 or less parts per million.

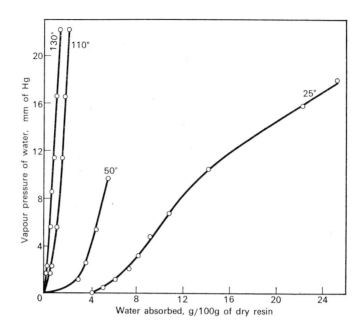

FIG. 2.5. Vapor pressure of water at equilibrium with Dowex 50-X8, KR, with various amounts of absorbed water. (Reproduced by permission from *Journal of Chemical Engineering Data*.)

Figure 2.5 presents data[33, 39] on the pressure of water vapor in equilibrium with the potassium form of Dowex 50-X8 containing various amounts of water. The figure indicates, for example, that 100 g of the resin contains 1.0 g of water after it is equilibrated at 110° with air containing enough water to yield 5.0 mm of vapor pressure. This resin can now be used to dry an organic liquid; the resin can absorb 19.0 g of water from the liquid before the water content of the liquid rises above the level corresponding to 15.0 mm pressure of water vapor.

References

1. G. E. BOYD, B. A. SOLDANO and O. D. BONNER, *J. Phys. Chem.*, **58**, 456 (1954).
2. E. GLUECKAUF and G. P. KITT, *Proc. Roy. Soc. (London)*, **A228**, 322 (1955).
3. J. A. MARINSKY and W. D. POTTER:
   (a) Microfilm USAEC, WAPD-C-188 (1953).
   (b) Microfilm AECU-3348 (1954).
   (c) G. R. HALL and M. STREAT, *J. Chem. Soc.*, 5205 (1963).
4. E. W. BAUMANN, *J. Chem. Eng. Data*, **5**, 376 (1960).
5. W. C. BAUMAN, J. R. SKIDMORE and R. H. OSMUN, *Ind. Eng. Chem.*, **40**, 1350 (1948).
5a. F. X. POLLIO, *Anal. Chem.*, **35**, 2165 (1963).
6. N. W. FRISCH and R. KUNIN, *Ind. Eng. Chem.*, **49**, 1365 (1957).
7. Y. MIZUTANI and S. SASAO, *J. Electrochem. Soc. Japan*, **29**, E199 (1961).
8. H. P. GREGOR, J. I. BREGMAN, F. GUTOFF, R. D. BRADLEY, D. E. BALDWIN and C. G. OVERBERGER, *J. Colloid Sci.*, **6**, 20 (1951).
9. R. KUNIN, *Anal. Chem.*, **21**, 87 (1949).
10. R. KUNIN and R. E. BARRY, *Ind. Eng. Chem.*, **41**, 1269 (1949).
11. J. I. BREGMAN and Y. MURATA, *J. Am. Chem. Soc.*, **74**, 1867 (1952).

12. R. KUNIN and F. X. McGARVEY, *Ind. Eng. Chem.*, **41**, 1265 (1949).
13. R. KUNIN, *Ion Exchange Resins*, Wiley, New York, 1958, p. 58.
14. G. E. MYERS and G. E. BOYD, *J. Phys. Chem.*, **60**, 521 (1956).
15. J. KIELLAND, *J. Am. Chem., Soc.*, **59**, 1675 (1937).
16. C. CALMON, *Anal. Chem.*, **24**, 1456 (1952).
17. I. GAMALINDA, L. SCHLOEMER, H. S. SHERRY and H. F. WALTON, *J. Phys. Chem.*, **71**, 1622 (1967).
18. G. E. BOYD, S. LINDENBAUM and G. E. MYERS, *J. Phys. Chem.*, **65**, 577 (1961).
19. K. W. PEPPER, D. REICHENBERG and D. K. HALE, *J. Chem. Soc.*, 3129 (1952).
20. K. A. KRAUS and G. E. MOORE, *J. Am. Chem. Soc.*, **75**, 1457 (1953).
21. J. MILLAR, *J. Chem. Soc.*, 1311 (1960).
22. M. G. SURYARAMAN and H. F. WALTON, *Science*, **131**, 829 (1960).
23. E. GLUECKAUF and R. E. WATTS, *Proc. Roy. Soc. (London)*, **A268**, 339 (1962).
24. W. RIEMAN and R. SARGENT, Ion exchange, in W. G. Berl, *Physical Methods in Chemical Analysis*, Academic Press, New York, 1961, vol. 4, p. 136.
25. G. D. MANALO, R. TURSE and W. RIEMAN, *Anal. Chim. Acta.*, **21**, 383 (1959).
26. H. P. GREGOR, J. BELLE and R. A. MARCUS, *J. Am. Chem. Soc.*, **76**, 1984 (1954).
27. D. REICHENBERG and W. F. WALL, *J. Chem. Soc.*, 3364 (1956).
28. R. M. WHEATON and W. C. BAUMAN, *Ann. N.Y. Acad. Sci.*, **57**, 159 (1953).
28a. G. D. MANALO, A. BREYER, J. SHERMA and W. RIEMAN, *J. Phys. Chem.*, **63**, 1511 (1959).
29. R. S. TURSE, W. H. GERDES and W. RIEMAN, *Z. Phys. Chem. (Frankfurt)*, **33**, 219 (1962).
30. G. BODAMER and R. KUNIN, *Ind. Eng. Chem.*, **43**, 1081 (1951).
31. S. A. BERNHARD and L. P. HAMMET, *J. Am. Chem. Soc.*, **75**, 1798 (1953).
32. P. D. BOLTON and T. HENSHALL, *J. Chem. Soc.*, 1226 (1962).
33. Z. CSUROS, G. DEAK and J. SZOLNOKI, *Acta Chim. Acad. Sci. Hung.*, **33**, 341 (1962).
34. J. M. CHRISTENSEN, *J. Org. Chem.*, **27**, 1442 (1962).
35. M. S. NEWMAN, *J. Am. Chem. Soc.*, **75**, 4740 (1953).
36. G. E. BOYD and B. A. SOLDANO, *Z. Elektrochem.*, **57**, 162 (1953).
37. C. E. WYMORE, *Ind. Eng. Chem., Product Research Develop.*, **1**, 173 (1962).
38. R. E. ANDERSON, *J. Chem. Eng. Data*, **8**, 32 (1963).
39. M. H. WAXMAN, doctor's thesis, Polytechnic Institute of Brooklyn (1952).

CHAPTER 3

ION-EXCHANGE EQUILIBRIUM

A. The Equilibrium Distribution

A.I. THE SELECTIVITY COEFFICENT

The exchange of ions between a solid ion-exchanging material and a solution is a typically reversible reaction. Suppose a solution containing the cations B^+ is shaken with a solid exchanger AR which contains the cations A^+. Ions B^+ enter the exchanger, while ions A^+ take their place in the solution. After a time, which may range from a few minutes to several days, depending primarily on the solid exchanger, no further change will be observed, and an equilibrium will have been established.

$$AR + B^+ \rightleftharpoons BR + A^+$$

If the ions B are doubly charged the equilibrium will be represented

$$2AR + B^{2+} \rightleftharpoons BR_2 + 2A^+$$

To represent the final distribution of concentrations we write a *selectivity coefficient*, which for exchanges between ions of equal charge has the form

$$E_A^B = \frac{[A^+][BR]}{[B^+][AR]} \quad \text{or} \quad \frac{[A^+][\bar{B}^+]}{[B^+][\bar{A}^+]} \tag{3.1a}$$

The symbols $[A^+]$ and $[B^+]$ indicate molar or molal concentrations in the solution. Equivalent concentrations may be used, if desired, to describe exchanges between ions of unequal charge.

The barred symbols refer to concentrations in the exchanger phase. Again these concentrations are represented in moles or equivalents per liter or kilogram. Square brackets conventionally indicate molar or molal concentrations in the solution within the exchanger (see p. 27). If the exchanging ions have equal charges it makes no difference what concentration units are chosen, since the units of E_A^B cancel. For exchanges between singly and doubly charged ions, however,

$$E_A^B = \frac{[A^+]^2[BR_2]}{[B^{2+}][AR]^2} \quad \text{or} \quad \frac{[A^+]^2[\bar{B}^+]}{[B^{2+}][\bar{A}^+]^2} \tag{3.1b}$$

Here the concentration units do not cancel and it is essential to specify the units used. Molal concentrations (moles of ions per 1000 g of water inside the exchanger) are hard to determine, as it is difficult to measure the water content of the exchanger. Molal or molar concentration units must nevertheless be used in discussing the Donnan equilibrium (p. 27). For evaluating selectivity coefficients it is more usual to express the internal concentrations

36

as *equivalent fractions* X_A and X_B. These are the numbers of gram-equivalents of the ions A and B per gram-equivalent of fixed ions in the exchanger. If the Donnan penetration of electrolyte can be neglected, $X_A + X_B = 1$. Another way to express ionic concentrations within the exchanger is as *gram-equivalents per kilogram of dry exchanger*. The ionic form of the dry exchanger must then be specified. For cation-exchange resins the H^+-form is commonly used; for anion-exchange resins, the Cl^--form.

An advantage[1] of relating the ionic concentrations to the quantity of *exchanger*, rather than to the solvent in the exchanger, is that the value of E_A^B is more nearly constant, being less dependent on X_A and X_B.

It is customary, though not mandatory, to write eqns (3.1) so that E_A^B is a number greater than unity. The ion B, that is to say, is considered to be more strongly held than ion A.

A.II. PARTITION RATIOS

The ratio of the concentration of an ion B in the exchanger to its concentration in the solution is called the "partition ratio". In many applications of analytical interest, such as elution chromatography (Chapter 6), the amount of one ion is small compared to the amount of the other. Let the ion whose concentration is small be B; then the partition ratio

$$K_a = \frac{[\bar{B}]}{[B]} \tag{3.2a}$$

(ionic charges are omitted for simplicity) is almost independent of the concentration of B and depends only on the concentration of A. This is evident from eqns. (3.1a) and (3.1b). If the ions A and B are of equal charge K_a depends inversely on the first power of the concentration of A in solution; if B is divalent and A univalent, K_a depends inversely on the square of the concentration of A.

We use the symbol K_a to mean a ratio of *solution concentrations*, as explained on p. 31 where the distribution of dissolved nonelectrolytes was discussed. To treat the chromatography of ionic species it is better to define the partition ratio with respect to the quantity of exchanger, thus

$$D = \frac{\text{amount of B in unit quantity of exchanger}}{\text{amount of B in unit quantity of solution}} \tag{3.2b}$$

The unit quantity of solution is generally the milliliter: the unit quantity of exchanger may be the milliequivalent of fixed ions, or it may be the gram of solvent-free exchanger. Thus the units of D are customarily ml/meq or ml/g. This matter will be discussed more fully in Chapter 6, where another ratio, C, which includes the ratio of the volumes of fixed and mobile phases will be introduced.

A.III. ELECTROSELECTIVITY

In exchanges of ions of equal charge the ratio between the concentrations of A and B does not change with dilution, aside from small effects due to non-ideality. If the ions are of unequal charge, two effects are noted. First, the ion of higher charge is usually more strongly held by the exchanger, and second, the distribution shifts with dilution. The more the solution is diluted, the more strongly the ion of higher charge is held by the resin, and vice versa. These effects, sometimes called *electro-selectivity*, have been used to good advantage in water-softening for many years. A dilute solution of calcium and magnesium salts ("hard" water) is passed through a column of exchanger containing sodium ions. The low concentration favors the absorption of calcium and magnesium ions by the exchanger. Then, in the regeneration step, a concentrated solution of sodium chloride is passed. The high concentration favors the removal of calcium and magnesium ions, which is just what is wanted. This principle is exploited in chemical analysis to separate ions according to their charge (see Chapter 8) and to concentrate trace constituents from dilute solutions.

B. Thermodynamics of Ion Exchange

B.I. THE EQUILIBRIUM CONSTANT

A proper representation of the ion-exchange equilibrium must recognize that neither the external solution nor that in the exchanger is ideal in the thermodynamic sense. The partial molal free energies are not linear functions of the logarithms of concentrations, and this is especially true of the exchanger phase, where the ions are much closer together than they are in the external solution. We can take cognizance of nonideality by introducing activity coefficients, as follows:

$$\mathbf{K}_A^B = \frac{m_A \bar{m}_B}{m_B \bar{m}_A} \frac{\gamma_A \bar{\gamma}_B}{\gamma_B \bar{\gamma}_A}, \tag{3.3}$$

where m and γ mean molality and activity coefficient. For simplicity the charges on ions A and B are considered to be equal, and the charge superscripts are omitted.

The quantity $\mathbf{K}_A^B$ is the *thermodynamic exchange constant*. For any given pair of exchanging ions, temperature and solvent, it is a true constant. Instead of the internal molality $\bar{m}$ one may, and usually does, use X, the equivalent fraction of the ion in the exchanger. The symbol $\bar{\gamma}$ is still used to represent the activity coefficient, and the numerical values of γ reflect the standard state chosen for the ionic species in the exchanger, as will be discussed below.

A comparison of eqns (3.1) and (3.3) shows that

$$\mathbf{K}_A^B = E_A^B \frac{\gamma_A}{\gamma_B} \frac{\bar{\gamma}_B}{\bar{\gamma}_A}. \tag{3.4}$$

Another quantity, the *equilibrium coefficient*, K_A^B, is defined by the equation

$$K_A^B = E_A^B \frac{\gamma_A}{\gamma_B}. \tag{3.5}$$

This is not a true constant, but depends on the ionic ratio in the exchanger. It is much easier to evaluate than $\mathbf{K}_A^B$, however, and graphs of K_A^B against X_B show the variations of activity coefficients within the exchanger.

Equations (3.4) and (3.5) are written for pairs of ions of equal charge. Corresponding equations for ions of unequal charge may easily be written.

Data for the calculation of the selectivity coefficient, E_A^B, are readily obtained by equilibrating known quantities of a solution of BCl (or other salt of B) of known concentration with known quantities of AR and then determining the concentrations of either A^+ or B^+

in either of the two phases. It is, of course, preferable to establish a "material balance" by the determination of both concentrations in each phase.

The determination of the equilibrium coefficient K_A^B requires, according to eqn. (3.5), the evaluation of the ratio γ_A/γ_B. Activity coefficients of individual ionic species have no thermodynamic significance, but their ratios do have meaning. If the co-ion, that is, the ion in solution which has the same charge sign as the exchanger matrix (for example, the chloride ion if cations are exchanged in a solution of chloride salts), is represented by X

$$\frac{\gamma_A}{\gamma_B} = \frac{(\gamma_\pm)_{AX}^2}{(\gamma_\pm)_{BX}^2} \tag{3.6}$$

where $(\gamma_\pm)_{AX}$ is the experimentally measurable mean activity coefficient of the salt AX. For a first approximation, one may use the value of $\gamma_\pm$ for a solution of the pure salt AX having the same concentration as the equilibrium solution obtained in the ion-exchange reaction, but for greater accuracy the activity coefficients should be measured in the actual binary salt mixture. Often this is very difficult to do, and activity coefficients in mixed salt solutions are usually estimated with the help of Harned's rule:[2]

$$\log \gamma_{AX} = \log \gamma_{(0)AX} + a_{AB}\mu_{AX} = \log \gamma_{(0)BX} - a_{AB}\mu_{BX}, \tag{3.7}$$

where γ is the mean activity coefficient [the same as $\gamma_\pm$ in eqn. (3.6)]; $\gamma_{(0)AX}$ is the coefficient of AX at zero concentration of AX in the presence of BX; μ_{AX}, μ_{BX} are the ionic strengths of AX and BX; a_{AB} is a constant, and the total ionic strength of the solution is considered to be constant.

For dilute solutions of salts of singly charged ions the ratio γ_A/γ_B is close to unity and can easily be estimated. If ions A and B have different charges, however, this is not so; for uni-divalent exchanges γ_A^2/γ_B is about 0.5 for ionic strengths of 0.5 to 1, and it cannot be ignored. Formally one can account for solution activities by using the coefficient K_A^B defined in eqn. 3.5.

To evaluate the ratio γ_A/γ_B which is needed for determining K_A^B is not easy. The first problem is the choice of standard reference states. If these are defined as ideal one-molal solutions of AX and BX in water (or whatever solvent is being used), then the quantity K_A^B is ipso facto unity. It is more practical to consider as reference states for the exchanger phase the pure, homoionic exchangers AR and BR in equilibrium with the pure solvent. To evaluate K_A^B from experimental data it is then necessary to know the amount and activity of the solvent in the exchanger at different ionic compositions, for the exchange of ions is always accompanied by movement of solvent into or out of the exchanger. It is also necessary to consider the presence of co-ions in the resin.

A rigorous thermodynamic analysis along these lines was made by Gaines and Thomas.[3] An approximate form of their equations, which serves well for exchanges in dilute aqueous solutions, is

$$\ln K_A^B = (z_B - z_A) + (n_2 - n_1) z_A z_B \ln \frac{P}{P_0} + \int_0^1 \ln K_A^B dN_B \tag{3.8}$$

where z_B, z_A are the charges on ions B and A; n_2, n_1 are the numbers of moles of solvent per equivalent of BR and AR, respectively, in their standard states; P is the vapor pressure of the solutions of BX and AX (assumed to be equal), P_0 is the vapor pressure of pure solvent; N_B is the equivalent fraction of cation B in the exchanger, that is the number of equivalents of B per equivalent of exchanger matrix.

In eqn. (3.8) the middle term is generally insignificant, and the first term is zero for exchanges of ions of equal charge. The first term enters because Gaines and Thomas chose to express solution concentrations in molar units, considering the standard free energy change for the conversion of an ideal one-molar solution of AX to an ideal one-molar solution of BX.

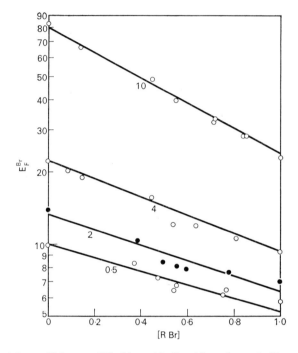

FIG. 3.1. Selectivity coefficients at 25° of bromide–fluoride exchange in Dowex 2 at crosslinkings indicated. (Reproduced by permission from *Journal of Physical Chemistry*.)

The selectivity coefficient for the exchange between fluoride and bromide ions[4] in anion-exchange resins is shown in Fig. 3.1. If similar plots of other uni-univalent anion-exchange reactions are examined, the following generalizations will be found to hold for almost all cases. (1) The selectivity coefficient is not constant, but depends strongly on the degree of crosslinking and on the fractions X_A and X_B of the ions in the resin; this shows that the activity coefficient ratio, $\bar{\gamma}_B/\bar{\gamma}_A$, depends on these variables. (2) The greater the crosslinking, the higher is E. (3) If B is the ion preferred by the resin, E_A^B decreases as X_B increases.

Figure 3.2 is a similar graph for cation-exchange resins.[5] The ordinate is the equilibrium coefficient, but in dilute aqueous solutions and for uni-univalent exchanges there is little difference between E_A^B and K_A^B. The generalizations made in the last paragraph apply to uni-univalent cation exchanges also, though more exceptions are found in cation exchange. One exception noted in Fig. 3.2 is that K_H^{Na} for the resin with 5.5% divinylbenzene *increased* slightly as X_{Na} from 0.10 to 0.46. Furthermore the second generalization fails when X_{Na} exceeds 0.67.

Figure 3.3 shows selectivity coefficients for a typical uni-divalent exchange, that of

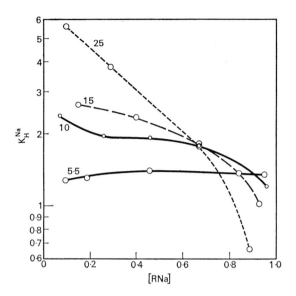

FIG. 3.2. Equilibrium coefficients of sodium–hydrogen exchange at 25° in sulphonated polystyrene resins of crosslinkings indicated. (Reproduced by permission from *Journal of the Chemical Society*.)

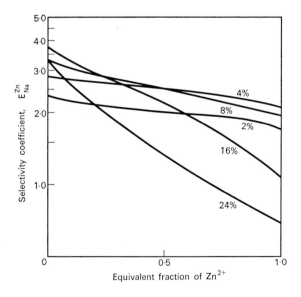

FIG. 3.3. Selectivity coefficients for $Z_n{}^{2+}$ — Na^+ exchange in Dowex 50 of different crosslinkings. (Reproduced by permission from *Journal of Physical Chemistry*.)

Zn^{2+} for $2Na^+$ in cation-exchange resins of different crosslinking.[6] There is more crossing of the graphs, and generalization (2) only holds for very small zinc loadings; generalization (3), however, appears to hold in all cases.

To compare equilibrium data for different pairs of ions and different exchangers it is not enough to quote one value of E for each exchange. One may, if he wishes, compare values of E for "zero loading", that is for $X_B = 0$; these data are useful for elution chromatography. Or one may use the thermodynamic constants $\mathbf{K}$ obtained from eqn. (3.8). Often only the last term in eqn. (3.8) is used.[7, 8]

One next asks whether it is possible to predict the constants $\mathbf{K}$ and the course of the curves shown in Figs. 3.1, 3.2, and 3.3. No theory has yet succeeded in predicting ion-exchange selectivities, though attempts have been made which will be discussed below. The variation of E with X_B and with the water content of the exchanger has, however, been explained in thermodynamic terms by Boyd et al.[4, 9] for resinous exchangers of the sulphonated polystyrene and quaternary-amine-substituted polystyrene types and for simple singly charged ions. They start by making a set of equilibrium measurements of E for the pair of ions under discussion (e.g. Na^+ and Li^+) for various ionic ratios with a resin of very low crosslinking, say 0.5% divinylbenzene. Then they take a series of resins of higher crosslinking (4, 8, and 12%, and so on) containing different proportions of the two cations, and for each resin and each cation composition they measure the water uptake over a range of partial pressures of water vapor up to saturation. It is also necessary to measure the equivalent volume of the dry resinates.

The free energy of ion exchange in an elastic material like a resin consists of two parts—a chemical part and a mechanical part. If the resin expands during exchange, because the entering ions are larger or more hydrated, or for any other reason, mechanical work is done. This contributes to the free energy and, therefore, to the equilibrium constant. From the Gibbs–Donnan theory of membrane equilibrium we write

$$\ln K_A^B = \ln \frac{\bar{\gamma}_A}{\bar{\gamma}_B} + \frac{P}{RT}(\bar{v}_{AR} - \bar{v}_{BR}), \tag{3.9}$$

where $\bar{v}_{AR}$, $\bar{v}_{BR}$ are the partial equivalent volumes of the swollen resinates at the particular resin composition considered, and P is the swelling pressure of the resin. Now P, $\bar{v}_{AR}$, and $\bar{v}_{BR}$ can be calculated from the experimental data, and it should be noted that this term in eqn. (3.9) accounts for roughly one-tenth of value of $\ln K$ for an 8% crosslinked resin. Thus the term $\ln(\bar{\gamma}_A/\bar{\gamma}_B)$ is much the more important.

The values of $\bar{\gamma}$ in eqn. (3.9) are not the same as those in eqn. (3.5), and this is the reason that the pressure–volume term appears in one equation and not in the other. They are referred to a different standard state: a hypothetical "infinitely dilute resin" having zero swelling pressure and no selectivity with respect to an infinitely dilute solution. The nearest practical approach to this is a resin of very weak crosslinking (0.5%), and this already shows some selectivity, even though its swelling pressure can be neglected. This selectivity must be found experimentally. Thus Myers and Boyd divide the term $\ln(\bar{\gamma}_A/\bar{\gamma}_B)$ into two terms, $\ln(\bar{\gamma}_A^*/\bar{\gamma}_B^*)$ to take account of this experimentally observed selectivity, and another term which can be calculated from the measurements of water activity. The final equation reads:

$$\ln K_c = \ln \frac{\gamma_A^*}{\gamma_B^*} + 55.51 \int_0^{\ln a_W} Y d\ln a_W + \frac{P}{RT}(\bar{v}_{AR} - \bar{v}_{BR}), \tag{3.10}$$

where

$$Y = \left[\frac{\partial(1/m)}{\partial X_{\text{B}}}\right] a_w = -\frac{1}{55.51}\left[\frac{\partial \ln(\bar{\gamma}_{\text{A}}/\bar{\gamma}_{\text{B}})}{\partial \ln a_w}\right],$$

m is the resin molality, X_{B} the equivalent fraction of ion B, and a_w the activity of the water.

In deriving these equations, Myers and Boyd use the cross-differentiation relations for ternary systems discussed by McKay.[10] The significance of this work is that it shows the importance of the water content of the resin as a factor modifying ion-exchange selectivity. Whatever affects the water content, be it crosslinking, ionic composition of the resin, or concentration of the external solution, will thereby affect the selectivity; changes in the ratio $\bar{\gamma}_{\text{A}}/\bar{\gamma}_{\text{B}}$ are determined by the water content alone. Correlation of data on resins of the same chemical type, but different crosslinking, is thus possible. In general, the smaller the water content of the resin, the greater its selectivity.

B.II. ENTHALPY AND ENTROPY

Enthalpy changes in ion exchange usually do not exceed 2–3 kcal per gram-equivalent and are often much smaller. This means that temperature does not have a great effect on ion-exchange equilibria and that close control of temperature in analytical applications is generally not needed. When ion exchange is coupled with other reactions in solution, however, the effect of temperature on these reactions must be considered. Metals are separated with the aid of complex-ion formation, and amino acids enter into acid–base reactions with the buffered eluting solutions. These reactions are affected by temperature, and in cases like these, careful regulation of temperature is needed to obtain efficient separations.

The entropy term $T\Delta S$ is often as great as, or greater than, the enthalpy term ΔH for ion-exchange reactions. This is particularly true in exchanges between univalent and divalent cations.[1, 6] The major entropy effect seems to be due to changes in hydration of the ions in the external solutions. Divalent ions are more strongly hydrated than univalent ions having the same radius, and hydration causes an orientation of water molecules with accompanying decrease in entropy. When a divalent ion leaves the water phase to enter the exchanger (where its hydration is much less) and two univalent ions are released to take its place, there is in general a gain in entropy which favors the exchange process. The enthalpy change is also positive, as a rule, and this opposes the exchange; the free energy change, which determines the preferred direction of exchange, is the resultant of the two.

C. Ionic Selectivity

C.I. EXPERIMENTAL DATA

We come now to the most important and most perplexing question concerning ion-exchange equilibrium: how can we predict ionic selectivity? The discussion in the preceding section did not answer this question; it merely related the selectivities of exchangers having the same chemical type. The "ideal" exchangers with nearly zero crosslinking show selectivity, and this is especially true of the anion exchangers.

Figures 3.4 and 3.5 summarize the selectivities of sulphonated polystyrene resins for singly charged cations and of quaternary ammonium polystyrene resins for singly charged anions. We notice immediately that the differences in binding strength for halide ions

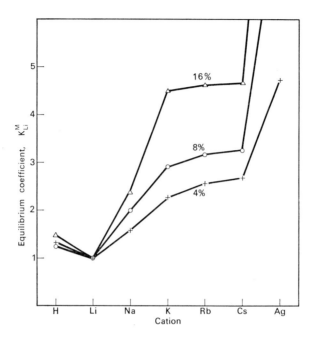

FIG. 3.4. Equilibrium coefficients for univalent cations in Dowex 50 of crosslinkings indicated. (Reproduced by permission from *Journal of Chemical Education*.)

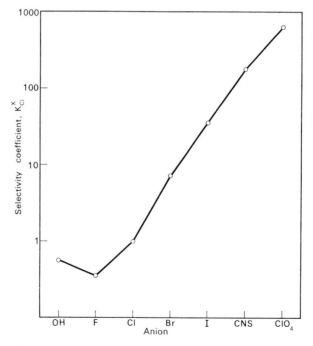

FIG. 3.5. Equilibrium coefficients for univalent anions in Amberlite IRA-400. (Chloride ion chosen as reference. Note logarithmic scale.) (Data used by permission from *Journal of Chemical Education*.)

on anion exchangers are far greater than the differences for alkali-metal ions on cation exchangers. This fact is of great importance to the experimenter, as anyone who has tried to remove iodide, or even chloride ions from a column of strong-base anion-exchange resin by passing sodium hydroxide solution will know. Large selectivity coefficients in ion exchange are a mixed blessing. On the other hand, the differences in binding of alkali metal ions by the cation-exchange resins are not great, and the separation of cesium from rubidium, for example, on a column of resin of this type would be tedious and nearly impossible.

The selectivity order for alkali-metal cations on sulphonated resins follows the well-known lyotropic series; the most strongly hydrated ion, Li^+, is held the most weakly, and the least hydrated ion, Cs^+, is held most strongly. With resins whose fixed ions are carboxyl ions, however, the selectivity order is just the reverse; Li^+ is held most strongly and Cs^+ the least,[11, 12] though the selectivity order is influenced by the degree of neutralization of the —COOH groups. Resins with fixed phosphonate groups show an inversion of selectivity with changing pH. At low pH values, where the ionic group is —PO_3H^-, Cs^+ is strongly preferred over Rb^+ and the other ions, but at high pH, where presumably the ionic group is PO_3^{2-}, Li^+ is preferred and the sequence is reversed.[13] A similar behavior is found with zirconium phosphate exchangers, and it is used in the atomic-power industry to separate the long-lived fission product, cesium-137. This element is preferentially absorbed with a very high degree of selectivity at low pH.

With divalent cations of inert-gas structure (Be^{2+}, Mg^{2+}, Ca^{2+}, Sr^{2+}, Ba^{2+}, Ra^{2+}) similar effects are observed. On sulphonate type exchangers, Ba^{2+} is held most strongly, and Mg^{2+} most weakly, of the common alkaline-earth cations. An exaggerated selectivity for radium ions is shown by inorganic exchangers of the zirconium phosphate type at low pH.

Other divalent ions are held strongly:[14, 15] see Table 3.1. It is of doubtful propriety to

TABLE 3.1. SELECTIVITY DATA FOR DIVALENT IONS

Cation	Selectivity data for crosslinking of		
	4%	8%	16%
UO_2	2.36	2.45	3.34
Mg	2.95	3.29	3.57
Zn	3.13	3.47	3.78
Co	3.23	3.74	3.81
Cu	3.29	3.85	4.46
Cd	3.37	3.88	4.95
Mn	3.42	4.09	4.91
Be	3.43	3.99	6.23
Ni	3.45	3.93	4.06
Ca	4.15	5.16	7.27
Sr	4.70	6.51	10.1
Pb	6.56	9.91	18.0
Ba	7.47	11.5	20.8

The resin was the polystyrene sulphonate, Dowex 50.
Selectivity data are referred to the lithium ion, are corrected for solution activities (eqn. (3.5)), and are for *one gram-equivalent* of ions, that is, they correspond to the *square root* of eqn. (3.1b). Solution concentrations are in moles per liter; resin concentrations are in equivalent fractions; temperature, 25°. (Used by permission from *Journal of Physical Chemistry*.)

compare the binding of ions having different charges, because the exchange equilibria shift with dilution; but as a rough rule one can say that in cation exchange, the higher the charge on an ion, the more strongly it is absorbed from solutions of 0.1 M concentration and below. Lhis generalization has been used as a basis for group separations of cations according to their charge.[16] With anions, such a generalization is harder, in part because most anions are derived from weak acids.

Special selectivity effects are shown by chelating resins, such as Dowex A-1. Data are

TABLE 3.2. SELECTIVITY COEFFICIENTS FOR THE CHELATING RESIN
DOWEX A-1
(R. Rosset, *Bull. Soc. Chim. France*, **1966**, 59)

Cation	Log E	Cation	Log E
Hg	4.18	Zn	1.29
UO_2	2.69	Co	1.18
Cu	2.65	Mn	0.69
VO	2.30	Ca	0.00
Pb	2.15	Ba	−0.17
Ni	1.67	Sr	−0.21
Cd	1.41	Mg	−0.32

E is the selectivity coefficient defined in eqn. (3.1a).

Rosset compared successive pairs of ions, e.g. Hg with UO_2, UO_2 with Cu, etc. In this table all ions are referred to Ca as standard.

(Used by permission of the Société chimique de France.)

shown in Table 3.2. Metal ions that form strong complexes with iminodiacetate ions in solution are also bound strongly by the resins, but no quantitative parallel can be drawn,[17] and the precise condition of coordination in such resins is still in doubt. Nevertheless, the great attraction of these resins for divalent ions can be put to good use in analysis, as, for example, in the removal of traces of calcium and manganese from caustic soda.[18]

From the practical standpoint one must recognize that ion exchange intrinsically is not a very selective process when compared with the action of many organic precipitants and complex-formers. It can be made much more selective, however, when it is coupled with the action of complex-formers in the solution. Addition of a complex-forming agent may work to take metal ions *off* the exchanger if the complexes formed are uncharged or have the same charge sign as the functional groups of the exchanger. If the complexes have charges opposite to the exchanger groups, however, the complexing agent will help to put the metal ions *into* the exchanger. The great majority of applications of ion exchange to the separation of metals use selective complex formation as an adjunct to ion exchange, the most striking example being the anion-exchange separation of metal-chloride complexes (see Chapter 8). One must never forget, when working with solutions of salts of metals other than the alkali and alkaline-earth metals, that complexes in solution are the rule rather than the exception. Lead(II) and mercury(II), for example, will behave very differently in chloride and nitrate solutions, and metals in high oxidation states, such as zirconium(IV), will form hydrolytic complexes unless a complexing agent stronger than water, such as F^-, is present. Some of these hydrolytic complexes have high molecular weights and cannot enter the pores of the exchanger.

C.II. THEORETICAL TREATMENT

It is evident that many factors influence ionic selectivity, and that a quantitative theory can cover only the simplest cases. The exchange of the alkali-metal cations is the obvious place to start. A convincing explanation of the selectivity of cation exchangers for the alkali metals and hydrogen has been given by Eisenman.[19]

Eisenman started by investigating the response of glass electrodes to various alkali-metal cations. Glasses act as ion exchangers, and glass electrodes function as ion-exchanging membranes. This has been shown by many investigators and in particular by the recent work of Doremus,[20] who measured diffusion coefficients of ions in glasses. Electrical potentials are easier to measure than ion-exchange distributions, but membrane potentials depend on two factors—the ion-exchange selectivity and the ratio of diffusion coefficients or mobilities. By measuring potentials of glass electrodes in solutions containing two ions, such as Na^+ and K^+ (in addition to the hydrogen ion, which is always present in aqueous solutions) an "electrochemical selectivity factor" is found *which is chiefly dependent on the ion-exchange selectivity*. The mobility ratio is in general only a tenth of the ion-exchange selectivity ratio. Eisenman investigated a great many glasses of widely different chemical compositions as well as a number of biological membranes. He found that *once the selectivity for a given pair of ions, for example K^+ and Na^+, was measured*, the selectivity *for any other pair*, for example K^+ and Cs^+, could be predicted with fair accuracy. The order of selectivity was not always the same, but the *number of different selectivity orders found was far smaller than the number statistically possible*. For the five cations, Li^+, Na^+, K^+, Rb^+, Cs^+, there are $(5!) = 120$ possible sequences if the order is random. Only 10 or 11 sequences were actually found. For a soda–lime glass containing no more than a trace of aluminum, selectivity is greatest for H^+, then decreases in the order $Na^+ > Li^+ > K^+ > Rb^+ > Cs^+$. A lithium–barium silicate glass gave an order $Li^+ > Na^+ > K^+ > Rb^+ > Cs^+$. For glasses high in aluminum the selectivity order was reversed: cesium ions affected the potential the most, and lithium ions the least: $Cs^+ > Rb^+ > K^+ > Na^+ > Li^+$. This is the order of selectivity for sulphonated polystyrene cation exchangers.

In Eisenman's work the selectivity factor for *one* exchange (Na^+—K^+ was used for reference) fixed the selectivity factors for all the other exchanges and hence the *selectivity sequence*. (There are some deviations; the points on the graphs show some scatter, but the statistical picture is very persuasive, including, as it does, collodion and biological membranes as well as glasses.) The inference is drawn that selectivity is determined by one basic cause. This is identified as the *electrostatic field strength at the fixed negatively charged site*. A silicate "site" in glass has a relatively concentrated charge and a high field strength compared to an aluminosilicate "site", where the excess negative charge is smeared over a greater volume:

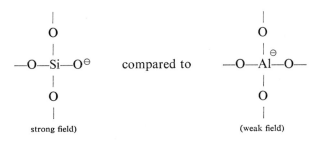

These may be likened to functional groups in ion-exchange resins:

$$-C \underset{O^{\ominus}}{\overset{O}{\diagdown}} \qquad \text{compared to} \qquad -S\overset{O}{\underset{O}{\updownarrow}}-O^{\ominus}$$

(strong field)   (weak field)

or, indeed, to simple halide ions:

$$F^- \qquad \text{compared to} \qquad I^-$$

(strong field)   (weak field)

As the simplest possible model for the ion-exchange process we may use the exchange of ions between an aqueous solution and solid, crystalline halide salts:

$$Na^+F^- \text{ (solid)} + K^+ \text{ (aqueous, 1 M)}$$

$$\rightarrow K^+F^- \text{ (solid)} + Na^+ \text{ (aqueous, 1 M)}$$

The standard free energy for such a reaction can be calculated from readily available data. A summary of the calculated free energies appears in Fig. 3.6, with Cs^+ as reference, i.e., the ordinates show $\Delta F°$, first, for $CsF(\text{solid}) + M^+(\text{aq.}) \rightarrow MF(\text{solid}) + Cs^+(\text{aq.})$, then for $CsCl(\text{solid}) + M^+(\text{aq.})$, and so on. The fluoride lattices prefer Li^+ over the other four cations; the iodide lattices prefer Cs^+; and in between, if the anion radius could be continuously varied along the straight lines shown in the graph, there would be exactly eleven different orders or sequences of selectivity. This is just what Eisenman observed in his glasses.

The halide crystal model for an ion exchanger is absurdly simple, yet essentially the same selectivity orders are predicted if the exchanger is likened to a *saturated solution* of the halides. One can also calculate *a priori* the energies of association of cations of any desired radius with various assemblies of negatively charged ions, and again one comes to the same conclusion, using the *bare, crystal-lattice radii of the positive ions*. Essentially, the exchange of alkali-metal cations presents a problem in electrostatics; there is competition between the fixed charges and the water dipoles. Fixed ionic sites with *high* field strengths win the contest and attract cations away from the water dipoles. The smallest cations, Li^+, are attracted the most strongly. Fixed ionic sites with *low* field strengths will lose the contest; the smallest cations are now attracted to the water dipoles, leaving the largest, Cs^+, to be bound selectively by the exchanger.

To apply this idea quantitatively one must be able to calculate the energies and entropies of hydration of the ions, which is virtually impossible *a priori*. These quantities may, however, be evaluated from experimental data.

It is now easy to see why different selectivity orders are found with different types of organic exchangers. Those with sulphonate ions follow the lyotropic series with Cs^+ the most strongly absorbed; those with carboxylate and phosphonate give other sequences, usually preferring Na^+ or Li^+ over the other alkali-metal ions, and the sequence depends on the degree of neutralization, i.e. upon the spacing of ionic sites. With phosphonate ions the selectivity sequence depends on whether the ions are singly or doubly charged; the reason for this is easily seen.

We have not discussed the relatively strong binding of the singly charged ions, Ag^+ and

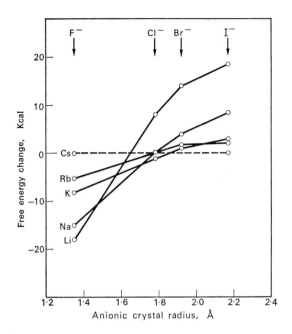

Fig. 3.6. Standard free energies for the hypothetical reactions $M^+(aq) + CsX(solid) \rightarrow Cs(aq) + MX(solid)$. The abscissae are the radii of the anions X^-. If X^- is the fluoride ion, LiF is the preferred solid; see text. (Reproduced by permission from *Biophysical Journal*.)

Tl^+. In the case of Ag^+ it seems that non-Coulombic forces (polarization of the Ag^+ ion) play an important part in enhancing the strength of binding to the exchanger. This effect is also evident in the large crystal-lattice energy of the silver halides and in the stability of silver-halide complex ions.

Divalent ions of the alkaline-earth series show increasing binding with increasing radius for the sulphonic acid exchangers, as would be expected from Eisenman's theory. In applying the theory, however, it appears that the spacing of the anionic sites is very important. This is an area that needs further investigation. As to divalent ions of the transition and post-transition metals, these are bound more strongly than would be expected from their radii alone, and, as we have already noted, it seems that nonelectrostatic forces play an important part. The state of solvation of such ions depends on the anionic field strength of the exchanger (see Chapter 11).

Turning now to the selectivity order of *anions*, the binding of halide and hydroxyl ions by strong-base anion exchangers follows the expected selectivity order, but the magnitude of the selectivity is much greater than the simple model would predict. Non-Coulombic forces evidently play a part, for the selectivity is very marked even with weakly crosslinked exchangers.[4] The anions cannot simply be treated as rigid spheres. Many phenomena, from those of the electrical double layer to the stability of metal-halide complex ions, give evidence of the deformability of chloride, bromide, and especially iodide ions. Nevertheless, deformation or polarization does not explain why perchlorate ions are held more strongly in ion exchange than iodide ions, or permanganate much more strongly than chromate.[21, 22] The question of anionic selectivity has been discussed with special reference to oxygen-containing anions by Chu et al.[23] These authors note that those anions are most strongly absorbed which are largest, have the smallest charge, and are the weakest bases. Thus ClO_4^- is more strongly held than I^- or ClO_3^-, MnO_4^- than $(CH_3)_3CCOO^-$. They attribute these differences to the interaction of the ions with the *water* outside the resin. The larger the ion, the more it disrupts the water structure, and this increases the free energy of the systems and destabilizes it. This effect would, however, be offset by high charge, which attracts the water dipoles, and by basic character, which also attracts water dipoles by hydrogen bonding and hydrolysis.

If an ion is very large it will not be able to enter the pores or framework of the ion exchanger. A resinous exchanger is not like a rigid wire screen, however; the crosslinked hydrocarbon matrix is flexible, and a large ion may be able to enter the resin if sufficient time is allowed for diffusion. In discussing the "screening" effect of a resin for large ions, therefore, one must be very careful to distinguish equilibrium effects from rate effects.

The foregoing discussion of theories of ionic selectivity is not exhaustive; there have been many other approaches to the problem, of which the ion-association models of Pauley[24] and of Rice and Harris[25] are among the most noteworthy. Future developments are likely to take account of different kinds of "cavities" with different degrees of solvent organization within the exchanger, and the inorganic exchangers, with their relatively well-defined and rigid frameworks, lend themselves well to such studies.[26] Obviously ion-exchange selectivity is a very complex phenomenon.

A comprehensive treatment of ion-exchange selectivity has recently been given by Reichenberg.[27]

C.III. RESINS AS NONAQUEOUS SOLVENTS

The attraction between ion exchangers and simple, nonpolarizable ions can be interpreted fairly well by the principles of electrostatics, and it is the ionic sites of the exchanger, their field strengths and their spatial arrangement, that determine selectivity orders. The molecular framework, or matrix, that links the fixed ions together is of minor consequence.

With large organic ions this is no longer the case. A resinous exchanger with a polystyrene matrix will strongly bind organic ions with aromatic rings. The strong binding of benzylammonium cations, $C_6H_5CH_2NH_3^+$, by sulphonated polystyrene resins is particularly noteworthy,[28] as is the binding of toluene-sulphonate ions, $CH_3C_6H_4SO_3^-$, by anion-exchange resins based on polystyrene. Cations like those of 1,10-phenanthroline are bound so strongly by sulphonated polystyrene resins that it is practically impossible to remove them. One can attribute these effects to interactions between π-electrons. If the resin does not have aromatic rings, the binding of aromatic ions is much weaker.

Comparing aliphatic ions, the cations of amines, or the anions of carboxylic acids with one another one finds that the longer the chain length and the more hydrocarbon-like the ion is, the more strongly it is bound, provided, of course, it is not so big as to be excluded for steric reasons. One can attribute this effect to exclusion from the water phase, following Chu et al.[23] just as logically as one can attribute it to the solvent action of the resin; however, in ligand exchange (Chapter 8) the polystyrene-base resins are much better "solvents" for long-chain amines than a carboxylic (polymethacrylic) resin with an aliphatic matrix. The clearest evidence of "solvent" action of polystyrene-base ion-exchange resins is, of course, their behavior in solubilization and salting-out chromatography (Chapter 9).

Another area where ion-exchange resins behave very much like nonaqueous solvents is in the absorption of metal-chloride complexes from aqueous hydrochloric acid solutions by strong-base anion-exchange resins. This phenomenon, which is of great importance in chemical analysis, will be discussed at length in Chapter 8. Iron(III), for example, is extracted from 8 to 9 M aqueous hydrochloric acid by a strong-base polystyrene resin with a distribution coefficient of 10^5. Now this very powerful absorption cannot be attributed to the stability of the complex anion $FeCl_4^-$, for this complex is actually very unstable in water. This anion may be stabilized within the resin because of its repulsion by the water, as suggested by Diamond et al.,[23] but it would seem, rather, that the resin was exerting a positive solvent effect, comparable to the extraction of the ion pair $H^+ FeCl_4^-$ by ethers, ketones, esters, and other oxygen-containing solvents. The effect of hydrochloric acid concentration is qualitatively similar in both phenomena. One can also point to the very strong binding of iron and other trivalent metals by cation-exchange (sulphonated polystyrene) resins at HCl concentrations of 6 M and above. This is a fact which the analytical chemist should remember when he wants to remove iron(III) from such resins; 2 M acid is much more effective than 6 M acid. Whitney and Diamond[29] explain this effect by the loss of water from the hydrated $Fe(OH_2)_6^{3+}$ cation at high acid concentrations, since $HClO_4$ shows the effect as well as HCl.

D. Experimental Methods

D.I. SHAKING

The measurement of distribution coefficients and selectivity coefficients is in principle very easy. One takes a weighed quantity of solid ion exchanger containing a known amount of exchangeable ions, places it in a bottle, adds a measured volume of solution containing a known amount of the other ionic species to be exchanged, and one shakes until equilibrium is reached (within the experimental limits of measurement), then one withdraws a measured volume of solution and analyzes it.

The points to be watched in this apparently very simple procedure include the following.

D.I.a. *Initial State of the Exchanger*

The moisture content at the time of weighing must not be allowed to change during a series of measurements. It need not be accurately known so long as the ionic content of the exchanger is known. If the exchanger ions are hydrogen ions, they are determined by titration of a separate sample preferably potentiometrically. This will reveal the presence of any weak-acid groups in a strong-acid exchanger. If the cation or anion exchanger is *not* a simple strong-acid or strong-base material, one must recognize that hydrogen or hydroxyl ions will remain in the exchanger except at extreme pH values, and that the pH of the solution will affect the distribution of the other ions.

D.I.b. *Approach to Equilibrium*

At least one experiment should be made in which samples of solution are withdrawn at different times so as to follow the progress toward equilibrium. As a rule no change will be observed after an hour or two of shaking, but weak-acid or weak-base resins, chelating resins, and many inorganic exchangers will react much more slowly (see Chapter 4). It is seldom that an exchanger is truly monofunctional, and if there is a small proportion of slowly reacting groups the last few per cent of the exchange may be quite slow. It is well, therefore, to test the attainment of equilibrium by approaching the distribution from two sides. A way to do this is to add to the bottle, after "equilibration", some of the ions that have just been removed from the exchanger, shake again, and measure the new distribution. Thus if a metal salt has been added to a hydrogen-form resin, one withdraws some solution for analysis, then adds some *acid* and re-equilibrates, withdrawing metal ions from the resin. If equilibrium is being attained, one should obtain a selectivity coefficient in this second experiment which is consistent with those obtained by direct reaction of resin with metal salt.

D.I.c. *Temperature*

Ion-exchange distributions are not greatly shifted by temperature unless complex-ion equilibria in solution are taking part. Temperature control $\pm 1°$–$2°$ is usually quite sufficient.

D.I.d. *Phase Separation*

With ion-exchange resins the boundary between the exchanger and the solution is well-defined and it is easy to withdraw solution for analysis without entraining any resin particles. A coarse glass-wool plug on the end of a pipette may be used if necessary, or solutions may be centrifuged if there is any possibility of fine suspended particles.

If the *resin* is to be analyzed (and this is always desirable) the phase separation is more difficult. One cannot simply filter off the resin and wash it with water; the washing causes a shift in the equilibrium particularly in exchanges between ions of unlike charge (p. 37). A good technique is to suck as much as possible of the solution away from the resin with a filter stick, then weigh the container, filter stick, and wet resin, find the weight of adherent solution, then measure the ionic contents of the resin and adherent solution combined. Then one has to know the amounts of water and electrolyte imbibed by the resin, and this is another problem. If the ambient solution is sufficiently dilute, however, the correction for the ion content of the adherent solution will be small.

D.I.e. *Donnan Exclusion*

When the air-dried resin is placed in the solution at the start of the experiment, it imbibes water and swells. If the solution is sufficiently dilute the amount of salt entering the resin with the water is negligibly small. Thus the electrolyte concentration of the solution *rises* when the resin and solution are mixed. Whether this effect is significant or not depends on the volume ratio of resin to solution.

D.I.f. *Analytical Errors*

It is tempting, especially if a radioactive tracer is used, to measure the concentration of one component only, and this only in the solution. All other concentrations are then found by difference. This may be a satisfactory procedure if the proportion of the measured component is small, but not when it is large. Especially in uni-divalent exchanges where the concentration of univalent ions is found by difference and is then squared, experimental errors can mount up rapidly. In such cases the concentration of *both* components should be measured.

D.II. TRACER-PULSE AND CONCENTRATION-PULSE METHODS

These methods for measuring selectivity coefficients were suggested by Helfferich and Peterson.[30] They depend on measurements of volumes of solutions passed through chromatographic columns, and it will be convenient to postpone a detailed discussion until Chapter 6, where the theory of chromatographic columns is treated. Ordinarily one uses ion-exchange equilibrium data to predict chromatographic elution volumes; the pulse methods do the reverse, and use elution volumes to evaluate equilibrium data. An advantage of this approach is that one does not have to confirm the attainment of equilibrium as noted in section D.I.b. above. Another advantage is that the methods lend themselves to automatic recording.

The tracer-pulse method gives the distribution ratio D (section A.II): the concentration-pulse method gives the rate of change of D with the proportions of exchanging ions. Thus the two methods complement one another. This matter will be explained more fully in Chapter 6.

References

1. I. Gamalinda, L. A. Schloemer, H. S. Sherry and H. F. Walton, *J. Phys. Chem.*, **71**, 1622 (1967).
2. H. S. Harned and B. B. Owen, *The Physical Chemistry of Electrolytic Solutions*, 3rd ed., p. 600, Reinhold, New York, 1958.
3. G. L. Gaines and H. C. Thomas, *J. Chem. Phys.*, **21**, 714 (1953).
4. G. E. Boyd, S. Lindenbaum and G. E. Myers, *J. Phys. Chem.*, **65**, 577 (1961).
5. D. Reichenberg and D. J. McCauley, *J. Chem. Soc.*, 2741 (1955).
6. G. E. Boyd, F. Vaslow and S. Lindenbaum, *J. Phys. Chem.*, **71**, 2216 (1967).
7. W. J. Argersinger, A. W. Davidson and O. D. Bonner, *Trans. Kansas Acad. Sci.*, **53**, 404 (1950).
8. E. H. Cruickshank and P. Meares, *Trans. Faraday Soc.*, **53**, 1289, 1299 (1957); *ibid.*, **54**, 174 (1958).
9. G. E. Myers and G. E. Boyd, *J. Phys. Chem.*, **60**, 521 (1956).
10. H. A. C. McKay, *Trans. Faraday Soc.*, **49**, 237 (1953).
11. H. P. Gregor, M. J. Hamilton, R. J. Oza and F. Bernstein, *J. Phys. Chem.*, **60**, 266 (1956).
12. C. E. Marshall and G. Garcia, *J. Phys. Chem.*, **63**, 1663 (1959).
13. H. Ti-Tien, *J. Phys. Chem.*, **68**, 1021 (1964).
14. O. D. Bonner and L. L. Smith, *J. Phys. Chem.*, **61**, 326 (1957); H. F. Walton, *J. Chem. Educ.*, **42**, 111 (1965).
15. H. F. Walton, D. E. Jordan, S. R. Samedy and W. N. McKay, *J. Phys. Chem.*, **65**, 1477 (1961).
16. J. S. Fritz and S. K. Karraker, *Anal. Chem.*, **31**, 921 (1959).
17. R. Rosset, *Bull. Soc. Chim. France*, 1964, 1845; *ibid.*, 1966, 59
18. A. J. Van der Reyden and R. L. M. Van Lingen, *Z. Anal. Chem.*, **187**, 241 (1962).
19. G. Eisenman, The electrochemistry of cation-sensitive glass electrodes, in *Advances in Analytical Chemistry and Instrumentation* Vol. 4 (ed. by C. N. Reilley), Wiley–Interscience, New York, 1965.
20. R. H. Doremus, *J. Phys. Chem.*, **68**, 2212 (1964).
21. J. Aveston, D. A. Everest and R. A. Wells, *J. Chem. Soc.*, 1958, p. 231.
22. H. P. Gregor, J. Belle and R. A. Marcus, *J. Am. Chem. Soc.*, **77**, 1713 (1955).
23. B. Chu, D. C. Whitney and R. M. Diamond, *J. Inorg. Nucl. Chem.*, **24**, 1405 (1962).
24. J. L. Pauley, *J. Am. Chem. Soc.*, **76**, 1422 (1954).
25. S. A. Rice and F. E. Harris, *Z. Phys. Chem.*, NF, **8**, 207 (1956).
26. H. S. Sherry, *J. Phys. Chem.*, **70**, 1158 (1966); H. S. Sherry and H. F. Walton, *ibid.*, **71**, 1457 (1967).
27. D. Reichenberg, chap. 7 of *Advances in Ion Exchange*, Vol. I (ed. by J. Marinsky), Marcel Dekker, Inc., New York, 1966.
28. S. R. Watkins and H. F. Walton, *Anal. Chem. Acta.*, **24**, 334 (1961).
29. D. C. Whitney and R. M. Diamond, *J. Phys. Chem.*, **68**, 1886 (1964).
30. F. Helfferich and D. L. Peterson, *Science*, **142**, 661 (1963).

CHAPTER 4

ION-EXCHANGE KINETICS

A. The Rate-controlling Step

An ion-exchange process that is ordinarily represented by a simple equation such as

$$LiR + Na^+ \rightarrow NaR + Li^+$$

actually occurs in five steps:[1]

(1) The sodium ion must reach the surface of the resin bead. Agitation of the solution or the passage of the solution through an ion-exchange column helps to bring the sodium ion to the interface. Nevertheless, regardless of how great the agitation may be, there is always a film of unstirred solution surrounding the resin bead; at least, the system behaves as if such a film existed. Therefore the sodium ion must diffuse through this film if the exchange is to occur. The thickness of the film is of the order of 10^{-3} or 10^{-2} cm, depending on the rate of stirring; of course, rapid stirring produces thin films.

(2) After the exchange has been in progress a short time, the lithium ions are depleted from the exchange sites at the surface of the resin. Therefore, the sodium must diffuse through the resin to an exchange site.

(3) The actual chemical exchange must occur.

(4) The liberated lithium ion must diffuse to the surface of the resin.

(5) The lithium ion must then diffuse through the stationary film into the bulk of the solution.

However, the principle of electroneutrality requires that steps (1) and (5) occur simultaneously at equal rates; i.e. for every sodium ion that diffuses through the stationary film toward the resin, a lithium ion must diffuse in the opposite direction. Analogously, diffusion of sodium and lithium ions inside the resin must occur at equal rates in opposite directions. Thus the number of steps is in reality only three: film diffusion, bead diffusion, and chemical exchange. The slowest of these governs the overall rate of exchange.

The rate-controlling step depends primarily on the concentration of the external solution. Under ordinary conditions, i.e. moderate stirring and a moderately or highly swollen resin, diffusion through the film is the slow step with solutions of the order of 0.01 N or less.[1] Diffusion in the resin is ordinarily the slow step with solutions of the order of 0.1 N or more.[1] No validated example has been found of an exchange controlled by the chemical reaction. One such report[2] has not been confirmed by later work.[2a]

B. Experimental Methods

B.I. SHALLOW-BED METHOD

A thin layer of resin, usually less than 1 cm deep, is supported in a small tube on a sintered-glass disc or a platinum screen.[1] It is very important to use resin particles of uniform size. A rapid stream

54

of the exchanging ion is passed through this bed. After an accurately noted time interval, the resin bed is rinsed with water. Analysis of the resin then indicates the extent to which the exchange occurred in the given interval. Under these conditions, the composition of the solution is virtually constant.

If the exchange is governed by bead diffusion in the shallow-bed method, the rate equation is rather complicated.

$$F = \frac{Q_t}{Q_\infty} = 1 - \frac{6}{\pi} \sum_{n=1}^{n=\infty} \frac{1}{n^2} \exp\left(\frac{-\bar{D}\pi^2 n^2 t}{r^2}\right). \tag{4.1}$$

Q_t is the extent of the exchange reaction after t seconds. Q_∞ is the extent of the exchange reaction after an infinite time interval. F is simply the ratio Q_t/Q_∞. The series of integers 1, 2, 3, 4, – – – is represented by n. $\bar{D}$ is the diffusion coefficient (cm²/sec) inside the resin, and r is the radius of the resin beads in centimeters. If B is defined as

$$B = \bar{D}\pi^2/r^2 \tag{4.2}$$

it follows that

$$F = 1 - \frac{6}{\pi} \sum_{n=0}^{n=\infty} \frac{1}{n^2} \exp\left(-Btn^2\right). \tag{4.3}$$

Thus F can be calculated for any value of Bt, and tables of F as a function of Bt are given in the literature.[1, 3] From experimentally determined values of F, values of Bt are found in these tables; then Bt is plotted against t. Throughout any given ion-exchange process, $\bar{D}$, r, and hence B are nearly constant. Therefore a linear plot of Bt vs. t is evidence that diffusion inside the resin is the slow step.

Figure 4.1 shows the plot of Bt vs. t for the exchange at 30° of [24]NaR with [23]Na+ when the total normality of electrolyte was 0.11. The resin contained both sulphonate and phenolic hydroxy groups. The linearity of the graph indicates that bead diffusion was the slow step. The slope of the graph is equal to B and has the value 0.0972. By substituting this and the value of the mean radius (0.0178) in eqn. (4.2) we find that $\bar{D} = 3.1 \times 10^{-6}$.

The value of the diffusion coefficient of sodium ion in aqueous solution[4] at 30° is 15×10^{-6}. The explanation of this large difference is that the diffusion of sodium ion through the resin is impeded both by the hydrocarbon part of the resin and by the electrostatic attraction between it and the sulphonate ions.

The rate equation[1] for exchange controlled by film diffusion in the shallow-bed method is

$$1 - F = \exp\left(\frac{-3Dt}{r\kappa\Delta r}\right), \tag{4.4}$$

where D is the diffusion coefficient in the films, κ is the equilibrium concentration in the resin of the exchanging ion (originally in the solution) divided by the concentration of the same species in the solution, and Δr is the thickness of the film. A plot of $-\log(1-F)$ vs. t should be linear under these conditions. The slope is $3D/2.30r\kappa\Delta r$.

Figure 4.2 shows a plot of $-\log(1-F)$ vs. t for an exchange[1] similar to the one shown in Fig. 4.1 except that the total concentration of the solution was only 0.001 M. Within the experimental error the points lie on straight lines, indicating that the slow step in the process was film diffusion.

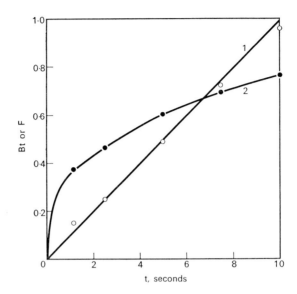

FIG. 4.1. Graph of an exchange controlled by bead diffusion in the shallow-bed method.[1] Concentration = 0.11 M. Graph (1): Bt vs. t. Graph (2): F vs. t. (Data used by permission from *Journal of American Chemical Society*.)

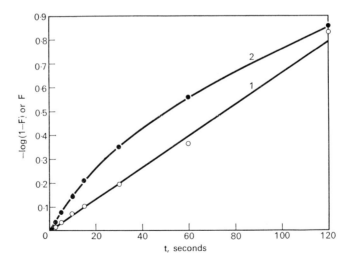

FIG. 4.2. Graph of an exchange controlled by film diffusion in the shallow-bed method.[1] Concentration = 0.001 M. Graph (1): $-\log(1-F)$ vs. t. Graph (2): F vs. t. (Data used by permission from *Journal of American Chemical Society*.)

B.II. LIMITED-BATH METHOD

A known quantity of resin (q_0 meq) is added suddenly to a solution containing an amount of electrolyte exactly equivalent to the added resin. The mixture is stirred for the desired time interval Then the resin is quickly removed from the solution.[5] Subsequent analysis of either the resin or the solution furnishes the data for the calculation of the extent of the exchange.

The equation for an exchange controlled by bead diffusion is

$$F = \frac{6Q_0}{r(Q_0 - Q_\infty)} \sqrt{\frac{\bar{D}t}{\pi}}. \tag{4.5}$$

Thus a plot of F vs. $\sqrt{t}$ is linear under these conditions, and $\bar{D}$ can be calculated from the observed slope.

If the exchange is controlled by film diffusion, a plot of $-\log(1 - F)$ vs. t is linear, as in the case of the shallow-bed method.

B.III. INDICATOR METHOD

Hale and Reichenberg[6, 3] studied the exchange between hydrogen-form sulphonic resins and sodium ion as follows. They prepared a solution containing a small known amount of sodium hydroxide, a much larger known amount of sodium chloride, and a little bromo-cresol green. At zero time they added to this solution a known quantity of hydrogen-form resin. They provided continuous uniform stirring. The quantity of resin (meq) was always greater than that of the sodium hydroxide but less—usually much less—than that of the sodium chloride. When the exchange had progressed to the point of liberating a quantity of hydrogen ion equivalent to the amount of sodium hydroxide taken, the indicator changed color suddenly. The experimenters noted this time with a stop-watch. They performed other experiments in any one series with the same amount of resin, the same total amount of sodium ion, the same volume of solution, but with different amounts of sodium hydroxide.

The apparatus required for this method is much less elaborate than that for the other two. On the other hand, this method is limited to those reactions for which a suitable indicator can be found.

If the sodium chloride is taken in large excess over the resin, as is usually the case, the resin is almost entirely in the sodium form when equilibrium is reached. Then Q_∞ can be taken as the quantity of resin. Q_t is the quantity of sodium hydroxide taken. Because of the large excess of sodium, the rate equations are the same as for the shallow-bed method.

In applying eqn. (4.2) to the exchange between hydrogen and sodium ions, the question is sure to arise whether $\bar{D}$ is the diffusion coefficient in the resin of the sodium ion or of the hydrogen ion. In fact, in all exchanges except those between two isotopes of the same element, $\bar{D}$ is the *interdiffusion coefficient* and has values intermediate between the $\bar{D}$ values of the two ions concerned.

Plots of Bt vs. t for the exchange between hydrogen-form resins and sufficiently concentrated solutions of sodium ion are linear within the experimental error (Fig. 4.3), indicating that bead diffusion controls the rate and that B (and hence $\bar{D}$ and r are essentially constant for any given set of conditions. Nevertheless, it has been proved both theoretically[7, 8] and experimentally[9, 10] that $\bar{D}$ depends on the composition of the resin and hence changes appreciably as the exchange progresses under any given set of conditions. If the exchanging ions

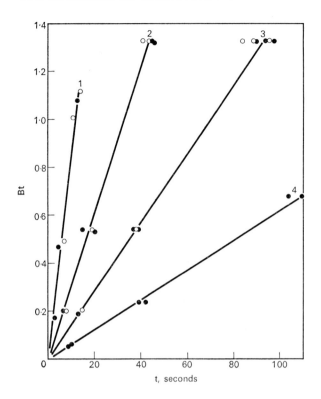

FIG. 4.3. Graphs of the exchange of hydrogen-form sulphonated polystyrene resins with sodium ion. ○: 0.91 M Na⁺. ●: 1.82 M Na⁺. (See Table 4.1.) (Reproduced by permission from *Journal of American Chemical Society*.)

have different diffusion coefficients, an electrical potential gradient is produced within the resin bead, and this influences the velocity of both ions.

C. Conditions that Influence the Rate

C.I. PARTICLE SIZE

Small particle size favors rapid exchange regardless of which method is used to study the rate and regardless of whether film diffusion or bead diffusion is the slow step. If the rate is controlled by diffusion through the film, the smaller particles react more quickly because they have greater specific surface and hence more diffusion occurs per unit time per unit quantity of resin. If bead diffusion is the slow step, the finer resin again reacts more rapidly because the exchanging ions have a shorter average distance to diffuse through the small particles.

A comparison of eqns. (4.1) and (4.4) indicates that particle size is more important when the slow step is bead diffusion rather than film diffusion because r appears in these equations to the second and first powers, respectively.

Graphs 3 and 4 in Fig. 4.3 and the corresponding data[6] in Table 4.1 illustrate the effect of particle size. The rate (proportional to the slope B) is greater for the smaller particle size, but $\bar{D}$, calculated by eqn. (4.2), is the same within the experimental error for both particle sizes (Table 4.1).

TABLE 4.1. KINETIC DATA OBTAINED BY THE INDICATOR METHOD WITH SULPHONATED POLYSTYRENES FOR THE EXCHANGE HR + Na⁺ → NaR + H⁺

Graph no. in Fig. 4.3	DVB (%)	r (cm)	Temp.	B	$\bar{D} \times 10^6$	Half-life* (sec)
1	5	0.0272	25°	0.082	6.1	3.7
2	17	0.0273	50°	0.29	2.2	10.4
3	17	0.0273	25°	0.0143	1.08	21.0
4	17	0.0446	25°	0.0016	1.23	49

* The time required for the conversion of one-half of the resin from the hydrogen to the sodium form. (Data used by permission from *Journal of American Chemical Society*.)

C.II. DIFFUSION COEFFICIENT INSIDE THE RESIN

The bead-diffusion coefficient $\bar{D}$ influences the rate of exchange only if the concentration of the solution is large enough so that diffusion inside the resin is the slow step. This diffusion coefficient depends on the following factors.

C.II.a. *The Degree of Swelling of the Resin*

Graphs 1 and 3 of Fig. 4.3 represent experiments done under essentially identical conditions except for the percentage of DVB in the polystyrene resin. The resin with 5% crosslinks was much more swollen than the one with 17% and had a value of $\bar{D}$ almost 6 times as great and consequently an exchange rate almost 6 times as great.

Ion-exchange reactions in nonaqueous media, particularly in nonpolar solvents, are very slow, sometimes only 1/1000 as fast as those in water. One reason is that the resins are very slightly swollen in these liquids. Another reason is that the resins are less dissociated in nonaqueous media, i.e. a larger fraction of the exchangeable ions are bound as ion pairs with the matrix ions and are thus not able to diffuse. Similarly, weak-acid resins in the free-acid form and weak-base resins in the free-base form are very slightly swollen and perform exchange reactions very slowly.[10a]

C.II.b. *Temperature*

Graphs 2 and 3 of Fig. 4.3 show the importance of temperature. A rise in temperature from 25° to 50° doubles the diffusion coefficient and hence the exchange rate.

C.II.c. *Nature of the Exchanging Ions*

Two aspects of the exchanging ions are important—the *valence* and the *size* of the hydrated ions. The diffusing ions are subject to coulombic attraction by the fixed ions of the resin. The greater the valence of the ion the greater is this retarding force. This phenomenon is illustrated by the data on the self-diffusion coefficients[11] (determined with radioactive tracers) of sodium, zinc, and yttrium ions at 25° in a sulphonated polystyrene with 10% DVB. The values are respectively 2.76×10^{-7}, 2.89×10^{-8}, and 3.18×10^{-9}.

Ionic size is important because the large ions have more difficulty than small ones in diffusing through the crosslinked hydrocarbon chains of the resin matrix. Table 4.2 shows the interdiffusion coefficients[5] of the indicated ions, paired with ammonium ion in each case in a phenolsulphonate resin. These substituted ammonium ions are essentially unhydrated in aqueous solution.

TABLE 4.2. INTERDIFFUSION COEFFICIENT
WITH AMMONIUM ION

Cation	$\bar{D} \times 10^9$
NMe_4^+	24
NEt_4^+	5
$NMe_3n\text{-}Amyl_2^+$	3
$PhNMe_2Et^+$	1
$PhNMe_2CH_2Ph^+$	0.06

(Data used by permission of the Faraday Society.)

C.III. DIFFUSION COEFFICIENT IN THE AQUEOUS FILM

The coefficient of diffusion in the film affects the exchange rate only at low concentrations of the aqueous solution. Values of some of these diffusion coefficients are given[4] in Table 4.3. The effect of the radius of the hydrated ion is seen in the values for Cs^+, K^+, Na^+, and Li^+, but this effect is less than in bead diffusion. Valence has only a minor influence. These diffusion coefficients increase about 2% per degree, a smaller temperature effect than in bead diffusion.

TABLE 4.3. DIFFUSION COEFFICIENTS IN AQUEOUS SOLUTION

Cation	$D \times 10^6$	Anion	$D \times 10^6$
H^+	93.4	OH^-	52.3
Cs^+	21.1	Cl^-	20.3
Tl^+	20.0	NO_3^-	19.2
K^+	19.8	BrO_3^-	14.4
Na^+	13.5	$C_2H_3O_2^-$, IO_3^-	10.9
Li^+	10.4	SO_4^{2-}	10.8
Pb^{2+}	9.8	CrO_4^{2-}	10.7
Cd^{2+}, Zn^{2+}, Cu^{2+}	7.2	$Fe(CN)_6^{3-}$	8.9
Ni^{2+}	6.9	$Fe(CN)_6^{4-}$	7.4

(Used by permission of Interscience Publishers.)

C.IV. STIRRING

According to eqn. (4.4) the rate of an exchange controlled by film diffusion increases as the thickness of the stationary film is decreased. This thickness, in turn, depends on the agitation of the mixture. In the limited-bath and indicator methods, the solution is stirred. In the shallow-bed method, the flow of the solution between the resin particle serves as the method of agitation. With increasing rate of stirring, the rate of exchange is increased up to a limit beyond which further increases in the rate of stirring have no effect on the rate of exchange. Table 4.4 illustrates this phenomenon with data[5] for the exchange between an ammonium-form resin and sodium ion under conditions where film diffusion is the slow step.

TABLE 4.4. EFFECT OF STIRRING ON RATE
OF EXCHANGE

Rate of stirring (rev/min)	Half-life (min)
470	5.9
660	3.8
750	2.9
860	1.2
990	1.0
1100	1.0

(Data used by permission of the Faraday Society.)

It seems likely that any system where the exchange is controlled by bead diffusion can be changed to one controlled by film diffusion simply by a sufficient decrease in the rate of agitation.

C.V. CONCENTRATION OF THE SOLUTION

An exchange between a resin and a 0.001 N solution is generally controlled by film diffusion. If the concentration is increased in a series of kinetic experiments where all the other conditions are kept unchanged, the rate of exchange increases linearly with concentration. However, a concentration is reached, usually about 0.01 N, where the increases in rate with further increases in concentration are less than linear. In this region, both film diffusion and bead diffusion play roles in determining the rate. With continuing increases in concentration, the rate reaches a limit. This is the region where bead diffusion is the rate-controlling factor. It is seen in Fig. 4.3 that 0.91 M and 1.82 M sodium ion gave the same rate within the experimental error.

The situation may be compared to a very simple electrical circuit consisting of a constant e.m.f., an ammeter, a constant resistor, and a variable resistor in series. If the conductance of the variable resistor is much less than that of the constant resistor, the current is directly proportional to the former. When the conductance of the variable resistor approaches that of the fixed instrument, the proportionality no longer obtains. When the variable conductance is much greater than the constant conductance, the current is constant in spite of variations in the former.

References

1. G. E. BOYD, A. W. ADAMSON and L. S. MYERS, Jr., *J. Am. Chem. Soc.*, **69**, 2836 (1947).
2. R. TURSE and W. RIEMAN, *J. Phys. Chem.*, **65**, 1821 (1961).
2a. A. VARON and W. RIEMAN, *J. Phys. Chem.*, **68**, 2716 (1964).

3. D. REICHENBERG, *J. Am. Chem. Soc.*, **75**, 589 (1953).
4. I. M. KOLTHOFF and J. J. LINGANE, *Polarography*, 2nd ed., vol. 1, p. 52, Interscience, London, 1952.
5. T. R. E. KRESSMAN and J. A. KITCHENER, *Discussions Faraday Soc.*, **7**, 90 (1949).
6. D. K. HALE and D. REICHENBERG, *Discussions Faraday Soc.*, **7**, 79 (1949).
7. F. HELFFERICH and M. S. PLESSET, *J. Chem. Phys.*, **28**, 418 (1958).
8. M. S. PLESSET, F. HELFFERICH and J. N. FRANKLIN, *J. Chem. Phys.*, **29**, 1064 (1958).
9. F. HELFFERICH, *J. Phys. Chem.*, **66**, 39 (1962).
10. F. HELFFERICH, *J. Phys. Chem.*, **67**, 1157 (1963).
10a. D. E. CONWAY, T. H. G. GREEN and D. REICHENBERG, *Trans. Faraday Soc.*, **50**, 511 (1954).
11. B. A. SOLDANO, *Ann. N.Y. Acad. Sci.*, **57**, 116 (1953).

CHAPTER 5

NONCHROMATOGRAPHIC APPLICATIONS

THIS chapter is devoted to relatively simple separations and to other applications of ion exchange to analytical chemistry. Most of these separations are actually accomplished by the use of a column and therefore might properly be classified as *chromatographic*. However, the separations of this chapter are so easy that they could be accomplished by batchwise processes without an intolerable number of repetitions. In contrast, the separations discussed in the next five chapters would require so many steps in a batchwise process that chromatography is the only practical method of performing the separation. Therefore, for purposes of convenience, the relatively easy separations treated in this chapter are classified rather arbitrarily as *nonchromatographic*.

Ringborn[1] considers seven truly batchwise separations. In each case he states the quantities of the ions to be separated, the quantity of the resin and the selectivity coefficient. Then he gives the detailed calculations which reveal whether the separation is quantitative.

A. Preparation, Care, and Use of an Ion-exchange Column

A.I. THE TUBE

The quantity of resin, and hence the size of the tube in which it is to be contained, depend chiefly on the nature and quantities of the substances to be separated. For most of the separations discussed in this chapter, the dimensions of the column are given in the respective references. The container may be a simple glass tube tapered at the lower end. The resin rests on a plug of glass wool or a sintered-glass disk near the lower end. The tube may be provided, as in Fig. 5.1, with a reservoir at the upper end to contain the solution that is to pass through the resin.

The presence of air bubbles in the resin bed interferes with the uniform flow of liquids through the tube and hence with the efficient use of the resin. To avoid the entrance of air into the resin bed, the meniscus of the liquid should not be allowed to fall below the upper level of the resin. A pinch-clamp or stopcock is usually provided at the lower end of the tube to stop the flow of liquid when the meniscus is 1 mm or more above the resin. Alternatively, the column may have a delivery tube such as is illustrated in Fig. 5.1. The flow of liquid through this tube will stop when the meniscus reaches the level L of the outlet of the delivery tube D. If the upper level of resin is below this point, air can enter the bed only by the slow evaporation of water.

If air is accidentally admitted to the bed, it can sometimes be removed by adding liquid to cover the bed and stirring the resin thoroughly with a long glass rod. This is difficult with long, narrow columns and with resins of small particle size. Air bubbles can also be removed by backwashing (see p. 66).

A.II. THE RESIN

The separations discussed in this chapter are generally performed with polystyrene resins containing either cation-exchange or anion-exchange groups, as the occasion requires.

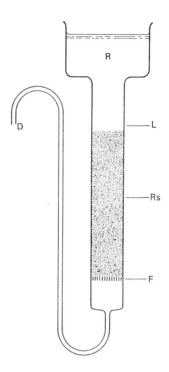

FIG. 5.1. A common type of ion-exchange column. *R*, reservoir; *Rs*, resin; *F*, sintered-glass filter disk; *D*, outlet; *L*, level of outlet.

Average crosslinking, about 8%, is satisfactory. Particle size is not critical, and 50–100 mesh is generally satisfactory.

A.II.a. *Removal of Fine Particles*

The commercially available resins frequently contain an appreciable quantity of particles very much smaller than the limits indicated on the label. These fine particles may pass through or clog the filter of the column and should therefore be removed before the resin is put into the tube. The removal is accomplished very simply by slurrying the resin in water, letting the mixture stand a few minutes until the large particles have settled, and then pouring off the turbid supernatant liquid. The process should be repeated a few times until all the particles settle quickly. The resin is then slurried again with water or an appropriate solution and poured into the tube to give a bed of the desired height.

A.II.b. *Conditioning*

Most commercial resins contain impurities both organic and inorganic, resulting from the manufacturing process. These should be removed by a conditioning procedure.

In the case of a sulphonated polystyrene, this consists of passing through the resin successively 1 M sodium hydroxide, water, 1 M hydrochloric acid, water, 95% ethanol, and water. The quantity of each liquid should be sufficient to displace nearly completely the previous solvent or cation. Three bed volumes suffice in all cases except the hydrochloric acid, where six bed volumes should be used because of the unfavorable value of E_H^{Na}. The cycle should be repeated about 3 times.

In the case of a carboxylic resin the same procedure may be used except that it is best to treat the resin as a thin bed on a Buchner funnel to avoid shattering a glass tube when the resin expands on being converted from the hydrogen to the sodium form. Three bed volumes of hydrochloric acid suffice.

The same sequence of liquids serves also for the conditioning of the strong-base anion exchangers. Since these resins are readily converted from the hydroxide to the chloride form, three bed volumes of hydrochloric acid suffice. On the other hand, larger volumes of 1 M sodium hydroxide are needed to convert these resins to the hydroxide form, 30 bed volumes for resins of type 1 and 6 for resins of type 2. It is best not to interrupt the conditioning when the resin is in the hydroxide form because of the relative instability in this form. Washing with alcohol is especially important with strong-base resins because the commercial grades contain organic impurities that seriously interfere with metal separations.

Weak-base anion exchangers may be conditioned by the same procedure. Three bed volumes of all liquids should suffice. Because of the large expansion that occurs when the resin is converted from the hydroxide to the chloride form, it is advisable to perform the conditioning on a Buchner funnel.

If a cation-exchange resin is to be used in either the sodium form or the hydrogen form or if an anion exchanger is to be used in the chloride or hydroxide form, the conditioning cycle can be stopped at such a point as to leave the resin in the desired form. If the conversion to this form must be quantitative, as is usually the case, the foregoing estimates of the required volume of the last electrolytic solution should not be trusted, but the passage of this solution through the column should be continued until negative qualitative tests are obtained for the ion being displaced from the column. If the resin is to be used in some other form, a suitable solution is passed through the column until qualitative tests on the effluent indicate completion of the exchange. Finally, about three bed volumes of water should be passed through the column to rinse out the last solution. The completeness of rinsing should be verified by a qualitative test. Rinsing a hydrogen-form resin with water, for example, should be continued until the effluent is basic to methyl orange.

If the column is to be used with a water-soluble organic solvent or with a mixture of such a solvent with water, this solvent or mixture should be used to rinse out the interstitial water and to bring the internal liquid of the resin into equilibrium with the external liquid. If the solvent to be used is insoluble in water (benzene, for example) or a mixture of such a solvent and a water-soluble solvent such as ethanol, the column must be rinsed first with the water-soluble solvent and finally with the nonpolar solvent or mixture. If a large volume change has occurred, the column should be backwashed again.

A.II.c. *Regeneration*

If, for example, a cation exchanger in the hydrogen form is used to remove metallic ions from solution, the resin sooner or later contains so many metal ions that they start to leak through the column into the effluent. Before this point is reached, it is generally desirable to regenerate the resin, i.e. to convert it again to its original form. This is usually accomplished simply by passing hydrochloric acid, 0.5 to 2 M, through the resin until qualitative tests reveal the absence of the metal ions in the effluent. In the case of cations that are tenaciously held by the resin, it is sometimes necessary to regenerate the resin by passing through it first a solution of a complexing anion, such as diammonium citrate, and then hydrochloric acid.

Of course, regeneration can be avoided by discarding a spent resin. However, this is an unnecessary expense. Furthermore, the preparation of a new column is generally more troublesome than the regeneration of a used one.

A.II.d. *Backwashing*

This consists in passing a stream of water upwards through the column at a velocity sufficient to break up the bed and form a suspension of the resin. A stopper with a tube leading to the sink is inserted in the upper end of the tube. The velocity of flow should not be sufficient to carry the resin particles out of the tube.

A.II.e. *Flow Rate*

The separations discussed in this chapter do not require rates of flow as slow as those necessary in Chapters 7, 8, and 9. It should be remembered, nevertheless, that a procedure that gives a satisfactory separation at the recommended flow rate may fail at a much higher rate. The appropriate references should be consulted to find the recommended flow rate for any particular procedure.

In many cases the column offers sufficient resistance to flow so that no special method of controlling the rate is necessary. In other cases, a partly closed stopcock or pinchclamp may be attached to the exit end of the column to control the velocity.

B. Preparation and Purification of Reagents
B.I. DEIONIZATION
B.I.a. *Deionization of Water*

Purified water is used in the laboratory in larger amounts than any other substance. In the first half of the present century, water was purified for laboratory use by distillation. Today the choice between distillation and deionization with ion exchangers depends on several factors that are discussed below.

The first deionization procedures consisted in passing the tap water first through a bed of hydrogen-form sulphonate or phenol-sulphonate resin, then through a bed of hydroxide-form anion exchanger. The effluent from the cation exchanger contained the acids corresponding to the salts in the original water. The extent of the removal of these acids by the anion-exchange resin depended on the basic strength of the latter. A strong-base resin would remove all acids very nearly completely; a weak-base resin would fail to remove very weak acids such as silicic, boric, and carbonic. If these acids are tolerable in the deionized water or if their salts are absent from the raw water, a weak-base resin is preferable because it is much easier and less expensive to regenerate it than a strong-base resin. This follows from the fact that the selectivity coefficient E_{Cl}^{OH} is large for a weak-base resin and small for a strong-base resin.

In the mixed-bed method of deionization, the raw water is passed through a mixture of hydrogen-form cation exchanger and hydroxide-form anion exchanger, which may be of either the strong-base or weak-base type. The advantage of the mixed-bed method is readily understood by considering the removal of sodium chloride. In the two-bed method, the absorption of sodium ion by the cation exchanger is not quite complete because of the reversibility of the exchange reaction

$$Na^+ + Cl^- + HR \rightleftharpoons NaR + H^+ + Cl^- \tag{5.a}$$

The sodium ion in the effluent from the cation exchanger and an equivalent amount of anion (mostly chloride) appear also in the effluent from the anion exchange. In the mixed-bed method, on the other hand, the removal of hydrogen and chloride ions by the anion exchanger

$$H^+ + Cl^- + ROH \rightarrow RCl + H_2O$$

drives reaction (5.a) to virtual completion.

The disadvantage of the mixed-bed method is that the regeneration of the resins is more difficult. The cation exchanger is usually regenerated with sulphuric acid, and the anion exchanger with sodium hydroxide (strong-base resin) or sodium carbonate (weak-base resin). Obviously, the resins must be separated from each other before they can be regenerated. Although it is possible to perform this separation by a flotation process, the cation exchanger being the denser, it is common practice to throw away the used cartridges of mixed resin in the ordinary laboratory installation.

In choosing between distillation and deionization, the most important considerations are the purity of the product and cost. Deionization, especially if the mixed-bed method and a strong-base anion exchanger are used, yields water of lower electrolyte content than the ordinary laboratory still but fails to remove nonelectrolytes.

The cost of fuel for the distillation of a liter of water is virtually independent of the salt content of the water. On the other hand, the cost of removal of one milliequivalent of salt by deionization is independent of the volume of water in which the salt was originally dissolved. Thus the cost of deionizing a liter of water is directly proportional to the normality of salt in the raw water. It follows that distillation is more economical for waters of large salt content and more expensive for waters of small salt content. When the raw water contains a large salt concentration and a product of very small concentration is desired, it is advantageous to apply distillation first and then deionization. A cartridge of mixed-bed resins serves to purify a very large volume of distilled water. Organic materials present in some water supplies of surface origin may coat the particles of resin and interfere with their action.

B.I.b. *Deionization of Nonelectrolytes*

Nonelectrolytes such as acetone and the lower alcohols can also be deionized by the same processes as used for water. These methods are applicable both to the nearly anhydrous organic liquids and to their aqueous solutions. In the former cases, the uptake of ions by the resins is slower than from aqueous solutions (p. 26). The first portions of effluent from the ion-exchange column may contain the nonelectrolyte in a different concentration from the influent because the resins may absorb the nonelectrolyte in preference to water. The chief electrolytic impurity in the lower primary alcohols and in aqueous formaldehyde is the acid formed by the slow atmospheric oxidation of these compounds. In many cases it suffices to remove the acid by passage through a column of anion exchanger in the free-base form. Since a strong base catalyzes the polymerization of formaldehyde, a weak-base resin should be used to remove acid from formaldehyde.[1a]

Ion-exchange resins are used extensively in the sugar industry to remove the electrolytes from the cane syrup. The methods used for the deionization of water, especially the usual two-bed method, have the disadvantage that the sucrose is subject to catalytic hydrolysis to glucose and fructose both by contact with the strong-acid resin and by the hydrogen ion in the effluent from this column. This difficulty is partly overcome by the "reverse deionization", in which the syrup is passed through the anion exchanger before the cation exchanger. A still better method employs a mixture of weak-acid and strong-base exchangers. The resins also remove most of the coloring matter present in the raw syrup.

B.I.c. *Deionization of Chromatographic Paper*

Filter paper often contains traces of soluble salts adsorbed so tenaciously that they cannot be removed by washing. These impurities often cause erratic behavior in paper chromatography. Schwartz and Pallansch[2] devised a method for removing these trace ions from the paper. They suspended a package of papers with a string in a stirred mixture of water, strong-acid cation exchanger and strong-base anion exchanger.

B.I.d. *Removal of Ordinary Electrolytes from Polyelectrolytes*

If a solution containing both ordinary electrolytes and polyelectrolytes is subjected to any of the foregoing deionization methods, a nearly complete separation is achieved. The polyelectrolytes are unable to penetrate into the particles of resin and hence are adsorbed only on the surface of the beads. The loss of polyelectrolyte is negligible if resins of high crosslinkage and large particle size are used. The ash content of albumin[3] was decreased from 4.28 to 0.0003% by this method.

B.II. MISCELLANEOUS ION-EXCHANGE METHODS FOR THE PREPARATION AND PURIFICATION OF REAGENTS

B.II.a. *Carbonate-free Sodium or Potassium Hydroxide*

It is well known that solutions of these bases are usually contaminated with the respective carbonate. Three methods have been proposed for the preparation of essentially carbonate-free base by means of ion-exchange resins.

In the first method,[4] an ordinary solution of approximately 0.1 M sodium hydroxide is passed through a 33·ml bed of chloride-form strong-base anion exchanger. The following reactions occur:

$$RCl + OH^- \rightleftharpoons ROH + Cl^- \tag{5.b}$$

$$2RCl + CO_3^{2-} \rightleftharpoons R_2CO_3 + 2Cl^- \tag{5.c}$$

Since the resin has a much greater affinity for chloride ion than for hydroxide ion, several liters of sodium hydroxide are required to displace the chloride ion from the column. The resin has a greater affinity for carbonate (at least in dilute solutions) than for hydroxide. Therefore the carbonate moves down the column more slowly than the hydroxide, and a portion of the effluent following the last of the chloride is 0.1 M sodium hydroxide free from carbonate. The column can be regenerated by the passage of hydrochloric acid and used repeatedly. This method is not recommended.

In the second method,[5] a concentrated solution of the sodium hydroxide is prepared. Sufficient barium hydroxide is added to precipitate all the carbonate. The solution is filtered with precautions against the absorption of carbon dioxide and diluted to the desired concentration with recently boiled water. It is then passed through a column of sulphonated polystyrene resin in the sodium form. Since barium has a large selectivity coefficient relative to sodium, it moves down the column slowly. Therefore a large volume of pure sodium hydroxide can be collected before barium ion emerges from the column. Of course, precautions must be taken to avoid the absorption of carbon dioxide by the effluent. By the use of potassium-form resin, a solution of carbonate-free potassium hydroxide can be prepared. The column can be regenerated by the use of sodium or potassium chloride.

In the third method,[6] 40 g of air-dried Amberlite IRA-400 in a column of 3.5 cm^2 cross-sectional area is converted to the hydroxide form by the passage of 2 l. of 1 M carbonate-free sodium hydroxide (prepared by dilution of the 18 M solution with carbonate-free water). The column is then washed with 2 l. of carbonate-free water. An accurately weighed quantity of about 50 meq of pure, dry sodium or potassium sulphate is weighed, dissolved in about 50 ml of carbonate-free water, and passed through the column. The effluent is caught in a 500-ml volumetric flask. The column is rinsed with carbonate-free water at a flow rate of about 8 ml/min until 500-ml has been collected; precautions are taken against the contamination of the effluent by carbon dioxide of the air. The concentration of the effluent is now calculated from the weight of the sulphate taken on the assumption that the exchange was quantitative.

$$SO_4^{2-} + 2\,ROH \rightarrow R_2SO_4 + 2\,OH^-$$

Obviously this method does not avoid the preparation of a carbonate-free solution of sodium hydroxide. Nevertheless, it has two advantages. (1) It serves to prepare carbonate-free potassium hydroxide which cannot be prepared by dilution of the concentrated solution. (2) It avoids the necessity of standardizing the base by titration. On the other hand, the agreement between the normality calculated from the weight of salt and that found by titration of a primary standard leaves something to be desired. On the assumption that the titration yields correct results, the relative standard error of the ion-exchange method was $\pm 0.33\%$. For microanalysis, the method has been adapted[7] to the preparation of 0.001 M sodium hydroxide.

B.II.b. *Potassium Nitrite*

According to the specifications[8] of the Committee of Analytical Reagents of the American Chemical Society, reagent-grade potassium nitrite should have a minimum assay of 94%. The major impurity is potassium nitrate, which cannot be removed by recrystallization. Ray[9] has devised the following method for obtaining potassium nitrite in a high state of purity. He recrystallized commercial sodium nitrite, originally containing about 3% of the nitrate, to a high degree of purity. He passed a 1 M solution of this nitrite through a bed of 5 lb of Amberlite IRA-400, thus converting it from the chloride to the nitrite form. He then passed 1 M reagent-grade potassium nitrite through the column. Because of the favorable value of $E_{NO_2^-}^{NO_3^-}$, the resin retained the nitrate quantitatively until it was almost completely converted to the nitrate form. The effluent was potassium nitrite at least 99.9% pure. The column can be regenerated with pure sodium nitrite and used repeatedly.

B.II.c. *Supporting Electrolytes in Polarography*

In the polarographic determination of very small concentrations of transition metals, the supporting electrolyte sometimes contains enough of these metals to introduce an appreciable error. In these cases, it is suggested[10] that the supporting electrolyte be passed through a column of cation exchanger before use. A chelating resin should be especially useful for this purpose.

C. Removal of Interfering Constituents
C.I. REMOVAL OF INTERFERING CATIONS

When an anion is to be determined, certain cations that are present in the sample or were added in the preliminary steps may interfere with the desired method of determination. In these numerous cases, ion exchange offers a very simple and elegant method of removing the undesired cations. It is necessary only to pass the solution through a column of sulphonated polystyrene, usually in the hydrogen form, and to rinse the column with a suitable volume of water or dilute acid. Three times the volume of the resin bed usually suffices. The combined effluent normally contains all the anions as the corresponding acids free from all cations except hydrogen.

The solution that is passed into the column is always slightly acidic to prevent the precipitation of hydroxides or salts of the metals present, such as ferric hydroxide or calcium phosphate. The hydrogen ions oppose the absorption of the cations according to equilibria such as

$$2HR + Ca^{2+} \rightleftharpoons CaR_2 + 2H^+$$

Fortunately, the selectivity coefficients of almost all metals with respect to hydrogen are rather large. Furthermore, as the solution flows down the column, the upper layers of resin absorb more metal than the lower layers. Thus the amount of resin needed for a quantitative removal of the metals is much less with the column procedure than it would be with a batch-wise process. The amount of resin needed depends on a large number of considerations: the nature and quantities of the metals in the sample, the volume and hydrogen-ion concentration of the solution, the nature of the resin, the ratio of height to cross-sectional area of the column, the temperature, and the flow rate. In general, the quantity of resin (in milliequivalents) should be about twice as great as the number of milliequivalents of all cations (including hydrogen ion) in the solution.

A few of the many applications of this type of separation are discussed below.

C.I.a. *Determination of Phosphate*

Whether phosphate is to be determined gravimetrically, photometrically, chelometrically, or acidimetrically, almost all cations except those of the alkali metals interfere. In the analysis of phosphate rock, the iron, aluminum, calcium, etc., of the sample are readily removed by passage through the hydrogen form of Dowex 50. If silica and fluoride are removed before the treatment with resin, the effluent contains only hydrochloric and phosphoric acids. Then the latter is very easily determined by titration with sodium hydroxide between the end points corresponding to the formation of the primary and secondary phosphates.[11]

This method is subject to a small negative error if the quantity of iron in the sample is very large. Then some phosphate is held on the resin by a reaction such as

$$FeR_3 + H_3PO_4 \rightarrow R_2FeH_2PO_4 + HR$$

This error is utterly negligible with the quantities of iron found in phosphate rocks.[12]

The method has been applied to a wide variety of samples containing either macro or micro quantities of phosphorus. An example of the latter is the indirect photometric determination of phosphate as chloranilate after the reaction

$$6PO_4^{3-} + 3La_2(C_6Cl_2O_4)_3 \rightarrow 6LaPO_4 + 9C_6Cl_2O_4^{2-}$$

The interfering cations were previously removed by ion exchange.[13] Hesse and Böckel[14] determined the phosphorus in nucleic acids by igniting the sample, passing a solution of the ash through a column of cation exchanger, and, finally, titrating the phosphate in the neutralized effluent with cerium(IV). Qualitative as well as quantitative purposes are served by the removal of interfering cations. For example, Wood[15] passed biological liquids through a column of Zeo-Karb 225 in the ammonium form to remove calcium and magnesium prior to the identification of phosphorus compounds by paper chromatography.

C.I.b. *Determination of Arsenic*

Although comparatively few cations interfere in the iodimetric determination of arsenic, elements such as iron, copper, and lead may not be present. The passage of the solution in dilute hydrochloric acid through a column of sulphonate resin removes the cations, whereas arsenic, both (III) and (V), is in the anionic form and passes into the effluent.[16]

C.I.c. *Determination of Borate*

Many cations interfere in the determination of boron, both by the acidimetric and photometric methods; and several authors have used ion exchange to remove these cations. One such method[17] for the determination of boron in steel consists in the photometric determination with azomethine-H after the solution has been passed through Dowex 50-X4. Other investigators[18] pass the solution through a mixed bed of strong-acid cation exchanger and a weak-base anion exchanger. This results in the removal of almost all electrolytes except boric acid (p. 66), leaving the solution in an ideal condition for the acidimetric[19] determination.

C.I.d. *Determination of Sulphate*

The well-known gravimetric determination of sulphate is subject to a rather serious error, usually negative, because of the coprecipitation of whatever cations may be present. For an accurate determination, all cations except hydrogen should be removed by ion exchange,[20] and the quantity of the latter ion should be decreased to a suitable value by evaporation. Volumetric determinations of sulphate by titration with precipitating agents such as barium chloride or lead nitrate are also subject to the error of coprecipitation. In addition, some cations may interfere by reacting with the indicator (rhodizonic acid or dithizone[21]). Again ion exchange is a very efficient method of removing the unwanted cations.

C.I.e. *Determination of Selenium*

The gravimetric method for selenium consists in reducing a solution of selenious acid to elemental selenium and weighing the latter. As in the case of sulphate, many metals are extensively coprecipitated and should be removed previously by cation exchange.[22]

C.I.f. *Determination of Fluoride*

Fluoride concentrations below 10^{-3} M are usually measured photometrically, although special selective electrodes are now coming into use. Until recently all photometric methods depended on the bleaching of complexes of zirconium ions (or other ions of high charge) with colored organic ligands such as alizarin. Many metal ions interfere with this process by combining with the organic ligand or with fluoride. The commonest interferences are from iron(III) and aluminum. Phosphate ions also interfere by combining with the zirconium.

The photometric method most used today depends on the formation of a blue complex of the reagent "alizarin complexone", fluoride ion, and the ion La^{3+} or Ce^{3+}. Again iron, aluminum, phosphate, and other ions interfere, though the interferences are less serious than in the older methods.

Fluoride can be separated from all interferences by the Willard and Winter distillation procedure. A more convenient and faster separation can be accomplished by ion exchange. Glasoe[22a] determined fluoride in iron ore and apatite by dissolving the minerals in 10 M hydrochloric acid and passing the solution into a bed of quaternary ammonium anion-exchange resin, rinsing with 10 M hydrochloric acid. Iron and phosphate were retained by the resin bed, while fluoride, being the most weakly held of all the anions, passed through it. Some aluminum passed also, but the interference which this caused was small and could be compensated. Where the chief interferences are from cations, as in water analysis, the sample may be passed through a cation-exchange resin before treatment with the alizarin-complexone reagent.[22b, 22c] If significant concentrations of phosphate were also present, it would be an easy matter to pass the solution through a short column of anion-exchange resin, which retains phosphate strongly, and wash with a dilute acid or salt solution to prevent retention of fluoride ions.

C.I.g. *Determination of Ruthenium*

The method of Zachariasen and Beamish[23] for the determination of ruthenium in ores involves melting the ore in the presence of an alloy of copper, nickel, and iron. When the melt solidifies, all the ruthenium is found in this metallic "button". It is dissolved in hydrochloric acid plus a little nitric acid, and the solution is passed through Dowex 50. The copper, nickel, and lead, which would interfere in the subsequent photometric determination of ruthenium, are retained by the resin, whereas the ruthenium passes into the effluent as $RuCl_6^{-2}$.

C.I.h. *Determination of Uranium*

The method of Fisher and Kunin[24] for the removal of interfering elements from uranium differs from the previous examples in that they use an anion exchanger. They treat the solution of the ore in dilute sulphuric acid with sulphurous acid to reduce iron(III) and vanadium(V). Then they pass the solution through Amberlite XE-117, type 2, in the bisulphate form. Uranium is retained by the resin as a sulphate complex while the other metals pass through the column. Finally, they pass 1 M perchloric acid through the resin to elute the uranium and determine it either photometrically or volumetrically.

C.I.i. *Determination of Indium by Activation Analysis*

Růžička and Starý introduced into activation analysis a new principle which they call substoichiometric determination.[25] They irradiated under identical conditions the unknown sample and a standard containing a known quantity of the element to be determined. Thus equal fractions of this element were converted to radioactive isotopes. They dissolved the samples and added the same quantity of carrier to each. Next they added to each solution the same quantity of some precipitating agent for the element to be determined, the quantity being less than equivalent to the added carrier. After filtration, they determined the radioactivity of both precipitates. The quantity of element in the unknown sample y is given by the equation

$$y = \frac{y_s a}{a_s},$$

where a denotes the measured radioactivity and the subscripts s refer to the standard sample.

A short time later,[26] they determined indium in germanium dioxide by substoichiometric activation analysis but replaced the precipitation by substoichiometric addition of a complexing agent and subsequent ion-exchange separation. After the irradiation and dissolution of the samples, they added 4.00 μmoles of indium carrier and 2.00 μmoles of ethylenediaminetetraacetate. Thus 2.00 μmoles of indium complex was formed in each solution. On passage of the solutions through columns of Dowex 50, the uncomplexed indium remained on the resin. They measured the radioactivity of the complexed indium in the effluents. In two experiments with 2.75 and 5.50 μg of indium added to 100 mg of germanium dioxide, they recovered 2.74 and 5.65 μg, respectively.

C.I.j. *Determination of Neutral Impurities in Amines of Large Molecular Weight*

Nelson et al.[27] dissolved a 5-g sample in 75 ml of propanol-2 and passed the solution through a column of hydrogen-form Dowex 50W-X4. Thus the amines were converted to cations and retained by the resin by reactions such as

$$C_{16}H_{33}NH_2 + HR \rightarrow C_{16}H_{33}NH_3R$$

while the neutral impurities, such as nitriles and amides, passed into the effluent. The authors evaporated the eluate to dryness and weighed the residue. If these impurities contain less than 16 carbon atoms per molecule, some of them are lost by volatilization.

C.II. REMOVAL OF INTERFERING ANIONS

When the determination of a certain cation is subject to interference by anions, the latter can generally be removed very simply by passage of the solution through a column of anion exchanger. The resin is usually used in the chloride form, although the nitrate and acetate also find application. The hydroxide form is generally unsuitable because the hydroxide ions liberated in the exchange would precipitate the cations inside the column.

By an extension of this principle, certain cations can be separated from other cations that form complexes more readily. A complexing anion is added to the solution to convert the easily complexed metals to anionic complexes. These are retained by the resin while the other cations pass through. Alternatively, the resin may be used in the form of the complexing anion (see below).

C.II.a. *Removal of Phosphate*

Phosphate interferes in the determination of many cations whether the determination is performed by gravimetry, chelatometry, or atomic-absorption spectrophotometry. Methods for the removal of phosphate by ion exchange prior to the determination of calcium and magnesium by titration with ethylenediaminetetraacetate have been described.[28, 29] Hinson[30] determined calcium in plants by atomic-absorption spectrophotometry after removing phosphate by ion exchange. David[31] determined strontium similarly.

C.II.b. *Determination of Methylpentoses*

When a mixture of methylpentoses and ordinary sugars is treated with periodate, the former are oxidized to acetaldehyde, the latter to formaldehyde. Wardi and Stary[32] removed the excess periodate by anion exchange and then determined acetaldehyde polarographically at an e.m.f. at which formaldehyde does not interfere.

C.II.c. *Determination of Alkali Metals in the Presence of Other Metals*

An elegant method for the determination of either sodium or potassium in the presence of vanadium(IV), iron(III), copper, nickel, and cobalt was devised by Samuelson and Schramm.[33] They poured the solution through a column (14 cm $\times$ 0.64 cm²) of Dowex 2 in the citrate form. The transition metals were retained on this column by reactions such as

$$R_3Ci + Cu^{2+} + 2NO_3^- \rightarrow RCuCi + 2RNO_3$$

The effluent from this column, containing only salts of the alkali metals, flowed directly into a column of the same dimensions containing the hydroxide form of Dowex 2. The effluent from the second column contained an amount of sodium (or potassium) hydroxide equivalent to the alkali metal in the sample.

$$ROH + Na^+ + NO_3^- \rightarrow RNO_3 + Na^+ + OH^-$$

The authors titrated this base with hydrochloric acid. They analyzed ten known mixtures containing from 2.0 to 3.2 mmol of alkali metal and from 0.5 to 1.2 mmol of another metal; the maximum relative error was 0.2%.

Samuelson *et al.*[34] extended the method to samples containing aluminum, manganese, zinc, calcium, and magnesium in addition to the metals mentioned above. The upper column, containing $R_2C_2O_4$, $RC_2H_3O_2$, and R_2H_2EDTA, removed all cations except those of the alkali metals while the lower column of ROH (Dowex 2 in all cases) changed the sodium or potassium salt into the free base. It was necessary to add some ethanol in order to retain all the metals except the alkalis on the first column. It was also necessary to pretreat the solution with a mixture of R_2H_2EDTA and $RC_2H_3O_2$ to remove most of the transition metals lest they be precipitated by the addition of alcohol. In spite of the more involved procedure, they obtained excellent results.

C.II.d. *Determination of Calcium and Magnesium in the Presence of Other Metals*

In the determination of calcium and magnesium in plant tissues, Carlson and Johnson[35] separated these metals from aluminum, copper, iron, manganese, zinc, and phosphate by a procedure similar in principle to the foregoing. They passed the properly buffered solution of the plant ash through a column of Dowex 21K previously treated with a solution of di-ammonium cyclohexanediaminetetraacetate. The resin retained all the metals except the calcium and magnesium, which they determined by chelatometry.

C.II.e. *Determination of Yttrium and Rare Earths in Steel*

Bornong and Moriarty[36] separated these elements from iron by precipitating them as fluorides, using thorium fluoride as a collector. They dissolved the precipitate and pre-cipitated the yttrium, rare earths, and thorium again as oxalate. Since thorium interferes in the final spectrographic determination of the rare earths and yttrium, they dissolved the mixed oxalates in nitric acid and passed the solution through a column of Dowex 1-X10. Thorium was retained as a nitrato complex while the yttrium and the rare earths passed into the effluent.

C.II.f. *Removal of Uranium from Alkali and Alkaline-earth Metals*

Graphite impregnated with uranium is used in some atomic reactors. Uranium inter-feres with the spectrographic determination of traces of alkali and alkaline-earth metals in this material. After ashing the sample and dissolving the ash in 9 M hydrochloric acid, the solution may be passed through Dowex 1-X10. This retains the uranium as a chloro com-plex, while the alkalis, alkaline earths, aluminum, and nickel go into the effluent.[37]

C.III. MISCELLANEOUS ANALYTICAL SEPARATIONS

C.III.a. *Analysis of Soaps and Greases*

Vamos and Simon[38] dissolved a 1-g sample in a mixture of equal volumes of benzene and methanol (or ethanol or propanol in case the sample fails to dissolve in the mixture of benzene and methanol). They passed this through a column of hydrogen-form strong-acid resin. Directly below this column was another of hydroxide-form strong-base anion ex-changer. The first column converted the soaps to free fatty acids, and the second absorbed these acids. The nonelectrolytes of the sample (high alcohols and hydrocarbons) ran through both columns. They evaporated the effluent from the lower column to dryness and weighed the residue of nonelectrolytes. They detached the columns and washed the fatty acids out of the lower column with an alcoholic solution of hydrochloric acid. By evaporat-ing this solution and weighing the residue, they determined the fatty acids. When an analysis of the metallic constituents of the sample was desired, they treated the upper column with hydrochloric acid and determined the various metals in the eluate. See p. 65 for preparation of the column for use with nonaqueous solvents.

C.III.b. *Determination of the Individual Fatty Acids in Fats*

The method of Hornstein *et al.*[39] follows. They dissolved the sample in petroleum ether and added to the solution some hydroxide form of Amberlite IRA-400 that had been previously treated with anhydrous ethanol and then with petroleum ether. After 5 min of stirring the reaction

$$3ROH + (AlkCOO)_3C_3H_5 \rightarrow 3ROOCAlk + C_3H_5(OH)_3$$

was complete. They separated the resin from the solution and washed it with petroleum ether. Then they treated the resin with a solution of hydrogen chloride in anhydrous methanol. The following reactions occurred:

$$ROOCAlk + HCl \rightarrow RCl + AlkCOOH$$

$$AlkCOOH + MeOH \rightarrow AlkCOOMe + H_2O$$

They added water to the methanolic solution and then extracted the methyl esters with petroleum ether. Finally, they determined the relative amounts of the esters by gas chromatography.

C.III.c. *Determination of Betaine in Sugar-beet Juice*

Although betaine, $Me_3^+NCH_2COO^-$, is quantitatively precipitated by ammonium reinekate, $NH_4Cr(III)(NH_3)_2(SCN)_4$, this method alone is not satisfactory for the determination of betaine in sugar-beet juice because other constituents such as choline

$$HOCH_2CH_2NMe_3OH$$

are also precipitated. It has been recommended[40] to pass the sample through a mixture of the hydroxide form of De-Acidite FF (a strong-base resin) and the hydrogen form of Amberlite IRC-50. This is, of course, a deionizing procedure and removes choline and other ordinary electrolytes; but betaine, being essentially neutral, passes through the column. Some protein, unable to penetrate into the resin, accompanies the betaine but does not interfere in the precipitation.

C.III.d. *Determination of Ammonium Ion in Blood*

Ammonium ion (plus ammonia) is often determined in blood and other biological fluids by spectrophotometry after it has been separated from other constituents by volatilization from an alkaline solution in a microdiffusion cell. Fenton[41] regarded this separation as unsatisfactory because some constituents of blood decompose slowly with evolution of ammonia. The reaction is aggravated by high pH and high temperature but occurs at an appreciable rate at room temperature and at the natural pH of blood.

Fenton chilled the blood immediately after sampling and centrifuged it while still cold to remove the corpuscles. He then passed the serum through a very small (3.0 cm × 0.39 cm²) chilled column of Amberlite CG 120 in the sodium form. This retained all the ammonium ion and a small fraction of protein. After washing the column, he passed a solution of sodium chloride through it. This removed the ammonium ion and a trace of protein. He gathered the latter by precipitation of aluminum hydroxide and finally determined the ammonium ion spectrophotometrically with indophenol blue.

Hutchinson and Labby[42] describe a similar procedure for the determination of ammonium ion in whole blood.

C.III.e. *Removal of Salts from Amino Acids and Condensed Phosphates*

In the separation of amino acids by ion-exchange chromatography (Chapter 7), the appropriate volume of eluate generally contains a single species of amino acid, which is, however, contaminated with a large quantity of salt from the eluent. The ratio of salt to amino acid may be as large as 1000. If an unexpected peak is found in the chromatogram, paper chromatography offers the quickest method of identifying the unknown amino acid. It is first necessary to remove the large amount of salt.

This is readily accomplished by the method of Dreze et al.[43] They passed the eluate fraction of no more than 4 ml through a very small bed of Dowex 2-X8 in the hydroxide form. This absorbed the amino acid and all the anions of the sample. Then they passed 1 M acetic acid through the column. The first eluate is pure water. Later when all the resin is converted to the acetate form, the effluent is 1 M acetic acid. Just before the acid appears in the effluent, there is a fraction containing the amino acid and no other solute. The foregoing procedure is applicable to both neutral and acidic amino acids. The authors[43] describe a similar procedure for the basic amino acids.

Rothbart et al.[44] have described a similar method for removing buffer salts from condensed phosphates.

C.III.f. *Spectrophotometric Determination of Silicate*

The spectrophotometric determination of silicate as the yellow silicomolybdate is subject to interference by ferric ions and by anions such as phosphate that form similar heteropoly acids. The blue reduction compound of silicomolybdate is more intensely colored but is subject to additional interference from chloride and nitrate. Anderson[45] removed the interfering substances by passing the solution of the silicate through a column of hydrogen-form Dowex 50-X8 and then through a column of Amberlite IR-4B in the hydroxide form. Because of the extreme weakness of silicic acid, it was not retained by the column of weak-base resin.

D. Determination of Total Salt

Let us consider a solution containing ammonium sulphate, sodium nitrate, and potassium chloride in which it is desired to determine the total normality of salt. Obviously it would be possible to determine each one of the cations (or anions) individually and then to calculate the total normality. A much simpler and more accurate method, however, consists in passing the sample through a column of strong-acid cation exchanger in the hydrogen form, rinsing the column with water and titrating the acid in the effluent.[46] If the sample contains acid, an aliquot must be titrated without treatment with the resin to serve as a blank correction. If the sample contains base, no such correction is necessary because the base will be absorbed by the hydrogen-form resin.

It is often possible to reverse the process, i.e. to use a strong-base anion exchanger in the hydroxide form and to titrate the effluent base with standard acid. This method is generally less desirable for the following reasons. (1) It is necessary to protect the alkaline effluent against absorption of carbon dioxide from the air. (2) Because of the small affinity of hydroxide ion for strong-base resins, the regeneration is more expensive and time-consuming than in the case of the cation exchanger. Weak-base or weak-acid resins cannot be used because the desired exchange reaction would not go to completion. (3) If the sample contains a cation whose hydroxide is insoluble, the method fails because of the precipitation inside the column. (4) In the case of samples containing much ammonium salts, the volatilization of ammonia would introduce an error unless the effluent is caught in excess standard acid which is then countertitrated.

D.I. SOURCES OF ERROR

Although the method with the cation exchanger has been used very extensively with excellent results, its thoughtless application may result in errors in some cases. If the sample contains an anion whose acid has a pK much above 5, phenolate for example, it is impossible to titrate the effluent satisfactorily with sodium hydroxide. If the anion of a very volatile acid such as cyanide or carbonate is present in the sample, there will be a loss of this acid from the effluent or even liberation of the acid gas inside the column. In these cases, it is best to acidify the sample and expel the volatile acid before use of the ion-exchange column.

Benzoic acid is an example of an insoluble acid with a soluble sodium or potassium salt. If an aqueous solution containing this salt is treated with a hydrogen-form resin, the benzoic acid is precipitated inside the column. Other acids such as the hydroxybenzoic acids may be sufficiently soluble in water so as not to be precipitated but will be eluted from the column only after extensive washing because of their great attraction to the resin by London forces (p. 31). In both of these cases, the problem may be solved by rinsing the column with ethanol or a mixture of ethanol and water. Another approach is to use the resin in the magnesium form and to titrate the liberated magnesium ion chelometrically.[47]

The behavior of complex ions must be considered. Some are so stable that they pass through the hydrogen-form resin without detectable decomposition. For example, $K_3Fe(CN)_6$ and $K_4Fe(CN)_6$ liberate respectively 3 and 4 (not 6) mmol of hydrogen ion per mmol of salt. Other complex ions are quantitatively broken up by the resin. It is rather surprising that many complexes of cadmium fall into this class. For example, even from solutions containing iodide, orthophosphate or oxalate, cadmium[48] is removed by the resin with liberation of two hydrogen ions per cadmium atom. Other complexes of intermediate stability are partly decomposed by the resin. These include the oxalate complexes of iron(III), aluminum, and lead,[50] and the sulphate complexes of chromium(III).[20, 49] From these facts, it is clear that the analytical chemist should not apply this method to samples containing complex ions until he ascertains, either from the literature or by a few preliminary experiments, the behavior of these complexes in the resin column.

Care must be exercised in interpreting the results if the sample contains salts of any polyprotic acid except sulphuric. For example, a sample consisting of x mmol of potassium primary phosphate and y mmol of sodium secondary phosphate will yield $(x + y)$ mmol of phosphoric acid in the effluent. This will require $(x + y)$ or $2(x + y)$ mmol of sodium hydroxide depending on whether the titration is performed to the first or second end point.

Finally, a caution should be sounded in regard to the purity of the solvents. Especially if dilute solutions of salts are analyzed there may be sufficient salt in the distilled water or acetic acid in the ethanol (if it be used) to introduce appreciable errors.

D.II. APPLICATIONS

This method has been applied to a wide variety of samples including blood serum,[51] flour,[52] natural waters,[53] starch hydrolysates,[54] sugar,[55] and the sulphite liquor of the paper industry.[56]

Shah and Quadri[57] devised an ingenious application of the determination of total salt with an anion exchanger to the classical Kjeldahl determination of nitrogen. They digested samples of about 3 mg with 0.1 ml of concentrated sulphuric acid in a sealed tube at 425° for 30 min. After cooling and opening the tube, they heated it at 90° for 5 min to expel sulphur dioxide. At this point the tube contained only ammonium bisulphate and sulphuric acid. They diluted the contents of the tube and passed the solution through a bed of Amberlite IRA-400 in the hydroxide form. The effluent solution contained nothing but ammonia, which could have been determined by titration with an acid. However, the authors used an iodimetric determination that gave a sixfold increase in sensitivity. They passed the effluent from the first column directly into another column containing the resin in the iodide form. In spite of the unfavorable selectivity coefficient and the weakness of ammonia as a base, the reaction

$$NH_3 + H_2O + RI \rightleftharpoons ROH + NH_4^+ + I^-$$

went quantitatively to the right by virtue of a large excess of resin iodide. Thus the effluent from the second column contained only ammonium iodide. They oxidized the iodide ion to iodate, which they determined iodimetrically.

The authors analyzed in duplicate six pure organic compounds containing from 6 to 18% nitrogen. The standard error was 0.17%.

This method is not applicable to samples containing a metal whose hydroxide is soluble, such as sodium or calcium. The hydroxides of these metals would accompany the ammonia in the effluent from the first column and would be converted into iodides in the second, thus introducing a positive error. For the same reason, sodium or potassium sulphate may not be added to hasten the digestion by raising the boiling point. However, the digestion proceeds rapidly in the sealed tube without catalysts of any sort.

E. Dissolving Insoluble Salts

Honda et al.[58] reported in 1952 that barium sulphate gradually disappears if a small quantity of it is stirred with a suspension of a cation exchanger in the hydrogen form and an anion exchanger in the hydroxide form. A year later, Osborn,[59] apparently unaware of the previous work, dissolved 250 mg of barium sulphate completely by agitating it with 10 g of strong-acid cation exchanger and 100 ml of water at 80° for 12 hr.

The dissolution in Osborn's experiment depends on the exchange reaction between HR and the small amount of barium ion in solution. The removal of barium ions from the solution causes further dissolution of the salt until the process is eventually complete. In Osborn's experiment the sulphate and bisulphate ions were permitted to accumulate in solution, thus decreasing, according to the solubility-product principle, the amount of barium ion in solution and hence retarding the dissolution. If an anion exchanger is also present, the sulphate ions are removed from solution, and the salt is dissolved more quickly.

Other factors that influence the kinetics of the process include the solubility product of the salt, the selectivity coefficient(s) of the exchange reaction(s), the ratio of resin and of water to salt, the rate of stirring and the temperature. Helfferich[60] has derived an approximate rate equation that takes most of these factors into account. This equation indicates that the rate of dissolution even in the presence of a large excess of strong-acid hydrogen-form resin will be very slow unless

$$\sqrt[v]{S} \gg 10^{-7},$$

where S is the solubility product and v the number of ions formed. Thus we should expect silver chloride ($S = 1.8 \times 10^{-10}$) to dissolve readily; Helfferich calculated 3 hr for complete dissolution. We should expect silver bromide ($S = 5.6 \times 10^{-12}$) to require a much longer period, and we should expect that the method would be impracticable with silver iodide ($S = 1.1 \times 10^{-18}$). In the derivation there is the tacit assumption that the anion of the slightly soluble salt does not have a strong tendency to combine with hydrogen ion. Hence, Helfferich's conclusions should not be applied to salts such as ferrous sulphide or lead chromate. This type of salt will dissolve more quickly than the equation predicts.

In the case of the sulphonated polystyrene resins, the hydrogen form is more efficient than any other form in dissolving slightly soluble salts. There are probably three reasons for this. (1) The selectivity coefficient of the cation of the slightly soluble salt vs. hydrogen is greater than the selectivity for the exchange with sodium or any other common metal. (2) Since the process is controlled by diffusion, the large diffusion coefficient of hydrogen ion is an advantage. (3) In the case of carbonates, oxalates, phosphates, sulphides, and similar salts whose anions have basic properties, the hydrogen ion liberated in the exchange re-presses the concentration of the anion, thus expediting the dissolution processes. Resins with iminodiacetate groups as functional groups, however, are more effective in the sodium form. See p. 81.

The dissolution of insoluble samples by treatment with a hydrogen-form strong-base resin offers two important advantages to the analytical chemist. (1) It avoids the introduction of a large amount of foreign substances. For example, the decomposition of a sample by a carbonate fusion introduces a large quantity of sodium or potassium which complicates all the following determinations. (2) The ion-exchange method simultaneously separates cations from anions. The latter are found in the filtrate from the resin, whereas the former are readily obtained by eluting the resin with a suitable acid.

E.I. APPLICATIONS

Ahmad and Khundkar[61] dissolved samples of calcium borate, using the hydrogen form of Zeo-Karb 225.

Kniested and Wahle[62] determined sulphate in mixtures of calcium sulphate and calcium carbonate as follows: They mixed a 250-mg sample with 50 g of hydrogen-form resin and 300 ml of water at 90°. After 30 min of stirring, the dissolution was complete. Then they filtered and washed the resin and titrated the filtrate with sodium hydroxide to the end point of methyl orange.

In the separation of radium, present as the sulphate, from its radioactive disintegration products, Dedek[63] found that ion exchange offered the most advantageous method of dissolution. Although radium sulphate has a solubility product of 4×10^{-11} and hence should dissolve in a reasonable time on treatment with only a cation exchanger, Dedek actually used a mixture of hydrogen-form cation exchanger and hydroxide-form anion exchanger.

Schafer[64] analyzed phosphate rocks by the following method. Into a 100-ml plastic bottle he put 50 mg of sample, 5–10 g of hydrogen-form Zeo-Karb 225, and 35 ml of water at 80°. He sealed the bottle and shook it overnight without further heating (except for a few samples rich in aluminum). Then he filtered and washed the resin, using plastic apparatus. He determined phosphate and fluoride by spectrophotometric methods on separate aliquots of the filtrate. He eluted the cations from the resin with hydrochloric acid and determined calcium in the eluate. Of course, he could have determined the other metals also in the same solution. An indication of the accuracy of the method is seen in the comparison in Table 5.1.

Bricker et al.[65] recommend the ion-exchange method for dissolving insoluble samples in classical (wet) qualitative analysis. They treated 50–100 mg of unknown with 10 ml of water and 5 g of hydrogen-form resin for 15–20 min at 60–80°. Sufficient of the following com-

TABLE 5.1. ANALYSIS OF PHOSPHATE ROCK
Sample 56B of the National Bureau of Standards

	% P_2O_5	% F	% CaO
Certified value	31.55	3.4	44.06
Schafer	31.57	3.39	44.09
	31.61	3.38	43.98

(Used by permission from *Analytical Chemistry*.)

pounds dissolved to give positive tests for both the cation and the anion (except, of course, that no tests were applied for the oxides): $PbSO_4$, $BaCO_3$, CaF_2, $BaSO_4$, $CaCO_3$, Hg_2Cl_2, CuO, ZnS, BiOCl, As_2O_3, CdS, $PbCrO_4$, PbO, FeS, and HgI_2.

Instead of the usual strong-acid resin in the hydrogen form, Rich[66] used chelating resin, Dowex A-1, in the sodium form. Since the selectivity coefficients of most cations vs. sodium ion with this resin are much greater than the corresponding selectivity coefficients vs. hydrogen ion in a sulphonated polystyrene resin, it is not surprising that Rich succeeded in dissolving slightly soluble salts more rapidly than those who worked with strong-acid resins. This is shown in Table 5.2. It should be noted that he dissolved silver iodide in 1 hr whereas Helfferich's[60] equation indicates that this salt cannot be dissolved in any reasonable time by a sulphonated resin.

TABLE 5.2. APPROXIMATE DISSOLVING
TIMES AT ROOM TEMPERATURE WITH
DOWEX A-1 (SODIUM FORM)

Compound	Time
AgCl	30–45 min
AgI	1 hr
$Ba_3(PO_4)_2$	15 min
$BaSO_4$	1–6 hr
CaF_2	1 hr
HgI_2	30–40 min
$Mn_3[Co(CN)_6]_2$	10–15 min
$PbCrO_4$	30 min
PbI_2	5–10 min
$Pb_3(PO_4)_2$	30 min
$PbSO_4$	5–10 min
$Zn_3[Co(CN)_6]_2$	15 min

(Used by permission from *Journal of Chemical Education*.)

Most cations can be readily eluted from Dowex A-1 by 6 M nitric acid. Silver ion requires lengthy washing with this eluent. Rich used concentrated nitric acid to remove mercuric ion and either aqua regia or sodium peroxide to remove chromic ion. The last three reagents cause partial or complete destruction of the resin. Perhaps a chelating agent such as ethylenediaminetetraacetate would remove these cations without injury to the resin.

F. Concentrating Trace Constituents

The analyst frequently has the problem of determining a constituent whose concentration is so small that the usual methods of analysis are not sufficiently sensitive or accurate. Ion exchange offers a very useful method of overcoming these difficulties.

F.I. FROM SOLUTIONS CONTAINING
NO ELECTROLYTE IN LARGE CONCENTRATION

The problem is simplest when the sample is an aqueous solution free from large concentrations of electrolytes or can readily be dissolved to give such a solution, e.g. the determination of trace metallic impurities in sucrose. In these cases, the procedure is very similar to that of deionization. If the constituents to be determined include both cations and anions, the solution is passed through columns of cation exchanger and anion exchanger, usually in the hydrogen and hydroxide forms. The constituents to be determined are then eluted from the respective column with suitable electrolytes. The volume of eluate is often only 0.01 or less of the original volume.

In comparison with evaporation, the ion-exchange method has the advantages of being faster and of avoiding contamination by dust and by dissolution of the glass container during the long evaporation. Furthermore, in the analysis of natural waters sampled on a field trip, there is an obvious advantage in carrying back to the laboratory a few small tubes of resin rather than many liters of sample. Still another advantage of the ion-exchange method is that it separates the cations of the sample from the anions. The determination of sulphate, for example, is more accurate if extensively coprecipitated cations such as sodium are not present.

The chief source of error in the application lies in the existence of the constituent to be determined in some form other than that of a simple ion. For example, ferric ion in dilute solution undergoes a slow hydrolysis to yield colloidal ferric hydroxide or basic salts; these may readily pass through the cation exchanger. In such cases it is often possible to add some reagent to the sample that will convert the constituent to be determined to some readily absorbable state. See, for example, the determination of titanium in natural water (p. 83).

If two ions of the same sign, A and B, are to be determined in one sample and if A is much more readily absorbed by the resin than B, the dimensions of the column must be chosen so that B is completely removed from the sample. In this case, A will be retained near the top of the column; and the volume of eluent needed to remove A from the column will ordinarily be greater than the volume needed to remove B, perhaps so great as to diminish somewhat the advantage of the ion-exchange method of concentrating. In this situation it is advantageous to pass the eluent through the column in the opposite direction to the passage of the sample solution. Then A has to travel only a short distance through the column and will be quantitatively eluted in a small volume.

F.I.a. *Analysis of Lake Waters*

The classical example of concentrating a trace solute in aqueous solution is Nydahl's determination of sodium, potassium, calcium, magnesium, chloride, and sulphate in the waters of Swedish lakes.[67] He passed samples of 1–5 l. through a 30-ml bed of Amberlite IR-100 and then through a similar bed of amberlite IR-4B at about 5 l/hr. He removed the cations from the resin with 100 ml. of 2.5 M hydrochloric acid and the anions with 100 ml of 0.5 M ammonia in a slow overnight elution. Then he determined them by conventional methods. He has good results in the determination of the ions listed above, but very low results for iron and phosphorus. These errors are probably due to the existence of these elements as complexes with humic acid or protein.

F.I.b. *Determination of Titanium in Natural Waters*

Titanium(IV) is an example of a metallic element that is not quantitatively retained by a column of cation exchanger because of its colloidal nature. Korkisch[68] first added 10 ml of 12 M hydrochloric acid to a liter of sample. Then he added 10 g of ascorbic acid and enough ammonia to give a pH between 4.0 and 4.5. Under these conditions, titanium is present as an anionic ascorbate complex. He passed the solution through a column, 10 cm × 0.28 cm², of Dowex 1-X8. The column retained the titanium along with similar complexes of tungsten, vanadium, uranium, thorium, and zirconium. After converting the resin to the mixed sulphate–fluoride form, he eluted the titanium with 60 ml of 0.05 M sulphuric acid containing 6 ml of perhydrol. The other metals remained in the column. He finally determined the titanium spectrophotometrically. Within the range of 10–80 μg of titanium per liter, recovery experiments showed a standard error of 0.4 μg/l.

F.I.c. *Determination of Iodine(131) in Drinking Water*

After adding small amounts of sodium iodide as carrier and thiosulphate and hydroxylamine hydrochloride to keep the iodine in the oxidation number of -1, Bently *et al.*[69] passed 5-l. samples through 40-ml beds of Deacidite FF, a strong-base anion exchanger. They eluted the iodide from the resin and determined its radioactivity. Determinations of known samples indicated recoveries of 95%, and concentrations as small as 0.05 $\mu\mu$c/l (3.1×10^{-15} M) could be detected.

F.I.d. *Determination of Metals in Sucrose*

For this purpose Noguchi and Johnson[70] dissolved 140 g of sample in 1 l. of water and passed the solution through a column of Dowex A-1 in the ammonium form, which quantitatively retained the metals. The investigators eluted the metals from the column with 50 ml of 2 M hydrochloric acid. They evaporated this effluent to dryness and heated it to drive off the ammonium chloride. They finally determined the few micrograms of zinc, iron, manganese, and copper in the residue by spectrophotometric methods.

F.I.e. *Determination of Strontium(90) in Milk*

Since milk may contain other radioactive isotopes of this element, the determination of strontium(90) is actually a determination of its radioactive daughter, yttrium(90). Porter *et al.*[71] stored the milk for 2 weeks at 1° to establish radioactive equilibrium between ⁹⁰Sr and ⁹⁰Y. After adding yttrium citrate as carrier, they passed 1 l. of milk through a 140-ml bed of Dowex 50W-X8, NaR, and then through a 30-ml bed of Dowex 1-X1, RCl. The former retained strontium, potassium, and calcium, while the latter retained the yttrium as a citrate complex. After rinsing the columns with water, they passed 2 M hydrochloric acid through the anion exchanger. A 35-ml fraction contained all the yttrium. They precipitated the yttrium as oxalate and determined the radioactivity of the precipitate. It is also possible to elute the radioactive cations from the cation exchanger and to determine them.

F.I.f. *Determination of Copper in Lubricating Oil*

All of the foregoing examples of trace determination depended on ion exchange in aqueous medium. This example and the one following, on the other hand, depend on ion exchange in nonaqueous solvents.

Buchwald and Wood[72] investigated the stability of several lubricating oils for internal-combustion engines by determining their copper contents after various periods of use. Under operating conditions, the oils undergo slow oxidation to organic acids which attack the metallic parts of the engine to yield oil-soluble salts.

In order to have enough copper for a spectrophotometric determination, the authors had to take samples of 20 g. They dissolved the sample in 60 ml of propanol-2 and passed the solution through a bed of sulphonated polystyrene resin in the hydrogen form, previously equilibrated with propanol-2. The resin retained the copper. Then they passed water through the column to remove the sample and the propanol. Next they used dilute sulphuric acid to elute the copper from the column, and they determined the copper spectrophotometrically in this effluent. Finally, in order to prepare the column for the next determination, they treated it with water to remove the sulphuric acid and with propanol-2 to displace the water.

F.I.g. *Determination of Nitrogen Compounds in Gasoline*

The nitrogenous compounds in gasoline are largely pyridine, quinoline, and their alkyl derivatives. Therefore they have basic properties. Snyder and Buell[73] passed a sample of 500 ml (less if the gasoline contained more than 0.2 parts per million of nitrogen) through a bed of Duolite C-10 in the hydrogen form, previously equilibrated with methanol. The resin retained the nitrogen compounds. After rinsing the column with *iso*-octane and methanol, they removed the pyridine, quinoline, etc., from the column with 25 ml of a 1% solution of ethyl amine in methanol. This effluent contained all the nitrogen compounds of the sample. The investigators measured its absorbance at 260 mμ. They calculated the total nitrogen content of the sample on the assumption that all of the nitrogen compounds in the gasoline have the same absorbancy at this wavelength. In order to prepare the column for the next determination, they treated it with aqueous sulphuric acid and then with methanol.

The investigators[73] found a standard deviation of 0.008 parts per million in replicate analyses of a sample containing 0.041 parts per million. With larger nitrogen contents the relative standard deviation of the method was a little better than 10%.

F.II. FROM SOLUTIONS CONTAINING LARGE CONCENTRATIONS OF ELECTROLYTES

The difficulty encountered in concentrating a trace constituent from a solution that contains also a large concentration of another electrolyte is illustrated by the problem of concentrating copper from an aqueous solution containing a trace of copper salt and a much larger concentration of ammonium chloride. Regardless of the original form of the resin, it is quickly converted to the ammonium form on contact with the solution, and the large concentration of ammonium ion prevents the quantitative absorption of the copper.

$$2NH_4R + Cu^{2+} \rightleftharpoons CuR_2 + 2NH_4^+$$

Riches[74] tried to concentrate 0.0004 M copper salt from 0.10 M ammonium chloride with a column of sulphonated phenolic resin. He was able to recover only 87% of the copper. Similar experiments with other transition metals yielded recoveries between 87 and 96%.

The problem of concentrating by ion exchange a trace of one ion in the presence of a large concentration of another requires the use of a resin with an abnormally high selectivity coefficient for the trace ion with respect to the bulk constituent. Chelating resins are ideally suited to the problem provided that the trace element forms a chelate and that the bulk ion does not. For example, Turse and Rieman[75] recovered between 99.0 and 99.6% of the copper from a solution 0.000016 M with copper and 1.0 M with ammonium chloride using columns of Dowex A-1, 1.1–1.7 cm in length, with flow rates as fast as 3.75 cm/min. The absorbed copper was readily removed by 10 ml of 1.0 M hydrochloric acid.

Other examples of the use of a chelating resin include the removal of traces of manganese[76] and copper[77] from sodium chloride and traces of calcium from lithium salts.[78]

The problem of separating a trace constituent from a bulk constituent is more difficult if the trace solute does not form a stable chelate. Nevertheless, Smales and Salmon[79] succeeded in determining the cesium in ocean water where the concentrations of cesium and sodium are respectively 4×10^{-9} M and 0.6 M. They used Zeo-Karb 215, a sulphonated phenolic resin, which has a much larger value of E_{Na}^{Cs} than the more common cation exchangers. More recently, a similar resin has been used to separate traces of cesium from ammonium salts.[80]

The separation of traces from bulk constituents is still more difficult if both the trace and bulk constituents form chelates. In these cases, ion-exchange chromatography (Chapter 8) is frequently the best method.

F.III. MICROQUALITATIVE SPOT TESTS

Extensive work has been done by several Japanese investigators—H. Kakihana, K. Kato, T. Murase, and M. Fujimoto—in the use of ion-exchange resins to increase the sensitivity of spot tests. Although Fujimoto has written a review[81] of this work, many papers in this field have been published after the review. The number of them is too great for complete listing here.

Some of the spot tests depend on the absorption by an appropriate resin of a highly colored complex ion of the metal to be detected. The resin should have a light color, and resins of low crosslinkage are preferred. An example of this type of test is the detection of iron as the red ferrous bipyridyl cation.[82] On a white spot plate put a few beads of a sulphonated polystyrene resin crosslinked with about 2% of DVB and cover them with one drop of the solution to be tested. Add 1 drop each of 3% hydroxylamine hydrochloride to reduce ferric to ferrous iron, 1 M sodium acetate to maintain the proper pH and 0.5% alcoholic bipyridyl. After waiting 20 min for the exchange reaction to occur, examine the resin with a magnifying glass. In the presence of iron, the beads show the red color of the ferrous bipyridyl cation. In the absence of other salts, 5×10^{-11} moles of iron can be detected; but many salts render the test less sensitive. By comparison with similar tests made with standard solutions of iron, a rough estimate of the quantity of iron can be made.

An example of the absorption of a colored anionic complex is the detection of titanium. When a strong-base anion-exchange resin is treated with 1 drop each of the unknown solution, 0.2 N sulphuric acid and 1% hydrogen peroxide, the resin turns yellow if titanium is present because of the absorption of the yellow peroxy anion of titanium.[83]

Another type of ion-exchange spot test involves the absorption by the resin of an anion that forms a highly colored precipitate with the cation to be detected. In the test for cobalt,[84] for example, a few beads of a strong-base anion-exchange resin are treated with 1 drop each of 0.05% nitroso R salt, 0.3% sodium acetate, and, finally, with 1 drop of the unknown solution. If cobalt is present, the resin turns red after 5 min because of the precipitation of the nitroso R salt of cobalt on the surface of the beads. Since the nitroso R salt itself is colored, the sensitivity is improved by heating the resin, after the 5-min period, with 1 drop of 2 M nitric acid to destroy the excess reagent.

A third type of ion-exchange spot test depends on the formation of a dye from a reagent previously absorbed by the resin. In the test for nitrite ion,[85] a few beads of strong-acid exchanger in the hydrogen form are treated with 1 drop of 0.1% m-phenylenediamine solution in 10% acetic acid. After 10 min, 1 drop of the unknown solution is added. In the presence of nitrite ion, the beads turn orange. The test may also be used for nitrate ion by previous reduction with a tiny amount of activated zinc dust.

References

1. A. RINGBORN, *Complexation in Analytical Chemistry*, pp. 209–19, Interscience, New York, 1963.
1a. G. A. CRISTY and R. E. LEMBCKE, *Chem. Eng. Progr.*, **44**, 417 (1948).
2. D. P. SCHWARTZ and M. J. PALLANSCH, *J. Agr. Food Chem.*, **10**, 82 (1962).
3. P. M. STROCCHI, *Ann. Chim. (Rome)*, **44**, 348 (1954).
4. C. W. DAVIES and G. H. NANCOLLAS, *Nature*, **165**, 237 (1950).
5. J. E. POWELL and M. A. HILLER, *J. Chem. Educ.*, **34**, 330 (1957).
6. J. STEINBACH and H. FREISER, *Anal. Chem.*, **24**, 1027 (1952).
7. B. W. GRUNBAUM, W. SCHONIGER and P. L. KIRK, *Anal. Chem.*, **24**, 1857 (1952).
8. COMMITTEE ON ANALYTICAL REAGENTS OF THE AMERICAN CHEMICAL SOCIETY, *Reagent Chemicals*, American Chemical Society, Washington, 1960.
9. J. O. RAY, *J. Inorg. Nucl. Chem.*, **15**, 290 (1960).
10. W. D. COOKE, private communication.
11. A. J. GOUDIE and W. RIEMAN, *Anal. Chem.*, **24**, 1067 (1952).
12. A. HOLROYD and J. E. SALMON, *J. Chem. Soc.*, 947 (1957).
13. K. HAYASHI, T. DANZUKA and K. UENO, *Talanta*, **4**, 244 (1960).
14. G. HESSE and V. BÖCKEL, *Mikrochim. Acta*, 939 (1962).
15. T. WOOD, *J. Chromatog.*, **6**, 142 (1961).
16. J. T. ODENCRANTZ and W. RIEMAN, *Anal. Chem.*, **22**, 1066 (1950).
17. R. CAPELLE, *Anal. Chim. Acta*, **25**, 59 (1961).
18. J. D. WOLSZON and J. R. HAYES, *Anal. Chem.*, **29**, 829 (1957).
19. M. HOLLANDER and W. RIEMAN, *Ind. Eng. Chem., Anal. Ed.*, **18**, 788 (1946).
20. O. SAMUELSON, *Svensk Kem. Tidskr.*, **52**, 115 (1940).
21. D. C. WHITE, *Microchim. Acta*, 254 (1959).
22. H. SCHUMANN and W. KÖLLING, *Z. Chem.*, **1**, 371 (1961).
22a. O. S. GLASOE, *Anal. Chim. Acta*, **28**, 543 (1963).
22b. P. G. JEFFERY and D. WILLIAMS, *Analyst*, **86**, 590 (1961).
22c. V. L. ZOLOTAVIN and V. M. KAZAKOVA, *Zavodsk. Lab.*, **31**, 297 (1965); *Chem. Abstr.*, **62**, 12900 (1965).
23. H. ZACHARIASEN and F. E. BEAMISH, *Anal. Chem.*, **34**, 964 (1962).
24. S. FISHER and R. KUNIN, *Anal. Chem.*, **29**, 400 (1957).
25. J. RŮŽIČKA and J. STARÝ, *Talanta*, **10**, 287 (1963).
26. A. ZEMAN, J. STARÝ and J. RŮŽIČKA, *Talanta*, **10**, 981 (1963).
27. J. P. NELSON, L. E. PETERSON and A. J. MILUN, *Anal. Chem.*, **33**, 1882 (1961).
28. R. JENNESS, *Anal. Chem.*, **25**, 966 (1953).
29. G. BRUNISHOLZ, M. GENTON and E. PLATTNER, *Helv. Chim. Acta*, **36**, 782 (1953).

30. W. H. Hinson, *Spectrochim. Acta*, **18,** 427 (1962).
31. D. J. David, *Analyst*, **87,** 576 (1962).
32. A. H. Wardi and Z. P. Stary, *Anal. Chem.*, **34,** 1093 (1962).
33. O. Samuelson and K. Schramm, *Z. Elektrochem.*, **57,** 207 (1953).
34. O. Samuelson, E. Sjöström and S. Forsblom, *Z. Anal. Chem.*, **28,** 323 (1955).
35. R. M. Carlson and C. M. Johnson, *J. Agr. Food Chem.*, **9,** 460 (1961).
36. B. J. Bornong and J. L. Moriarty, *Anal. Chem.*, **34,** 871 (1962).
37. J. A. Goleb, J. P. Faris and B. H. Meng, *Appl. Spectroscopy*, **16,** 9 (1962).
38. E. Vámos and F. Simon, *Acta Chim. Acad. Sci. Hung.*, **27,** 347 (1961).
39. I. Hornstein, J. A. Alford, L. E. Elliott and P. F. Crowe, *Anal. Chem.*, **32,** 540 (1960).
40. A. Carruthers, J. F. F. Oldfield and H. J. Teague, *Analyst*, **85,** 272 (1960).
41. J. C. B. Fenton, *Clin. Chem. Acta*, **7,** 163 (1962).
42. J. H. Hutchinson and D. H. Labby, *J. Lab. Clin. Med.*, **60,** 170 (1962).
43. A. Dreze, S. Moore and E. J. Bigwood, *Anal. Chim. Acta*, **11,** 554 (1954).
44. H. Rothbart, H. W. Weymouth and W. Rieman, *Talanta*, **11,** 33 (1964).
45. L. H. Anderson, *Ark. Kemi*, **19,** 243 (1962).
46. O. Samuelson, *Z. Anal. Chem.*, **116,** 328 (1939).
47. E. Sjöström and W. Rittner, *Z. Anal. Chem.*, **153,** 321 (1956).
48. O. Samuelson, *Svensk Kem. Tidskr.*, **58,** 247 (1946).
49. K. H. Gustavson, *Svensk Kem. Tidskr.*, **58,** 274 (1946).
50. R. Djurfeldt, J. Hansen and O. Samuelson, *Svensk Kem. Tidskr.*, **59,** 14 (1947).
51. B. D. Polis and J. G. Reinhold, *J. Biol. Chem.*, **156,** 231 (1944).
52. Y. Pomeranz and C. Lindner, *Anal. Chim. Acta*, **15,** 330 (1956).
53. J. P. Hilfiger, *Chim. Anal.*, **31,** 226 (1949).
54. M. Maurer, *Stärke*, **13,** 161 (1961).
55. Y. Pomeranz and C. Lindner, *Anal. Chim. Acta*, **11,** 239 (1954).
56. O. Samuelson, *Svensk Papperstid.*, **48,** 55 (1945).
57. R. A. Shah and A. A. Quadri, *Talanta*, **10,** 1083 (1963).
58. M. Honda, Y. Yoshino and T. Wabiko, *J. Chem. Soc., Japan, Pure Chem. Sect.*, **73,** 348 (1952).
59. G. H. Osborn, *Analyst*, **78,** 220 (1953).
60. F. Helfferich, *Ion Exchange*, McGraw-Hill, New York, 1962.
61. S. J. Ahmad and M. H. Khundkar, *Proc. Pakistan Sci. Conf.*, **12,** Pt. 3, C6–C7 (1960); *Chem. Abs.*, **56,** 8000g (1962).
62. H. Kniested and J. Wahle, *Chem. Tech.*, Berlin, **13,** 10 (1961).
63. W. Dedek, *Z. Anal. Chem.*, **173,** 399 (1960).
64. H. N. S. Schafer, *Anal. Chem.*, **35,** 53 (1963).
65. P. Key, E. Sohl, R. Tiews and C. Bricker, *J. Chem. Educ.*, **40,** 416 (1963).
66. R. Rich, *J. Chem. Educ..* **40,** 414 (1963).
67. F. Nydahl, *Proc. Intern. Assoc. Theor. Appl. Limnology*, **11,** 276 (1951).
68. J. Korkisch, *Z. Anal. Chem.*, **178,** 39 (1960).
69. R. E. Bently, R. P. Parker, D. M. Taylor and M. P. Taylor, *Nature*, **194,** 736 (1962).
70. Y. Noguchi and M. J. Johnson, *J. Bacteriol.*, **82,** 538 (1961).
71. C. Porter, D. Cahill, R. Schneider, P. Robbins, W. Perry and B. Kahn, *Anal. Chem.*, **33,** 1306 (1961).
72. H. Buchwald and L. G. Wood, *Anal. Chem.*, **25,** 664 (1953).
73. L. R. Snyder and B. E. Buell, *Anal. Chem.*, **34,** 689 (1962).
74. J. P. R. Riches, *Chem. and Ind.*, 646 (1947).
75. R. Turse and W. Rieman, *Anal. Chim. Acta*, **24,** 202 (1961).
76. H. Imoto, *Bunseki Kagaku*, **10,** 124 (1961); *Chem. Abs.*, **55,** 24383b (1961).
77. H. Imoto, *Bunseki Kagaku*, **10,** 1354 (1961); *Chem. Abs.*, **57,** 413e (1962).
78. R. L. Olsen, H. Diehl, P. F. Collins and R. B. Ellestad, *Talanta*, **7,** 187 (1961).
79. A. A. Smales and L. Salmon, *Analyst*, **80,** 37 (1955).
80. A. G. Pratchett, Atomic Energy Research Estab. (Gt. Brit.), R3404, 1961.
81. M. Fujimoto, *Chemist-Analyst*, **49,** 4 (1960).
82. M. Fujimoto, *Bull. Chem. Soc. Japan*, **30,** 283 (1957).
83. M. Fujimoto, *Bull. Chem. Soc. Japan*, **29,** 833 (1956).
84. M. Fujimoto, *Bull. Chem. Soc. Japan*, **30,** 278 (1957).
85. M. Fujimoto, *Bull. Chem. Soc. Japan*, **29,** 600 (1956).

CHAPTER 6

THEORY OF
ION-EXCHANGE CHROMATOGRAPHY

IN CHROMATOGRAPHY, two or more constituents of a sample are separated from each other more or less completely by virtue of the differences in their distribution ratios between a stationary phase and a mobile phase in intimate contact with each other. Ion-exchange chromatography makes use of an ion exchanger as the stationary phase; the solutes to be separated are taken up by the solid to different extents by virtue of ion-exchange reactions. Other types of chromatography (ion-exclusion, ligand-exchange, salting-out, and solubilization) also use ion-exchange resins as the stationary phase but do not depend on ion-exchange reactions for the separation.

A. Subdivisions of Ion-exchange Chromatography

There are three types of ion-exchange chromatography, differing from each other in technique and applications. These are (1) elution, (2) frontal, and (3) displacement.

A.I. ION-EXCHANGE ELUTION CHROMATOGRAPHY

In this technique the ion-exchange column is first treated with the same electrolytic solution that is to be used as eluent until the exchangeable ion of the resin is completely replaced by the similarly charged ion of the eluent

$$R_zA + zEl^\pm \longrightarrow R_zEl + A^{\pm z}$$

Next a small amount of sample dissolved in a small volume (preferably of eluent) is added to the top of the column. Then more eluent is passed through the column. Fractions of eluate (or effluent) are collected and analyzed, or the eluate is analyzed automatically.

The elution graph of a mixture of chloride, bromide, and iodide[1] is shown in Fig. 6.1. All the chloride appeared in the first 55 ml of eluate, all of the bromide in the next 45 ml, and all of the iodide between $U = 140$ and $U = 350$. (U is the volume of eluate collected starting at the addition of the sample.) The separation of the halides from each other was quantitative because halide could not be detected in the fraction at $U = 45$ nor in the four fractions between $U = 100$ and $U = 150$. On the other hand, the halides were not obtained as their pure sodium salts. The chloride fractions were mixed with nearly 0.50 M sodium nitrate, and the fractions containing bromide and iodide were mixed with nearly 2.0 M sodium nitrate.

The reader may wonder if the change in concentration of eluent is necessary. If 2.0 M sodium nitrate had been used for the entire elution, all the elution curves would have been shifted to the left, and the curves of chloride and bromide would have partly overlapped, indicating incomplete separation. On the other hand, if 0.50 M sodium nitrate were used for the entire elution, the curve for iodide would have been shifted far to the right and would be spread out more than 1000 ml along the U axis. Then the iodide concentration would have been too small for satisfactory determination.

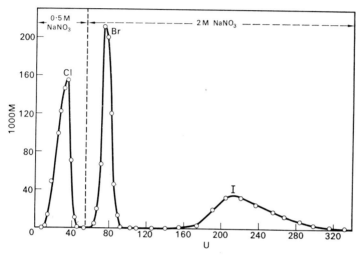

FIG. 6.1. Elution graph of halide mixture. Column $= 6.7$ cm $\times$ 3.4 cm² Dowex 1-X10, 100–200 mesh. Eluent $= 55$ ml 0.50 N sodium nitrate followed by 2.0 N sodium nitrate. (Reproduced by permission from *Analytical Chemistry*.)

A.II. ION-EXCHANGE FRONTAL CHROMATOGRAPHY

In this technique, the resin is first converted to the form of an ion with a smaller selectivity coefficient than any of the ions to be separated. Then the sample solution is passed through the column. No eluent is used.

Figure 6.2 shows the frontal graph obtained by passing a solution of sodium chloride, bromide, and iodide, each 0.10 N, through a column of 12 meq of acetate-form resin. Since acetate has a smaller selectivity coefficient than any of these halides, it is driven down the column in front of the halides. Almost all of the acetate is displaced from the column before any of the halides appears in the eluate. Chloride, having a smaller selectivity coefficient than bromide or iodide, precedes them down the column; and pure 0.30 N sodium chloride emerges from the column as soon as the last of the acetate has left. However, bromide and iodide are also moving down the column, the latter more slowly; and bromide appears in the effluent at $U = 60$. The concentration of bromide in the eluate rises rapidly to 0.15 N while the concentration of chloride falls rapidly to the same value. Eventually the iodide breaks through the column and rises rapidly to 0.10 N, while the concentrations of chloride and bromide decrease to this value. Note that the total concentration of eluate is always approximately equal to the total concentration of influent solution, 0.30 N, and that the final effluent has the same composition as the influent. The only constituent of the sample that is isolated in a pure condition by frontal chromatography is the one with the smallest affinity for the resin.

A.III. ION-EXCHANGE DISPLACEMENT CHROMATOGRAPHY

The procedure of displacement chromatography is a cross between those of elution and frontal chromatography. The resin is first quantitatively converted to the form of an ion with a selectivity coefficient smaller than that of any of the ions to be separated. Then the mixture to be separated is poured into the column; the quantity of sample is much larger than in elution chromatography, amounting to perhaps one-tenth of the total exchange capacity of the column. Finally, an eluent is passed through the column. The eluent must be the salt of an ion with a greater selectivity coefficient than any of the sample ions.

An example is found in the separation of lithium and sodium by Jouy and Coursier.[2] They passed a solution of lithium and sodium sulphates (containing much more of the former compound) into a column (110 cm × 0.79 cm²) of hydrogen-form Dowex 50, 200–400 mesh. When the total quantity of lithium and sodium added amounted to 18% of the total

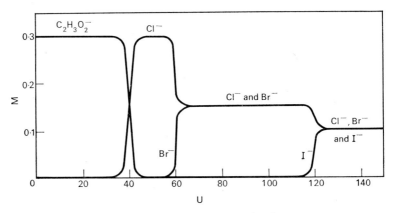

FIG. 6.2. Frontal graph of halide mixture.

exchange capacity of the column, they eluted with 1.0 N ammonium sulphate. The displacement graph is given in Fig. 6.3.

Although Dowex 50 normally prefers hydrogen ion to lithium ion, the situation is reversed in the presence of sulphate because the hydrogen ion in the aqueous phase is largely converted to the bisulphate ion. Therefore hydrogen ion is displaced from the resin by the descending band of lithium and sodium. Similarly, ammonium ion, with the largest selectivity coefficient of all these cations, drives the lithium and sodium in front of it. Thus pure sulphuric acid emerges first from the column at the same concentration as the influent ammonium sulphate, 1.0 N. The lithium and sodium in the mixed band are gradually separated because lithium, having the smaller selectivity coefficient, moves down the column faster than sodium. Thus, after the last of the hydrogen ion is displaced from the column, the lithium sulphate emerges, most of it free from both hydrogen and sodium ions. When all of the lithium has been displaced from the column, a small quantity of pure sodium sulphate emerges. Finally, ammonium sulphate breaks through, and its concentration rapidly rises to 1.0 N.

If the quantity of mixed sulphate added to the column had been much smaller, the ammonium ion would have appeared in the effluent before the last of the lithium was displaced. Then no *pure* sodium sulphate would have been obtained.

A.IV. RELATIVE ADVANTAGES OF THE THREE METHODS

Elution chromatography is the only one of these three procedures by which a quantitative separation of the constituents of the sample can be obtained. It is, therefore, the method most used in quantitative analysis. The contamination of the eluate fractions with large amounts of the eluent is not a serious disadvantage because the eluent can be chosen so as not to interfere with the subsequent determinations of the separated sample constituents. For example, sodium nitrate does not interfere with the titration of the halides with silver nitrate. Another advantage of elution chromatography is that no regeneration of the column is generally required between elutions of similar samples. For example, in the separation of the halides the resin is in the nitrate form at the end of the elution. It is

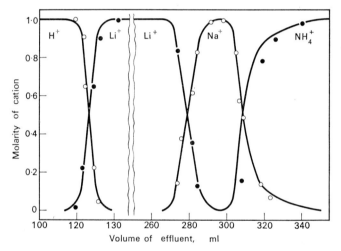

FIG. 6.3. Displacement graph of mixture of lithium and sodium sulphates. (Reproduced by permission of the *Société chimique de France*.)

necessary to pass only about 70 ml of 0.50 M sodium nitrate through the column to displace the more concentrated nitrate from the interstitial solution before adding the next sample of mixed halides. A minor disadvantage of elution chromatography is the requirement that the sample be small. Nevertheless, the concentration of sample constituents in the appropriate eluate fractions is generally large enough to permit good spectrophotometric determinations.

Displacement chromatography is the only one of the three methods by which a pure solution of each sample constituent can be obtained. Therefore it is used chiefly for preparative purposes. The ability to handle large samples adds to its usefulness for this purpose. Still another advantage is that greater flow rates may be used than in elution chromatography. A disadvantage is that the regeneration of the column is time-consuming. For example, after the separation of lithium and sodium by displacement chromatography, the resin is in the ammonium form. It must be converted to the hydrogen form before the next sample of lithium and sodium can be treated. Since the selectivity coefficient of hydrogen vs. ammonium is small, a large quantity of acid must be passed through the column to complete the conversion.

No quantitative separations can be achieved by *frontal* chromatography, and only one sample constituent can be obtained in pure form. For these reasons, ion-exchange frontal chromatography has very little use.

B. Importance of Theoretical Considerations

Let us consider the problem faced by an analyst who wishes to devise a method for the separation of two similarly charged ions (cations, for example) by ion-exchange elution chromatography. Let us further assume that no published literature is available on this particular separation. Unless theoretical guidance can be used, the analyst must make a large number of purely random choices.

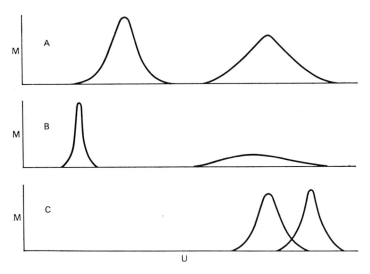

FIG. 6.4. Elution graphs of binary mixtures. (Reproduced by permission from *Record of Chemical Progress*.)

He must select an exchanger from the bewildering number of commercial products differing from each other in matrix, degree of crosslinking, ionogenic group, and particle size. He must decide on the dimensions of his column. Not only must he choose an eluent from the dozens of available electrolytes, but he has to select the concentration. He must decide whether to add a buffer to the eluent or to use it at the pH of its solution in pure water. Since the addition of a complexing agent often improves the efficiency of a separation, he faces the additional problem of which complexing agent, if any, to employ. He also has to select the temperature and flow rate of the elution.

Let us assume that he is sufficiently fortunate in his selections to obtain on his first elution of the mixture two Gaussian elution curves. These may fall into any one of three classes as illustrated[2a] in Fig. 6.4. Graph *A* represents a good separation. Graph *B* represents a quantitative but inefficient separation; there is too great a distance between the two curves, resulting in waste of time and eluent. The overlapping curves of graph *C* indicate incomplete separation.

It is very unlikely that a graph like *A* will be obtained in the first elution after a random selection of the elution conditions. After the first elution, the analyst must generally alter the conditions of elution until he gets a satisfactory graph. The possible number of combinations of elution conditions is so great that the analyst may quit the task before success is achieved unless he is guided by theoretical principles. Furthermore, the probability of success by trial and error decreases as the number of constituents in the sample increases.

Prior to 1954, when equations were first derived to express the effect of the concentration and pH of the eluent on the elution behavior of anions of weak inorganic acids,[3] two different sets of investigators in two different laboratories attempted to develop a method for the analysis of mixtures of the commercially important lower condensed phosphates. Both worked without benefit of theory, and both failed. By the application of theory to this problem, the separation of sixteen members of this group of anions has been achieved.[4]

Even the analyst who merely wishes to apply a previously published method of separation by ion-exchange chromatography will probably profit from a knowledge of the theory. The reason for this is that two different batches of ion-exchange resin bearing the same name, e.g. Dowex 1-X10, 100–200 mesh, will probably differ appreciably from each other in chromatographic behavior. This is understandable when one considers the many competing reactions that occur in the synthesis of the resin (Chapter 1). Therefore exact repetition of a published method does not guarantee a successful separation. For example, whereas a 6.7-cm column of Dowex 1-X10, 100–200 mesh, served to separate the halides as indicated in Fig. 6.1, a 7.3-cm column was needed when a different batch of the "same" resin was used.[1] In cases like this, a knowledge of the theory will generally indicate to the analyst what condition should be changed, in what direction and by what amount.

Two distinct theories may be applied to ion-exchange elution chromatography: (1) the plate-equilibrium or more simply the plate theory, and (2) the mass-transfer or rate theory. We shall consider the plate theory first.

C. Plate Theory of Ion-exchange Elution Chromatography

C.I. HISTORY OF THE PLATE THEORY

Martin and Synge originated the plate theory of chromatography in 1941 and applied it to liquid–liquid systems.[5] Mayer and Tompkins showed in 1947 that it is also useful in ion-exchange chromatography.[6] Further advances were made by Rieman and his coworkers, who derived equations for the effect of the concentration and pH of the eluent,[3] for the effect of both stepwise[7]† and continuous[8] changes in the composition of the eluent, and for calculating the optimum height of the column.[3]

C.II. ASSUMPTIONS OF THE PLATE THEORY

Two basic assumptions are necessary in the derivation of the equations of the plate theory.

(1) It is assumed that the column consists of a certain number of "theoretical plates" and that equilibrium is established between the stationary and mobile phases in each plate before the liquid flows on to the next lower plate. Since equilibrium is approached rather slowly, this assumption is reasonably valid only if the flow rate is slow and if a resin of small particle size and not too great crosslinkage is used. The plate theory *per se* fails to

---

† Although reference 7 deals with salting-out chromatography, the equations are equally valid for ion-exchange chromatography.

indicate the maximum allowable flow rate and particle size. Experimental evidence indicates that 100–200 or preferably 200–400 mesh resin of 4 or 8% DVB and flow rates of about 0.5 cm/min are satisfactory for most separations. This flow rate can be doubled if the ions to be separated are of moderate size and sufficiently different in selectivity coefficients. On the other hand, large or difficultly separable ions may require 2% DVB and flow rates as small as 0.2 cm/min.

(2) The second assumption is that in any plate of the column at any time during the elution the quantity of any counter ion of the sample is much less than the quantity of counter ion of the eluent. This assumption requires small quantities of sample. Again, the plate theory fails to indicate the maximum allowable sample for any given column. Empirical evidence indicates that the quantity of sample should not exceed 0.01 times the quantity of resin (both measured in milliequivalents) for most separations. This ratio can be increased to 0.1 for very easy separations and should be decreased to 0.005 for very difficult ones.

C.III. EQUATION FOR THE PEAK VOLUME OF AN ELUTION CURVE

The purpose of this section is to derive an equation expressing the volume of eluate at the peak of the elution curve U^* in terms of the interstitial volume of the column V and the distribution ratio C of the sample constituent in question. The distribution ratio is defined as the quantity of sample constituent in the resin of any plate divided by the quantity in the interstitial solution of the same plate after equilibration is attained. The experimental determination of the interstitial volume of a column is discussed on p. 133, where a useful table of values is also given.

We assume that only v ml of sample solution is added to the column, where v represents the interstitial volume of one plate. Although this condition is hardly ever realized in actual chromatography, the distortion of the curve caused by larger sample solutions is not serious provided that the ratio of sample solution to peak volume does not exceed 0.01.

Let $L_{n,r}$ denote the fraction of the sample constituent in the interstitial solution of the rth plate after nv ml of eluent has been added to the column. Similarly, let $S_{n,r}$ be the fraction in the resin phase of the same plate. By definition of C,

$$S_{n,r} = CL_{n,r}. \tag{6.1}$$

In the uppermost plate, which is designated plate 0, when $n = 0$, that is after the addition of the sample but before the addition of the eluent,

$$S_{0,0} = CL_{0,0}$$

$$S_{0,0} + L_{0,0} = 1.$$

Therefore

$$L_{0,0} = \frac{1}{1+C},$$

$$S_{0,0} = \frac{C}{1+C}.$$

With the addition of the first v volume of eluent, the solution from the zero plate moves to the first plate, but the sample constituent in the resin remains in plate zero.

$$S_{1,0} + L_{1,0} = S_{0,0} = \frac{C}{1 + C}.$$

By combination of this with eqn. (6.1),

$$L_{1,0} = \frac{C}{(1 + C)^2},$$

$$S_{1,0} = \frac{C^2}{(1 + C)^2}.$$

Similarly, in plate 1,

$$S_{1,1} + L_{1,1} = L_{0,0} = \frac{1}{1 + C},$$

$$L_{1,1} = \frac{1}{(1 + C)^2},$$

$$S_{1,1} = \frac{C}{(1 + C)^2}.$$

After the addition of $2v$ ml of eluent, the condition of plate 0 is

$$S_{2,0} + L_{2,0} = S_{1,0} = \frac{C^2}{(1 + C)^2},$$

$$L_{2,0} = \frac{C^2}{(1 + C)^3},$$

$$S_{2,0} = \frac{C^3}{(1 + C)^3}.$$

The first plate has two sources of sample constituent, the content of the resin before the last transfer of solution and the content of the solution transferred from plate 0. Therefore

$$S_{2,1} + L_{2,1} = S_{1,1} + L_{1,0} = \frac{C}{(1 + C)^2} + \frac{C}{(1 + C)^2} = \frac{2C}{(1 + C)^2},$$

$$L_{2,1} = \frac{2C}{(1 + C)^3},$$

$$S_{2,1} = \frac{2C^2}{(1 + C)^3}.$$

In the second plate,

$$S_{2,2} + L_{2,2} = L_{1,1} = \frac{1}{(1 + C)^2},$$

$$S_{2,2} = \frac{C}{(1 + C)^3},$$

$$L_{2,2} = \frac{1}{(1 + C)^3}.$$

The foregoing calculations can be continued to find the fraction of sample constituent in either phase of any plate at any value of n. The results of these calculations for $L_{n,r}$ up to $r = 5$ and $n = 7$ are given in Table 6.1. The general equation for the mass transfer is

$$S_{n,r} + L_{n,r} = S_{n-1,r} + L_{n-1,r-1}.$$

The general equations for $S_{n,r}$ and $L_{n,r}$ are

$$S_{n,r} = \frac{n!}{(n-r)!\,r!} \frac{C^{n-r+1}}{(1+C)^{n+1}},$$

$$L_{n,r} = \frac{n!}{(n-r)!\,r!} \frac{C^{n-r}}{(1+C)^{n+1}}.$$

These last two equations are valid only if $n \geq r$. Obviously both $S_{n,r}$ and $L_{n,r}$ are zero if $n < r$.

Let p denote the number of plates in the column exclusive of the top plate. Then in the last plate

$$r = p,$$

$$L_{n,p} = \frac{n!}{(n-p)!\,p!} \frac{C^{n-p}}{(1+C)^{n+1}}. \tag{6.2}$$

TABLE 6.1. VALUES OF $L_{n,r}$

	$r = 0$	$r = 1$	$r = 2$	$r = 3$	$r = 4$	$r = 5$
$n = 0$	$\dfrac{1}{1+C}$	0	0	0	0	0
$n = 1$	$\dfrac{C}{(1+C)^2}$	$\dfrac{1}{(1+C)^2}$	0	0	0	0
$n = 2$	$\dfrac{C^2}{(1+C)^3}$	$\dfrac{2C}{(1+C)^3}$	$\dfrac{1}{(1+C)^3}$	0	0	0
$n = 3$	$\dfrac{C^3}{(1+C)^4}$	$\dfrac{3C^2}{(1+C)^4}$	$\dfrac{3C}{(1+C)^4}$	$\dfrac{1}{(1+C)^4}$	0	0
$n = 4$	$\dfrac{C^4}{(1+C)^5}$	$\dfrac{4C^3}{(1+C)^5}$	$\dfrac{6C^2}{(1+C)^5}$	$\dfrac{4C}{(1+C)^5}$	$\dfrac{1}{(1+C)^5}$	0
$n = 5$	$\dfrac{C^5}{(1+C)^6}$	$\dfrac{5C^4}{(1+C)^6}$	$\dfrac{10C^3}{(1+C)^6}$	$\dfrac{10C^2}{(1+C)^6}$	$\dfrac{5C}{(1+C)^6}$	$\dfrac{1}{(1+C)^6}$
$n = 6$	$\dfrac{C^6}{(1+C)^7}$	$\dfrac{6C^5}{(1+C)^7}$	$\dfrac{15C^4}{(1+C)^7}$	$\dfrac{20C^3}{(1+C)^7}$	$\dfrac{15C^2}{(1+C)^7}$	$\dfrac{6C}{(1+C)^7}$
$n = 7$	$\dfrac{C^7}{(1+C)^8}$	$\dfrac{7C^6}{(1+C)^8}$	$\dfrac{21C^5}{(1+C)^8}$	$\dfrac{35C^4}{(1+C)^8}$	$\dfrac{35C^3}{(1+C)^8}$	$\dfrac{21C^2}{(1+C)^8}$

With any constant values of C and p, the value of $L_{n,p}$ is zero when $n = 0$. With increasing values of n, the value of $L_{n,p}$ increases to a maximum and then decreases to almost zero at sufficiently large values of n. Let L^* denote the maximum of $L_{n,p}$ and let n^* denote the corresponding value of the n. At sufficiently large values of p, such as pertain to practical chromatography, the maximum value of $L_{n,p}$ $(n = n^*)$ differs only slightly from the previous and the following values $(n = n^* - 1$ or $n = n^* + 1)$. Therefore we may write as a very close approximation

$$L_{n^*, p} = L_{(n^*-1), p}.$$

Substitution of eqn. (6.2) in this equation yields

$$n^* = p(1 + C). \tag{6.3}$$

From the definitions of the various symbols it is obvious that

$$nv = U, \tag{6.4}$$

$$n^*v = U^*, \tag{6.5}$$

$$vp = V. \tag{6.6}$$

Therefore

$$U^* = V(1 + C) \tag{6.7}$$

This is a very important equation in ion-exchange chromatography. It will be used later to explain the effect of the concentration and pH of eluent on the elution.

C.III.a. *Constancy of C*

In the derivation of eqn. (6.7) it was tacitly assumed that C is constant. We shall now test the validity of this assumption by a mathematical analysis. Incidentally, we shall derive a useful equation relating C to the concentration of eluent and to the selectivity coefficient of the sample ion vs. the eluent ion El (assumed to be univalent). For the sake of simplicity, we shall assume that neither the sample ion A^{+z} nor the eluent ion $El^{\pm}$ forms complexes with any constituent of the solution. Complexing in this sense is meant to include the reaction between a basic anion such as acetate and hydrogen ion.

The exchange reaction that occurs during the elution is

$$A^{\pm z} + z\text{REl} \rightleftharpoons \text{R}_z\text{A} + z\text{El}^{\pm}$$

$$E = \frac{[\text{El}^{\pm}]^z [\text{R}_z\text{A}]}{[\text{A}^{\pm z}][\text{REl}]^z}.$$

Square brackets denote the molarities of the dissolved ions or mole fractions of the resin constituents.

By definition of the distribution ratio

$$C = \frac{wQ[\text{R}_z\text{A}]}{v[\text{A}^{\pm z}]}, \tag{6.8}$$

where w denotes the weight in grams of dry resin in one plate and Q the specific exchange capacity of the resin. By combination of the last two equations,

$$C = \frac{wQE[\text{REl}]^z}{v[\text{El}^{\pm}]^z}.$$

Since the quantity of sample is much smaller than the amount of eluent, [REl] $\simeq$ 1. For the same reason, [El$^\pm$] represents not only the concentration of eluent ion in the interstitial solution but also the concentration of the eluent solution in the reservoir. Also $w/v = W/V$, where W is the weight of dry resin in the whole column. Thus

$$C = \frac{WQE}{V[El^\pm]^z}. \tag{6.9}$$

It should be noted that z is the valence of the sample ion, not of the eluent.

In any given elution, W, Q, and V are constant. Thus, the constancy of C depends on the constancy of E and [El$^\pm$]. Clearly if the eluent concentration is changed during the elution, the value of C will be altered, and eqn. (6.7) will not be applicable.

Since E depends on the composition of the resin (p. 40) and since the value of [REl] changes as the band of sample ion moves down the column, it may be argued that C can not be constant. Although this is strictly true, the variation in E is negligible in a properly conducted elution because [REl] is constant at a value very close to unity.

In summary, C is constant, and eqn. (6.8) is valid unless the composition of the eluent is changed during the elution.

C.IV. EFFECT OF CONCENTRATION OF ELUENT

The upper graph of Fig. 6.5 shows an unsatisfactory elution of 0.2 meq each of sodium oxalate and potassium bromide through a column, $6.7\,\text{cm} \times 3.8\,\text{cm}^2$, of Dowex 1-X10, 150–200 mesh, with 0.0500 M sodium nitrate as eluent.[8] The lower graph shows that the use of 0.100 M sodium nitrate in the same column gave an excellent separation.

Equations (6.8) and (6.9) hold the key to the explanation of the improvement caused by the change in concentration of eluent. By combination of these equations

$$U^* = \frac{WQE}{[El^\pm]^z} + V. \tag{6.10}$$

Since V is usually appreciably smaller than the first term on the right, we may write as a crude approximation

$$U^* = \frac{WQE}{[El^\pm]^z}. \tag{6.11}$$

Since WQE is usually independent of the concentration of the eluent, eqn. (6.11) indicates that doubling the value of [El$^\pm$] will decrease the peak volumes of bromide and oxalate respectively to one-half and one-quarter of their former values. Equation (6.10) gives a more accurate calculation of the effect of a change in eluent concentration but requires a knowledge of value of V. From the known values of U^* and [El$^\pm$] of the first elution, WQE is evaluated. Then U^* can be calculated for any other value of [El$^\pm$].

Table 6.2 shows a very satisfactory agreement between the observed and calculated values of U^*. Although agreement within 3% or less is generally found, there are a few combinations of sample ion, resin, and eluent for which eqn. (6.10) gives very poor agreement between calculated and observed values of U^*. For example, the elution of iodide through the same column with 2.00 M sodium nitrate gave $U^* = 220$, from which $E = 111$. From this value of E, it is readily calculated that the value of U^* with 1.00 M sodium nitrate should be 431. The observed value was 323. Discrepancies like these can be interpreted as resulting from variations in E with changes in eluent concentrations. Such variations are not

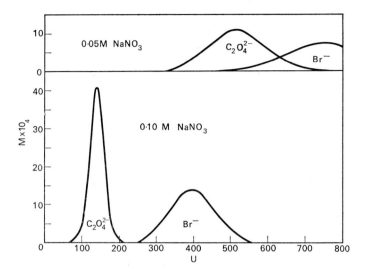

FIG. 6.5. Effect of concentration of eluent.

TABLE 6.2. EFFECT OF CONCENTRATION OF ELUENT ON PEAK VOLUMES ($V = 10$)

Ion	0.0500 M NaNO$_3$	0.100 M NaNO$_3$		
	Observed	Observed	Eqn. (6.10)	Eqn. (6.11)
Br$^-$	760	396	385	380
C$_2$O$_4$$^{2-}$	515	139	139	129

(Data used by permission from *Analytical Chemistry*.)

surprising in view of the fact that selectivity coefficients are defined in terms of concentrations, not activities.

Changing the concentration of the eluent is effective in improving a poor separation chiefly when the ions to be separated have different valences. The separation of ions of the same valence is improved by changing the eluent concentration only if the original eluent was so concentrated that both sample ions had small and nearly equal U^* values; then a decrease in concentration increases both peak volumes and may achieve a good separation.

If the elution curves of two ions overlap with the ion of smaller valence emerging first, a decrease in eluent concentration should be applied. For example, in an elution through a column of Dowex 50 ($V = 31$) with 1.00 M hydrochloric acid, the curves of potassium and magnesium overlapped slightly. Potassium ($U^* = 85$) preceded magnesium ($U^* = 120$). With 0.700 M hydrochloric acid, the separation was quantitative with peak volumes at 112 and 217 respectively.

C.V. EFFECT OF pH OF ELUENT

Equation (6.8) and hence also eqns. (6.9), (6.10), and (6.11), are not applicable to the elution of a weak acid on its salt with an eluent of low pH. Let us consider first the case of a weak monoprotic acid HA. It will exist in solution as both HA and A^-. Therefore instead of eqn. (6.8) we must write

$$C = \frac{wQ[RA]}{v[A^-] + v[HA]}. \tag{6.12}$$

By applying the equation for the ionization constant K of the weak acid, $[HA]$ can be eliminated from this equation.

$$C = \frac{wQ[RA]}{v[A^-]} \frac{K}{K + [H^+]}, \tag{6.13}$$

where $[H^+]$ is the hydrogen-ion concentration of the eluent or the interstitial solution. Then it follows that

$$C = \frac{WQE}{V[El^-]} \frac{K}{K + [H^+]}, \tag{6.14}$$

$$U^* = \frac{WQE}{[El^-]} \frac{K}{K + [H^+]} + V. \tag{6.15}$$

In the case of a triprotic acid, H_2A, eqn. (6.12) is replaced by

$$C = \frac{wQ[R_3A] + wQ[R_2HA] + wQ[RH_2A]}{v[A^{3-}] + v[HA^{2-}] + v[H_2A^-] + v[H_3A]}$$

which leads to

$$C = \frac{WQ}{V} \cdot \frac{\dfrac{E_1 K_1 [H^+]^2}{[El^-]} + \dfrac{E_2 K_1 K_2 [H^+]}{[El^-]^2} + \dfrac{E_3 K_1 K_2 K_3}{[El^-]^3}}{[H^+]^3 + K_1 [H^+]^2 + K_1 K_2 [H^+] + K_1 K_2 K_3}, \tag{6.16}$$

where E_1, E_2, and E_3 are the selectivity coefficients of the primary, secondary, and tertiary anions, respectively, and K_1, K_2, and K_3 are first, second, and third ionization constants.

Analogous equations for di- and tetraprotic acids are given elsewhere.[3]

Two characteristics of eqns. (6.14) and (6.16) should be noted. In the first place, at sufficiently low hydrogen-ion concentration, each of these equations can be simplified to eqn. (6.9). "Sufficiently low hydrogen-ion concentration" means that $[H^+] < 0.01\,K$ for a monoprotic acid or that $[H^+] < 0.01\,K_1$ for a triprotic acid. In the second place, the peak of the elution curve of a weak acid can be shifted simply by changing the pH of the eluent. Therefore, if two acids of different strengths give overlapping elution curves, the separation can be improved by changing the pH of the eluent.

This is illustrated in Fig. 6.6. Curve 3 represents the elution of the anion of a strong acid through an anion-exchange resin; it was calculated by eqns. (6.7), (6.10), (6.29), and (6.30) (p. 107) from the following assumptions: $WQ = 69.6$, $E_B = 2.00$, $H = 19.1$, $A = 3.80$, $V = 29.0$, $[El] = 1.00$, $p = 400$, and $J = 0.500$ (the number of millimoles of sample ion). Curve 2 represents the elution of a weak acid; it was calculated with eqn. (6.15) and the additional assumptions that $E_A = 19.0$ and that pH $=$ p$K - 1.00$. It is seen that the curves overlap. However, by decreasing the pH 0.20 units, the curve of the weak acid is shifted to position 1. Since the change in pH (other conditions remaining constant) does not affect the elution curve of the strong acid, a good separation is obtained at pH $=$ p$K - 1.20$. Curve 4 indicates that a good separation is also obtained at pH $=$ p$K - 0.70$.

Since eqn. (6.15) was used to calculate the positions of curves 1, 2, and 4 in Fig. 6.6, this figure can not serve as an indication of the reliability of equations such as this. The following experiment was performed[3] to test the validity of equations such as (6.15) and (6.16) that predict the effect of pH on the elution behavior of weak polyprotic acids.

Orthophosphoric acid was eluted three times with 0.15 M potassium chloride through a column of Dowex 50-X10 with known values of W, Q, and V. Each time the eluent was buffered at a different pH. From each elution curve, C was evaluated. These three values of C along with the corresponding values of $[\text{'H}^+]$ were substituted in eqn. (6.16) to yield three simultaneous equations that were solved for the values of E_1, E_2, and E_3. Then various values of hydrogen-ion concentration were assumed and eqn. (6.16) was solved for the value of C corresponding to each pH. Thus curve 1 of Fig. 6.7 was constructed. The black circles on

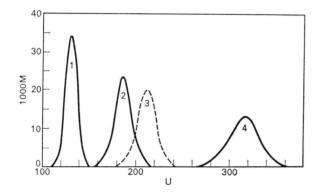

FIG. 6.6. Effect of pH of eluent. Curve 1, weak acid, pH $=$ pK–1.20. Curve 2, weak acid, pH $=$ pK–1.00. Curve 3, strong acid. Curve 4, weak acid pH $=$ pK–0.70. (Reproduced by permission of Interscience Publishers.)

this curve represent the three elutions that were used to evaluate the three selectivity coefficients. Then other elutions were done at other pH values, and the corresponding C values were calculated from the elution graphs. These points, represented by open circles, fit well on the curve and indicate that orthophosphoric acid follows eqn. (6.16) closely. Figure 6.7 also contains similar curves for pyrophosphoric and triphosphoric acid as examples of tetra- and pentaprotic acids. Although the complete equation for pyrophosphoric acid involves four selectivity coefficients, solutions of this acid and its salts within the pH range studied contain negligible amounts of the univalent ion. Hence only the other three selectivity coefficients need to be evaluated and three elutions (black circles in the figure) suffice. Similarly, with triphosphoric acid, only E_3, E_4, and E_5 are required from pH 3.5 to 11.2.

Equations like (6.14) and (6.16) for the effect of the hydrogen-ion concentration of the eluent on the distribution ratio are not applicable to organic acids except very polar organic acids such as oxalic. Indeed, the peak volume of many organic acids is *increased* by a decrease in the pH of the eluent—an effect directly opposite to that predicted by these equations. In the derivation of eqns. (6.14) and (6.16) it was tacitly assumed that the sorption of the nonionized acid by the resin is negligible. Although this assumption is acceptable for inorganic acids, it is not valid for moderately hydrophobic organic acids. In such cases, the resin often exerts a stronger attraction for the nonionized acid by van der Waals forces

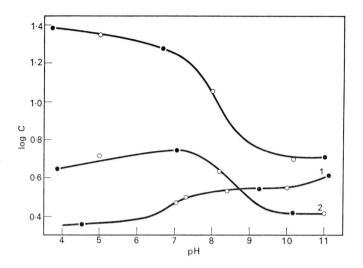

FIG. 6.7. Effect of pH of eluent. Curve 1, orthophosphate, eluent = 0.25 M Cl⁻. Curve 2, pyrophosphate, eluent = 0.25 M Cl⁻. Curve 3, triphosphate, eluent = 0.15 M Cl⁻. (Reproduced by permission from *Analytical Chemistry*.)

than for its anion by ion exchange. Then a decrease in pH converts more of the acid to the nonionized form and causes an increase in C.

A few words of advice about the buffering of the eluents are in order. No problem is encountered in buffering an eluent in the range where a mixture of ammonia and an ammonium salt is a satisfactory buffer. For example, a 0.15 M chloride eluent of pH 8.22 was prepared[3] by adding to a solution 0.14 M with potassium chloride and 0.01 M with ammonium chloride sufficient 15 M ammonia (about 0.46 ml/l) to give the desired pH. Chloride is the only anion in this eluent.

On the other hand, the preparation of an eluent buffered at a pH about 5 presents a problem. A mixture of acetic acid and an acetate is a good buffer in this range for most purposes. However, the addition of an acetate to a chloride buffer introduces another ion that competes for resin sites with both the chloride and the anions of the sample. In other words, the acetate functions also as an additional eluting ion and may invalidate the equations relating C to the concentrations of the eluents and of the hydrogen ion.

Two precautions are necessary in order to avoid serious trouble from this source. In the first place, the buffer can be chosen so that its anion has a small selectivity coefficient; acetate buffers are very good in this respect. In the second place, the concentration of buffer should be small in comparison with the major constituent of the eluent. On the other hand, excessively dilute buffers permit too great a variation of pH in the interstitial solution. For example,[3] when a sample of triphosphate was eluted with potassium chloride poorly buffered at pH 8.00, the pH of the eluate was originally 8.00, rose to 8.27, fell to 7.88, and finally rose to 7.97. At the pH of this eluent, the triphosphate exists principally as $HP_3O_{10}^{4-}$ and $P_3O_{10}^{5-}$. Since the resin at first absorbs more $HP_3O_{10}^{4-}$ than $P_3O_{10}^{5-}$, the equilibrium

$$HP_3O_{10}^{4-} \rightleftharpoons H^+ + P_3O_{10}^{5-}$$

is first displaced to the left with consequent rise in pH. Later, as the triphosphate is driven from the resin, the equilibrium is displaced in the opposite direction, and the pH falls. Finally, after all the triphosphate has been driven from the resin, the pH of the eluate approaches that of the eluent.

The pH of the eluent also affects the separation of cations if the eluent contains an anionic complexing agent. This is discussed in the next section.

C.VI. EFFECT OF COMPLEXING AGENTS IN THE ELUENT

Let us consider first the elution of two cations, A^{+z} and B^{+z}, through a cation-exchange resin with a noncomplexing eluent such as ammonium nitrate. Let us assume that A^{+z} emerges from the column before B^{+z} but that the two elution curves overlap badly. The addition of some anionic complexing agent such as ammonium lactate, tartrate, citrate, or ethylenediaminetetraacetate very frequently changes the poor separation into a good one.

The cation A^{+z} reacts reversibly with one or more of the ligands to give a complex whose formula may be written AL, AL_2, AL_3, AHL, $A(HL)_2$, ... where the charge on the complex will depend on the charges on the cation and the ligand but is always less than z and may be zero or negative. If the charge on the complex is zero or negative, it will ordinarily be absorbed by the resin to a negligible extent. Then the C value of A^{+z} is decreased in proportion to the fraction of A^{+z} converted to the complex, and C becomes zero when the complexation is complete. Therefore the addition of the complexing agent moves the elution graph of A^{+z} to the left. Even if the complex is positively charged, it will probably have a smaller affinity for the resin than the uncomplexed cation; and the effect of the addition of the complexing agent is still to shift the elution curve to the left.

Of course, the cation B^{+z} will probably also react with the ligand; and its elution curve will also be moved toward the left. However, if the stability constants of the two complexes are different, the effect of the ligand on the two elution curves will be different. Very fortunately, among any related group of cations, the selectivity coefficients of sulphonated polystyrene resins follow the reverse order of the stability constants of the complexes with any of the common anionic ligands. Thus the alkaline earths are eluted through Dowex 50 by ammonium chloride in the sequence Ca < Sr < Ba < Ra; and the stability constants of their complexes follow the opposite order. Thus the effect of the addition of the ligand is to increase the distance between the elution graphs and hence to improve the separation. The same is true of the rare earths and of the actinides.

If only one complex were formed between a given cation and a given ligand, it would be a simple task to derive an equation relating C to the selectivity coefficient of the cation, the concentration of the ligand, and the stability constant of the complex. Although such equations always predict qualitatively the effect of adding or changing the concentration of the ligand, the quantitative agreement between the predictions of the equation and experimental results is generally poor. The formation of more than one complex from a given cation and a given ligand may be the cause of the discrepancy, e.g. $CaLc^+$ and $CaLc_2$ in the case of calcium lactate.

Since most anionic ligands are the anions of weak acids, the concentration of ligand in a given solution may be decreased by decreasing the pH. Thus pH is also an important factor in separations involving complexation.

The cations of many transition metals can react with a sufficient number of anionic ligands to impart a negative charge to the complex, e.g. $FeCl_4^-$ or $CoCl_3^-$. Then the metal is absorbed by anion-exchange resins. The C value of lead(II), for example, with Dowex 1-X8 is zero if the concentration of hydrochloric acid in the eluent is zero (and if no other complexing agents are present). As the concentration of hydrochloric acid in the eluent is increased, C rises. However, the distribution coefficient reaches a maximum[9] of 26 in 1.2 M HCl. Further increases in the concentration of hydrochloric acid decrease the value of C; in 3.5 M HCl, C is 13.

No simple satisfactory equation has been derived for expressing the value of the distribution coefficient of a transition metal as a function of the concentration of hydrochloric acid. The following naïve discussion may serve to clarify the *qualitative* effects. The increases in the concentration of hydrochloric acid up to 1.2 M serve to convert increasing fractions of the lead into the uncharged ($PbCl_2$) or the anionic ($PbCl_3^-$ and $PbCl_4^{-2}$) complexes. Cationic lead (Pb^{2+} and $PbCl^+$) is excluded from the resin by Donnan's principle; the neutral complex is not excluded, and the anionic complexes may undergo exchange reactions. Therefore C increases with increasing concentration of hydrochloric acid until only a very small fraction of the lead remains as cations. This point is reached at about 1.2 M HCl. Further increases in the concentration of hydrochloric acid decrease C by driving the following equilibria to the right.

$$R_2PbCl_4 + 2Cl^- \rightleftharpoons 2RCl + PbCl_4^{-2}$$

$$RPbCl_3 + Cl^- \rightleftharpoons RCl + PbCl_3^-$$

A more rigorous treatment of this subject may be found elsewhere.[10-13]

Many other metals show maxima in the plots of C versus the concentration of hydrochloric acid.

Glycols with the hydroxyl groups on adjacent carbon atoms form complexes with borate ion, BO_2^-, and can be separated by anion-exchange chromatography.[14] The eluent is sodium borate $NaBO_2 \cdot H_2O$ containing sufficient sodium hydroxide to prevent appreciable hydrolysis of the borate ion. For this rather unusual type of ion-exchange chromatography, the following equation has been derived.[15]

$$\frac{WQ[RBO_2]}{CV} = \frac{K_1}{K_2}[BO_2^-] + \frac{1}{K_2}.$$

K_1 is the stability constant of the complex GBO_2^-, where G represents the glycol. K_2 is the equilibrium constant of the reaction

$$G + RBO_2 \rightleftharpoons RGBO_2$$

$[BO_2^-]$ is the molarity of sodium borate in the eluent, and $[RBO_2]$ is the mole fraction of RBO_2 in the resin (less than unity because of the presence of sodium hydroxide in the eluent). For any given eluent and glycol, all quantities in the equation can be determined readily except K_1 and K_2. A plot of $WQ[RBO_2]/CV$ vs. $[BO_2^-]$ should be linear, and K_1 and K_2 can be evaluated from the intercept and slope of the line. In accordance with the theory, the plots for the five glycols investigated[15] were linear up to $[BO_z^-] = 0.25$.

C.VII. EQUATION OF THE ELUTION CURVE

Heretofore, we have studied rather simple equations that relate the position of the peak of the elution curve to the concentration and pH of the eluent. Now we shall consider the equation that expresses the concentration of a sample constituent in any fraction of the eluate as a function of the volume of eluate.

Since the solution flows from the last plate into the eluate, the concentration of the sample constituent in the eluate is

$$M = \frac{JL_{n,p}}{v},\qquad (6.17)$$

where J is the number of millimoles of the constituent of the sample. Thus eqns. (6.2), (6.4), and (6.17) enable one to calculate M for any value of U if values of C, p, J, and v are known or assumed.

However, eqn. (6.2) is extremely awkward to use. Whereas very few mathematical tables give factorials beyond 200!, the values of n in almost all chromatographic separations run up to several thousand.

The mathematical labor is diminished somewhat by applying Stirling's approximation

$$y! = e^{-y}y^y \sqrt{(2\pi y)}$$

to eqn. (6.2). Then

$$L_{n,p} = \frac{1}{\sqrt{(2\pi)}} \frac{n^{n+\frac{1}{2}}}{(n-p)^{n-p+1/2} \cdot p^{p+1/2}} \frac{C^{n-p}}{(1+C)^{n+1}},\qquad (6.18)$$

$$\begin{aligned}
\log L_{n,p} = {} & (n+\tfrac{1}{2}) \log n + (n-p) \log C - 0.39909 \\
& - (n-p+\tfrac{1}{2}) \log (n-p) - (p+\tfrac{1}{2}) \log p \\
& - (n+1) \log (1+C).
\end{aligned}\qquad (6.19)$$

Equations (6.18) and (6.19) are still very inconvenient. To illustrate the difficulty, let us calculate M for $p = 600$, $C = 9.0000$, $n = 6100$, $v = 0.13300$, and $J = 1.0000$.

$$\begin{aligned}
\log L_{n,p} = {} & 6100.5 \log 6100 + 5500 \log 9.0000 - 0.39909 - 5500.5 \log 5500 \\
& - 600.5 \log 600 - 6101 \log 10.000,
\end{aligned}$$

$$\begin{aligned}
= {} & 23{,}092.405665 + 5248.32000 - 0.39909 - 20{,}573.85018 \\
& - 1668.27908 - 6101.00000,
\end{aligned}$$

$$= -2.76527,$$

$$L_{n,p} = 0.0017168,$$

$$M = 0.011843.$$

It should be noted that in order to have five significant figures in the final value of $L_{n,p}$, five decimal places must be carried in the computation of $\log L_{n,p}$; this requires carrying ten significant figures in some of the multiplications. Clearly eqn. (6.18) or (6.19) is not convenient for calculating the concentration of the sample constituent in any specified fraction of the eluate.

Actually, three significant figures in the value of $L_{n,p}$ would suffice. In the foregoing calculations, $L_{n,p}$ was computed to five significant figures in order that a more reliable comparison could be made between the results obtained by eqn. (6.19) and a much more convenient eqn. (6.25) to be derived shortly.

C.VII.a *The Gaussian Elution Equation*

An ion-exchange elution that is performed with sufficiently small sample, flow rate, and particle size gives an elution graph that closely resembles the Gaussian curve, or the curve of normal distribution. Furthermore, with a sufficiently large value of p, such as pertains in any satisfactory separation by ion-exchange chromatography, the value of M for any given values of p, U, V, C, and J, as calculated by eqns. (6.19) and (6.17) checks very closely the value calculated by the Gaussian equation with properly chosen parameters. Therefore a reliable and comparatively simple elution equation (M as a function of U) can be derived by combining eqns. (6.17) and (6.18) with the Gaussian equation.

The simplest form of the Gaussian equation is

$$y = \exp(-x^2)$$

or in our variables

$$M = \exp(-U^2). \tag{6.20}$$

This is not satisfactory for three reasons: (1) It predicts that the peak volume, U^*, is located at $U = 0$, a condition that never obtains in chromatography. (We define U as the volume of eluate collected from the beginning of the addition of the sample.) (2) It predicts that the peak value of M, i.e. M^*, will be unity whereas M^* is usually much smaller and depends on J, C, and p. (3) It predicts an invariant width of the graph, whereas the width actually depends on C and p. (To compare the width of two Gaussian curves, the width must be measured at some constant fraction of M^*; it is customary to measure the width at M^*/e.) The first two short-comings of eqn. (6.20) are eliminated by substituting $U - U^*$ and by inserting M^* as a factor on the right-hand side. Thus eqn. (6.21)

$$M = M^* \exp - (U - U^*)^2 \tag{6.21}$$

is better than eqn. (6.20) but still has invariant width. We obtain the correct width of the graph by inserting the appropriate factor in front of $(U - U^*)^2$. We shall now evaluate this factor.

Let the appropriate factor be a^2. Then

$$M = M^* \exp\{-a^2(U - U^*)^2\}. \tag{6.22}$$

Since the total amount of any sample constituent eventually appears in the eluate,

$$J = \int_{-\infty}^{+\infty} M dU = M^* \int_{-\infty}^{+\infty} \exp\{-a^2(U - U^*)^2\} dU$$

Since the value of this definite integral is $\sqrt{(\pi)}/a$,

$$J = \frac{M^* \sqrt{\pi}}{a},$$

$$a = \frac{M^* \sqrt{\pi}}{J}. \tag{6.23}$$

From eqn. (6.17),

$$M^* = \frac{JL^*}{v}, \tag{6.24}$$

where L^* is the maximum value of $L_{n,p}$. Therefore

$$a = \frac{\sqrt{(\pi)}\,L^*}{v}.$$ (6.25)

To evaluate L^*, eqn. (6.18) is rewritten with L^* in place of $L_{n,p}$ and n^* in place of n. Then this equation is combined with eqn. (6.3) to yield

$$L^* = \frac{1}{\sqrt{\{2\pi p C(1+C)\}}}.$$ (6.26)

Substitution of this expression in eqn. (6.25) yields

$$a = \frac{1}{v\sqrt{\{2pC(1+C)\}}}.$$

From eqn. (6.6)

$$a = \frac{1}{V}\sqrt{\left\{\frac{P}{2C(1+C)}\right\}}.$$ (6.27)

Substitution of this value of a and the value of V from eqn. (6.7) in eqn. (6.22) yields the Gaussian elution equation

$$M = M^* \exp\left\{\frac{-p}{2}\left(\frac{1+C}{C}\right)\left(\frac{U-U^*}{U^*}\right)^2\right\}.$$ (6.28)

It is more convenient in its logarithmic form:

$$\log M = \log M^* - 0.217p\left(\frac{1+C}{C}\right)\left(\frac{U-U^*}{U^*}\right)^2.$$ (6.29)

Another useful equation is obtained by combining eqns. (6.24), (6.26), and (6.6):

$$M^* = \frac{J}{V}\sqrt{\left\{\frac{p}{2\pi C(1+C)}\right\}}.$$ (6.30)

In order to compare eqns. (6.29) and (6.19) for both convenience and results, we shall now calculate M by eqn. (6.29) with the same assumed values as on p. 100: $p = 600$, $C = 9.0000$, $n = 6100$, $v = 0.13300$, and $J = 1.0000$. From eqn. (6.6), $V = 79.800$. From eqn. (6.30), $N^* = 0.012909$. From equation (6.3), $n^* = 6000$. From eqn. (6.5), $U^* = 798.00$. From equation (6.4), $U = 811.30$. Finally, from eqn. (6.28), $M = 0.011769$. This value checks satisfactorily with the value ($M = 0.011843$) calculated by eqn. (6.19).

The relative difference between the results of these two equations would be somewhat greater nearer the end of the graph, i.e. at larger values of U. It would also be greater for smaller values of p, i.e. for shorter columns. From a theoretical point of view, the values obtained by eqn. (6.18) are more nearly correct than those obtained by eqns. (6.28) or (6.29); but the differences are always negligible. In the calculations of eqn. (6.29) there is no serious loss of significant figures as in the case of eqn. (6.18); and the use of three significant figures in the multiplications and divisions gives sufficient accuracy in the answer.

C.VII.b. *Evaluation of the Number of Plates*

If the theoretical plate had a constant known height, it would be possible, of course, to calculate the number of plates in any given column from its height. Unfortunately this is not the case. The plate height depends on the nature of the resin, the flow rate, the temperature, the eluent, and even the migrant ion. Therefore it is necessary to evaluate p for any given set of conditions. The method consists of performing an elution, plotting the graph, and applying eqn. (6.31), which we shall now derive.

We define U_a as the value of U when $M = M^*/e$. Substituting these values for U and M in eqn. (6.28), we find

$$\exp{(-1)} = \exp\left\{\frac{-p}{2}\left(\frac{C+1}{C}\right)\left(\frac{U_a - U^*}{U^*}\right)^2\right\}$$

$$-1 = \frac{-p}{2}\left(\frac{C+1}{C}\right)\left(\frac{U_a - U^*}{U^*}\right)^2$$

$$p = 2\left(\frac{C}{1+C}\right)\left(\frac{U^*}{U_a - U^*}\right)^2. \qquad (6.31)$$

C.VII.c. *Experimental Test of the Gaussian Elution Equations*

A rather typical graph of the elution of a single solute through an ion-exchange column is plotted in Fig. 6.8. The column measured 6.2 cm $\times$ 3.8 cm². Its interstitial volume was 9.0 ml. The circles represent analyzed fractions of the eluate .The continuous graph is drawn through the circles and is the experimental elution curve. Careful observation reveals that the graph is not perfectly symmetrical but that it "tails" slightly, i.e. the descending slope is less steep than the ascending slope. This is a rather common characteristic of elution graphs.

In spite of this slight asymmetry of this graph we shall compare it with the ideal elution graph as expressed in eqn. (6.29). From the graph, $U^* = 133.9$, $M^* = 340 \times 10^{-6}$. From eqn. (6.7), $C = 13.9$. The broken horizontal line represents $M = M^*/e = 125 \times 10^{-6}$. This line intersects the experimental curve at $U = 121.8$ and $U = 148.5$. These are the two values of U_a. Substitution of 121.8 for U_a in eqn. (6.31) yields $p = 230$. Substitution of 148.5 for U_a yields $p = 172$. If the curve were perfectly symmetrical, the two values of p would be in perfect agreement. The average value of p is 201. Thus P, the number of plates per cm, is 32. Finally, by substituting $M^* = 340 \times 10^{-6}$, $U^* = 133.9$, $C = 13.9$, and $p = 201$ in eqn. (6.29) the value of M corresponding to any value of U can be calculated. Thus the broken curve of Fig. 6.8 is obtained. The agreement between the experimental and theoretical curve is reasonably good.

C.VII.d. *Equation for Calculating the Ideal Length of Column*

If two constituents of a sample give overlapping elution curves and if the previously discussed theoretical considerations indicate that a satisfactory separation cannot be achieved by adding a complexing agent or by changing the pH or concentration of the eluent, the last resort is to use a longer column. A simple equation based on eqns. (6.28) and (6.7) enables the analyst to calculate the minimum length of column that will give a

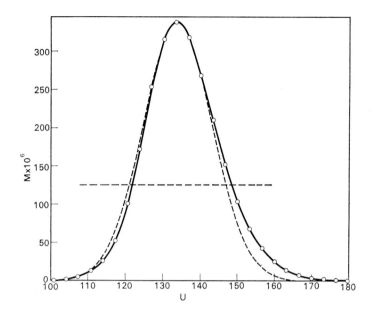

FIG. 6.8. Comparison of theoretical and experimental elution curves.

quantitative separation. Let us consider a separation to be quantitative if there is 0.05% of cross contamination.

Let U_i denote the value of U at which the two elution curves intersect when there is 0.05% cross contamination. Therefore the first constituent of the sample (denoted by subscript 1) has been 99.95% eluted when $U = U_i$. Comparing the usual form of the Gaussian equation,

$$y = \frac{1}{\sqrt{(2\pi)}} \exp\left(\frac{-x^2}{2}\right),$$

with eqn. (6.28), we see that

$$\left(\frac{U - U^*}{U^*}\right) \sqrt{\left\{\frac{p(1 + C)}{C}\right\}}$$

corresponds to x. From tables of probability integrals[16]

$$\int_{x=-\infty}^{x=3.29} y\,dx = 0.9995.$$

Therefore

$$\left(\frac{U_i - U_1^*}{U_1^*}\right) \sqrt{\left\{\frac{p_1(1 + C_1)}{C_1}\right\}} = 3.29$$

$$U_i = U_1^* + 3.29U_1^* \sqrt{\left\{\frac{C_1}{p_1(1 + C_1)}\right\}}.$$

But U_i is also the point where 0.05% of the second constituent has been eluted. We now use the negative value of the square root and find that

$$\left(\frac{U_2{}^* - U_i}{U_2{}^*}\right) \sqrt{\left\{\frac{p_2(1 + C_2)}{C_2}\right\}} = 3.29,$$

$$U_i = U_2{}^* - 3.29 U_2{}^* \sqrt{\left\{\frac{C_2}{p_2(1 + C_2)}\right\}}.$$

Now by equating the two values of U_i, substituting P_1H and P_2H for p_1 and p_2 respectively, and solving for $\sqrt{H}$, we find that

$$\sqrt{H} = \frac{3.29}{C_2 - C_1} \left[\sqrt{\left\{\frac{C_2(1 + C_2)}{P_2}\right\}} + \sqrt{\left\{\frac{C_1(1 + C_1)}{P_1}\right\}} \right]. \qquad (6.32)$$

Since $C(1 + C) \cong C + \frac{1}{2}$ unless C is small, the following simpler equation is usually satisfactory:

$$\sqrt{H} = \frac{3.29}{C_2 - C_1} \left(\frac{C_2 + \frac{1}{2}}{\sqrt{P_2}} + \frac{C_1 + \frac{1}{2}}{\sqrt{P_1}}\right). \qquad (6.33)$$

The relative error introduced in this approximation is 6.1% when $C = 1$, 1.0% when $C = 3$, and 0.11% when $C = 10$.

If some degree of cross contamination other than 0.05% is selected as the criterion of a satisfactory separation, eqns. (6.32) and (6.33) can be used simply by substituting the appropriate value of the probability integral for 3.29. Some values of these probability integrals are given in Table 6.3.

TABLE 6.3. PROBABILITY INTEGRALS

Degree of cross contamination (%)	Limit of probability integral
0.01	3.74
0.02	3.56
0.05	3.29
0.1	3.09
0.2	2.88
0.5	2.58
1.0	2.33

To illustrate the calculation of the ideal length of column, let us consider the separation of malic and tartaric acids by elution through Dowex 1-X8, 200–400 mesh, with an eluent 2.00 M with acetic acid and 0.400 M with sodium acetate. With a column 6.0 cm × 0.95 cm² and at a flow rate of 0.50 cm/min, the two elution curves overlapped. Separate elutions of the two acids under the same conditions gave curves from which the following values were calculated: for malic acid $C = 16.2$, $P = 20$; for tartaric acid $C = 25.2$, $P = 34$. Because of rather similar ionization constants ($pK_1 = 3.5$, $pK_2 = 5.1$ for malic acid, 3.0 and 4.4 for tartaric), changes in the pH of the eluent failed to yield a satisfactory elution. Therefore it was decided to use a longer column. Substitution of the foregoing values of C and P in eqn. (6.33) yielded

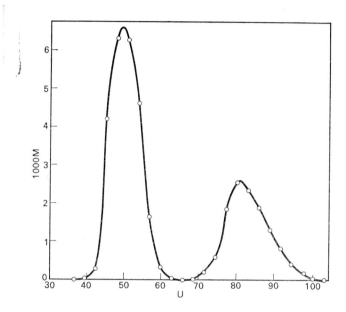

FIG. 6.9. Separation of malic and tartaric acids. (Reproduced by permission of Elsevier Publishing Co.)

$$\sqrt{H} = 3.10,$$

$$H = 9.60.$$

It is recommended that the calculated value of H be multiplied by 1.04 or 1.05 to allow for experimental errors in the evaluation of C and P, for inaccuracies in eqn. (6.33) caused by the several approximations in its derivation, and for the failure of experimental elution graphs to follow exactly the Gaussian elution equation. In this case the analyst prepared a column 10.0 cm in length, which gave the quantitative separation[17] shown in Fig. 6.9.

It should be emphasized that the resin used to lengthen a column (or to prepare a longer column) should come from the same batch as that used to determine the values of C and P that served to evaluate the ideal length. As stated on p. 93, resins of identical label, but taken from different batches, often differ appreciably in chromatographic properties.

If the elution graphs of the two ions to be separated are markedly asymmetric, the P value of the first ion should be calculated from the descending slope only, and the P of the second ion from the ascending graph only.

Equations (6.32) and (6.33) have been used many times in the Rutgers laboratory and have been found to be very reliable. This equation can also be used to calculate how much a column may be *shortened* if the original elution graph shows an unnecessarily large distance between the curves as in graph B of Fig. 6.4.

C.VII.e. *Equation for Calculating which Constituents of a Given Mixture Can Be Separated by a Column of Given Length*

To illustrate the application of this equation, we shall consider the determination of each compound in a mixture of the salts of the acids listed in Table 6.4. The method[18] developed

TABLE 6.4. VALUES OF $\Delta \log C$ FOR SOME ORGANIC ACIDS OF PHOSPHORUS

No.	Formula	C	$\log C$	$\Delta \log C$
1	H_3PO_4	0.96	−0.018	
				0.198
2	$MePO(OH)_2$	1.645	+0.216	
				0.260
3	$HOPO(OMe)_2$	2.99	0.476	
				0.003
4	$MePO(OH)OMe$	3.01	0.479	
				0.215
5	$EtPO(OH)OMe$	4.94	0.694	
				0.198
6	$EtPO(OH)OEt$	7.80	0.892	
				0.153
7	$EtPO(OH)OPr$	11.1	1.045	
				0.250
8	$MePO(OH)OBu$	17.8	1.250	

for the analysis of such mixtures depends on salting-out chromatography rather than ion-exchange chromatography, but the equation under consideration is applicable to both techniques. Separate elutions were performed on pure specimens of each compound in the mixture with a column of special cation-exchange resin and an eluent of 4 M lithium chloride plus 1 M hydrochloric acid. Values of C and P for each compound were calculated from the elution graphs. The average value of P for all of these compounds was 20. It was decided, somewhat arbitrarily, to use a 50-cm column. The question, then, was how many of these compounds could be separated from the others by elution through this column with the aforementioned eluent.

The equation that gives the answer to this question is derived as follows. Equation (6.32) is simplified by assuming that $P_1 = P_2 = P$ and that both values of C are much larger than unity so that $(C + 1)C = C^2$. Thus

$$\sqrt{(HP)}\,(C_2 - C_1) = 3.29(C_2 + C_1).$$

Dividing by C_1 and solving for C_2/C_1 yields

$$\frac{C_2}{C_1} = \frac{\sqrt{(HP)} + 3.29}{\sqrt{(HP)} - 3.29}.$$

$$\Delta \log C = \log C_2 - \log C_1 = \log\left(\frac{\sqrt{(HP)} + 3.29}{\sqrt{(HP)} - 3.29}\right). \tag{6.34}$$

Substitution of $H = 50$ and $P = 20$ yields 0.090 for the right side of this equation. This means that any pair of compounds with $\Delta \log C$ substantially greater than 0.090 will be separated quantitatively under the chosen conditions of elution and that any pair of compounds with $\Delta \log C$ substantially less than 0.090 will not. The approximations used in deriving eqn. (6.34) detract seriously from its accuracy. For example, the question of the separability under the foregoing conditions of a pair of compounds with $\Delta \log C = 0.090$ or 0.100 can not be answered reliably by eqn. (6.34).

Reference to Table 6.4 indicates that compounds 1, 2, 5, 6, 7, and 8 can be isolated under the selected conditions but that compounds 3 and 4 will overlap badly. Some other method must be used to determine these compounds in the given mixture. Actually the fraction of eluate containing both numbers 3 and 4 was divided into two aliquots; one was treated with persulphate to convert both acids to orthophosphoric while the other was treated with hydrochloric acid to convert only compound 3 to orthophosphoric acid. Then spectrophotometric determinations of this acid in both aliquots furnished the data needed to calculate the quantity of both compounds.

C.VII.f. *The Ideal Relationship between C_1 and C_2*

It is obvious that the ease of separation of two constituents of a sample depends on the ratio C_2/C_1. The absolute values of the distribution ratios are also important, however, as is illustrated in Table 6.5. Each column represents a chromatographic separation of two

TABLE 6.5. IDEAL RELATIONSHIP BETWEEN C_1 AND C_2

	a	b	c	d	e	f
C_1	0.125	0.250	0.500	1.00	2.00	4.00
C_2	0.250	0.500	1.00	2.00	4.00	8.00
C_1C_2	0.0312	0.125	0.500	2.00	8.00	32.0
H, ideal (cm)	59.7	35.4	22.6	15.5	12.9	11.4
V_b (ml)	119	70.8	45.2	31.0	25.8	22.8
V (ml)	46.4	27.6	17.6	12.1	10.1	8.90
p	597	354	226	155	129	114
$U_2{}^*$ (ml)	58.0	41.4	35.2	36.3	50.5	80.1
U_f (ml)	60.9	46.1	41.3	45.2	65.6	107
time (m in)	76.2	57.6	51.6	56.5	82.1	134

constituents whose distribution ratios are given in the first two rows of figures. Next are given the products of the two C values. The minimum heights required for a quantitative separation were calculated by eqn. (6.32) with the assumption that $P_1 = P_2 = 10.0$. The bed volumes V_b were calculated on the assumption that each column had a cross-sectional area of 2.00 cm². The interstitial volumes were calculated with the assumption that $V = 0.390\ V_b$. The numbers of plates p were calculated as 10.0 H. Equation (6.7) was used to calculate the values U^*. It was assumed that the elutions were terminated when $M_2 = M_2{}^*/1000$; the volumes of effluent U_f collected at this point were calculated by substituting this value in eqn. (6.29). Finally, the times required were calculated on the assumption that the flow rate was 0.400 cm/min or 0.800 ml/min.

In each case, the values of C_1 and C_2 were selected so that C_2/C_1 is 2.00. The values of C_1C_2 increase, on the other hand, from 0.0312 to 32. With increasing values of C_1C_2, we find decreasing values of H, and hence also of V_b, V, and p. On the other hand, the values of $U_2{}^*$ decrease at first with increasing values of C_1C_2, but then increase with further increases in C_1C_2. The reason for this is that the compounds of large C require large volumes of eluent to remove them even from short columns. The values of U_f and time show the same trends as those of $U_2{}^*$. The minimum time for a quantitative separation corresponds to a value of C_1C_2 between 0.500 and 2.00; i.e. roughly unity. If different values of C_2/C_1, P, A, V/V_b, and flow rate had been selected (constant for each column in the table), the times required to complete the separation would have differed from those of Table 6.5, but the same conclusion would had been reached; i.e. for a given ratio of C_2/C_1, the best value of C_1C_2 is approximately unity.

In selecting the conditions for a chromatographic separation of two compounds or ions, the analyst should bear in mind not only the two desirable relationships between C_1 and C_2 but also the fact that the values of C_1 and C_2 can be changed by changing the concentration of eluent and sometimes also by adjusting the pH of the eluent or adding a complexing agent.

If three constituents of a sample are to be separated by ion-exchange chromatography, it is obviously impossible to have the desirable large ratios of C_2/C_1 and C_3/C_2 and also to maintain values of C_1C_2 and C_2C_3 close to unity unless the composition of the eluent is changed in the course of the elution. In fact, such changes are often applied in the chromatographic separation of multicomponent mixtures. The theory of such elutions is discussed in the next section.

C.VIII. CALCULATION OF U^* IF THE ELUENT IS CHANGED DURING THE ELUTION

C.VIII.a. *Stepwise Changes*

Let us consider the elution of a multicomponent mixture in which the eluent is changed once during the elution. All constituents whose peak is eluted *before* the change in eluent follow eqn. (6.7). On the other hand, any constituent whose U^* has not been reached at the change of eluent undergoes a change in C when the eluent is changed and therefore does not follow eqn. (6.7). We shall now derive an equation for the U^* of such a constituent.

Let U_1 be the volume of the first eluent used. Let C_1 and C_2 denote the C values of *one* constituent corresponding to the first and second eluent.

If the first eluent were used until the peak of the solute in question is eluted, the value of U^* would be given by eqn. (6.7). The fraction h_1 of the column height traveled by the peak of this constituent under the influence of the first eluent is equal to the total volume of the first eluent divided by the volume of this eluent that would pass through its peak if this eluent were used exclusively.

$$h_1 = \frac{U_1}{U_1^* - V} = \frac{U_1}{C_1 V},$$

where U_1^* is the value of U^* that would obtain if the first eluent were used exclusively. The fraction h_2 of the column height still to be traversed under the influence of the second eluent is

$$h_2 = 1 - \frac{U_1}{C_1 V}. \tag{6.35}$$

If the second eluent were used exclusively the value of U^* would be

$$U_2^* = C_2 V + V.$$

However, when the first portion of the second eluent enters the column, the interstitial solution is occupied by the first eluent. Therefore, only $C_2 V$ ml of the second eluent is effective in moving the peak. Thus the volume U_2 of the second eluent that is needed to complete the elution of the peak is

$$U_2 = h_2 C_2 V.$$

Substitution of eqn. (6.35) in this equation yields

$$U_2 = \left(1 - \frac{U_1}{C_1 + V}\right) \quad C_2 V = C_2 \left(V - \frac{U_1}{C_1}\right).$$

The total volume of both eluents needed to elute the peak is then

$$U^* = V + U_1 + U_2 = V + U_1 + C_2 \left(V - \frac{U_1}{C}\right). \tag{6.36}$$

Analogously it can be proved that with three eluents

$$U^* = V + U_1 + U_2 + C_3 \left(V - \frac{U_1}{C_1} - \frac{U_2}{C_2}\right).$$

In general, for $(n + 1)$ eluents and n changes,

$$U^* = V + \Sigma U_n + C_{n+1} \left(V - \sum \frac{U_n}{C_n}\right), \tag{6.37}$$

the summations being taken from $n = 1$ to $n = n$.

Equations (6.36) and (6.37) were first derived by Freiling[19] and later independently by Breyer.[20]

When the changes in eluent involve only changes in concentrations, the values of U^* calculated by eqn. (6.36) or (6.37) agree with the observed values with an average relative error of 2.5%. It is probable that greater discrepancies will be encountered when changes of the pH of the eluent are made because the reaction of the buffer of the second (or subsequent) eluent with the buffer constituents of the first (or previous) eluent absorbed by the resin will prevent a truly stepwise change in pH.

A change in eluent concentration during the elution is desirable if the successive values of log C for any one eluent show a sudden jump. For example, if the log C values of the five constituents of a mixture with the first eluent are -0.25, $+0.15$, 0.54, 0.91, and 1.03, there is little advantage in making a change in the eluent concentration. On the other hand, the log C values -0.21, $+0.02$, 0.75, 1.02, and 1.92 show sudden jumps after the second and fourth constituents. In such a case, an increase in eluent concentration after the *first* compound has been completely eluted from the column and another change after the *third* compound has been eluted will improve the efficiency of the separation. (It is assumed in this paragraph that all the ions to be separated have the same valence.)

C.VIII.b. *Gradient Elution*

An elution in which the composition of the eluent is changed continuously is called a gradient elution. We shall consider only gradient changes in the *concentration* of the eluent.

The device with a constant-volume mixing chamber sketched in Fig. 6.10 delivers to the column an eluent whose concentration follows the equation[21]

$$[El] = M_2 - (M_2 - M_1) \exp\left(-\phi/V_R\right)$$

where ϕ denotes the volume of liquid delivered to the column, M_1 the concentration of solution originally in the mixing chamber, M_2 the concentration of solution in the upper reservoir, and V_R the volume of solution in the mixing chamber. When M_1 is zero, this equation becomes

$$[El] = M_2 \{1 - \exp\left(-\phi/V_R\right)\}. \tag{6.38}$$

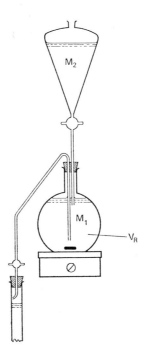

FIG. 6.10. Apparatus for exponential gradient. (Reproduced by permission from *Analytical Chemistry*.)

This is the equation for an exponential gradient.

The apparatus shown in Fig. 6.11 delivers to the column an eluent whose concentration follows the equation[21]

$$[El] = M_3 - (M_3 - M_2)\frac{\phi}{V_R}\exp(-\phi/V_R) - (M_3 - M_1)\exp(-\phi/V_R).$$

Both constant volume mixing chambers contain the same volume of solution and the original concentrations of the three solutions are M_1, M_2, and M_3 as indicated in the figure. If M_1 is zero and if $M_3 = 2M_2$, this equation becomes

$$[El] = M_2\left\{2 - \frac{\phi}{V_R}\exp(-\phi/V_R) - 2\exp(-\phi/V_R)\right\}. \qquad (6.39)$$

If the further restriction is applied that $\phi < V_R/2$, the values of ϕ calculated from eqn. (6.39) check those calculated from eqn. (6.40) within 3%,

$$[El] = \phi M_2/V_R. \qquad (6.40)$$

Although other types of apparatus have been devised to deliver gradients corresponding to other equations, the exponential and linear gradients expressed by eqns. (6.38) and (6.40) are the most important and are the only ones that will be discussed in this book.

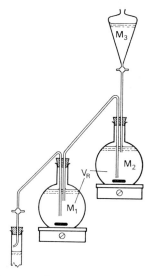

FIG. 6.11. Apparatus for linear gradient. (Reproduced by permission from *Analytical Chemistry.*)

The rate of movement of a peak of a solute in the column depends on the concentration of the eluent *at the peak*. This concentration may be calculated by eqn. (6.38) or (6.40), ϕ now denoting the volume of eluent that has passed through the peak. That is, $\phi = U - V'$, where V' is the interstitial volume of the column above the peak. From eqn. (6.10) it follows that

$$U - V' = \frac{W'QE}{[\text{El}]^z},\tag{6.41}$$

where W' is the weight of resin in the column above the peak. Therefore

$$\phi = \frac{W'QE}{[\text{El}]^z}.\tag{6.42}$$

Also

$$\phi^* = U^* - V,\tag{6.43}$$

where ϕ^* is the value of ϕ when the peak reaches the bottom of the column.

When an increment of eluent solution dU moves through any horizontal plane of the column, a smaller increment $d\phi$ moves through the peak because the peak moves downward through an increment dW' of resin. Therefore, from eqn. (6.42),

$$d\phi = dU - dV' = \frac{QE\,dW'}{[\text{El}]^z}.$$

For [El] we may substitute $f(\phi)$. Therefore

$$d\phi = \frac{QE\,dW'}{[f(\phi)]^z}.$$

We may now substitute for $f(\phi)$ the right-hand side of eqn. (6.38) or (6.40) according to whether the gradient is exponential or linear. Next the equation is integrated between the limits $\phi = 0$ to $\phi = \phi^*$ and $W' = 0$ to $W' = W$. Thus ϕ^* is found, and U^* is found by eqn. (6.43). The details of the integration may be found elsewhere.[8, 22]

The final equations follow.

Exponential gradient, $z = 1$,

$$\phi^* + V_R \exp\left(-\phi^*/V_R\right) - V_R = WQE/M_2.$$

Exponential gradient, $z = 2$,

$$\phi^* - V_R \left\{1 - \exp\left(-\phi^*/V_R\right)\right\}\left\{2 - \frac{1}{2}[1 + \exp(-\phi^*/V_R)]\right\} = WQE/M_2^2.$$

Exponential gradient, $z = 3$,

$$\phi^* - 3V_R \left\{1 - \exp\left(-\phi^*/V_R\right)\right\} + \frac{3}{2} V_R[1 - \exp\left(-2\phi^*/V_R\right)]$$

$$- \frac{1}{3} V_R [1 - \exp\left(-3\phi^*/V_R\right)] = WQE/M_2^3.$$

Linear gradient, $z = 1$,

$$\phi^* = \sqrt{\left(\frac{2WQEV_R}{M_2}\right)}.$$

Linear gradient, $z = 2$,

$$\phi^* = \sqrt[3]{\left(\frac{3WQEV_R^2}{M_2^2}\right)}.$$

Linear gradient, $z = 3$,

$$\phi^* = \sqrt[4]{\left(\frac{4WQEV_R^3}{M_2^3}\right)}.$$

Chloride and bromide as examples of univalent ions and oxalate as an example of a bivalent ion were eluted with sodium nitrate with both linear and exponential gradients. In each of 35 such elutions covering a range of U^* values from 54 to 222, the observed U^* was compared with the calculated value.[8, 22] The average ratio of the calculated to observed value was 1.01 with a standard deviation of 0.022. On the other hand, iodide ion failed to follow the foregoing equations, because, as stated on p. 98, $E_{NO_3}^I$ depends on the nitrate-ion concentration.

C.VIII.c. *Comparison of Gradient and Stepwise Changes*

The chief advantage of gradient elution over stepwise changes is that less of the operator's time is required to set a gradient elution in operation than is required for several stepwise changes of eluent concentration; at least, this is true once the apparatus for gradient elution has been assembled.

Theoretically, the elution curves in gradient chromatography should show a slight heading (ascending slope less steep than the descending slope). Actually, since the elution curves with constant composition of eluent show a slight tendency to tail, the effect of gradient elution usually balances this; and the curves in gradient elution are usually more nearly symmetrical than in nongradient elution.

Since stepwise changes provide greater flexibility in the composition of the eluent than do gradient changes, any required separation can usually be performed in less total time by stepwise changes than by gradient changes.

C.IX. WIDTH OF ELUTION CURVES

Let us define the width ω of an elution curve as the width when $M = M^*/e$. It follows then from eqn. (6.31) that

$$\omega = 2(U_a - U^*) = 2U^* \sqrt{\left\{\frac{2C}{p(1 + C)}\right\}}.$$

Combining this with eqn. (6.7) and substituting HP for p, we find that

$$\omega = 2V \sqrt{\left\{\frac{2C(1 + C)}{HP}\right\}}.$$

For any given resin and eluent, the ratio V/V_b is constant. For our present purpose, we shall assign a value of 0.39 to this ratio and write

$$V = 0.39V_b = 0.39\,AH, \tag{6.44}$$

where A is the cross-sectional area of the column.
Therefore

$$\omega = 0.78A \sqrt{\left\{\frac{2HC(1 + C)}{P}\right\}}. \tag{6.45}$$

Equation (6.45) indicates that the width of elution curves increases as C increases. This means that for any given column and any one eluent, each successive elution curve is wider than the preceding one because the constituents emerge in the order of increasing C values. On the other hand, a constituent that emerges shortly after a change in the eluent may be much narrower than its predecessors because it may have a very small C value as a result of the change in eluent.

A short-sighted consideration of eqn. (6.45) might lead to the conclusion that separations are improved by using short and narrow columns since ω is decreased by decreasing H and A. The error of this reasoning is obvious when eqn. (6.46) is also considered. From eqns. (6.7) and (6.44)

$$U^* = 0.39\,(1 + C)AH.$$

The distance between peaks of two constituents is therefore

$$\Delta U^* = U_2^* - U_1^* = 0.39\,(C_2 - C_1)AH. \tag{6.46}$$

By comparison of eqns. (6.45) and (6.46) it is seen that both the width of the curves and the distance between the peaks vary directly as the cross-sectional area of the column. Therefore, although narrow graphs are obtained with narrow columns, the advantage of the narrower graphs is compensated by the smaller distances between the peaks. Further comparison of these two equations reveals that the width increases as the square root of the height, whereas the distance between the peaks increases directly as the height. Therefore increases in height cause more rapid increases in the distance between peaks than in the width and hence facilitate separations.

C.X. PROCEDURE FOR DEVELOPING A METHOD OF SEPARATION BY ION-EXCHANGE ELUTION CHROMATOGRAPHY

The analyst who wishes to design a method for the quantitative separation of the constituents of a qualitatively known mixture (or series of such mixtures) will find the following discussion helpful. It is assumed that the constituents of the mixture are available in reasonable purity.

The first decision concerns the resin to be used. In most cases the strong-acid or strong-base polystyrene resins will serve best. Crosslinkages of 8 % are best in most cases, but very large ions may require 4 %. This entails the disadvantage of having the column shrink if the concentration of eluent is increased in the course of the elution. Then, when the dilute eluent is used again in the next elution, the expansion of the resin may cause a violent shattering of the chromatographic tube (see p. 25). Fine particle size of resin is desired to expedite the attainment of equilibrium in the column, which favors Gaussian curves. Resins larger than 100–200 mesh should very seldom be used for chromatographic separations. The use of narrower size ranges than those supplied commercially has been found to be advantageous in the separation of amino acids[23] and hydroxyacids.[24] However, in view of the time required for the size grading, the use of the narrower size range is probably justified only for very difficult separations that are to be performed on a routine basis.

Two points should be borne in mind in the selection of the electrolyte that is to be used as eluent. (1) It should not interfere in the determination of the sample constituents in the fractions of the eluate. For example, sodium nitrate is a satisfactory eluent for the separation of chloride, bromide, and iodide because the separated halides can be determined in the eluate fractions by titration with silver nitrate (or by iodimetry in the case of iodide).[1] Hydrochloric acid is a good eluent for the separation of the alkali metals because the separated alkali metals can be readily determined by flame photometry or more accurately by titration with silver nitrate after evaporation to dryness. In the latter case, a correction must be applied for the coprecipitation of hydrochloric acid by the alkali halides. (2) The affinity of the eluent ion for the resin should not differ too greatly from those of the ions to be determined. If the selectivity coefficient of the sample ion relative to the eluent ion is too great, a very large concentration or a very large volume of eluent will be required to elute the sample. If it is too small, the several C and U^* values will be too small for a good separation. The use of a very dilute eluent to avoid this difficulty entails violation of the second assumption of the plate theory (p. 94); then asymmetric elution graphs are obtained, and the plate theory is not applicable.

Two pure samples of each constituent of the mixture should then be eluted, each with a different concentration of eluent, at a flow rate of 0.4–0.6 cm/min. If approximately Gaussian graphs are obtained, the flow rate was satisfactory; indeed, a slightly higher flow rate might be tried if desired. If the curves are markedly asymmetric, the elutions should be repeated with slower flow rate, smaller resin particles, and/or higher temperature. Short columns (6–10 cm) may be used for these preliminary elutions to save time.

For each preliminary elution, U^*, C, and p are evaluated. (The determination of total column capacity WQ was described on p. 24; the interstitial volume V may be determined as described on p. 134 or estimated from Table 7.1 (p. 134).) For each compound eluted, two values of E can be calculated by eqn. (6.9). If these agree for any one constituent, it can be assumed to have a constant E; and C for any concentration of eluent can be calculated by eqn. (6.9). If any constituent of the mixture shows a variable E, it should be eluted with several other concentrations of eluent so that a plot of C vs. concentration of eluent can be made. If the nature of the mixture to be separated suggests that the use of eluents containing buffers or complexing agents would facilitate the separations, elutions of each compound of the mixture with several such eluents should be performed.

When sufficient data have been collected regarding the effect of the composition of the eluent on the C values of the several constituents of the mixture, the analyst is ready to plan the separation. If the mixture contains only two constituents, the eluent is chosen so as to give a maximum value of C_2/C_1 and a value of C_1C_2 near unity. The desired height of column is then calculated by eqn. (6.32) or (6.33).

If the mixture contains more than two constituents but not more than eight or ten, the height of column is calculated from the C values of the two compounds that give the minimum ratio of C's. Stepwise or gradient changes in the eluent may be applied if it seems desirable (see p. 114).

If the mixture contains more than ten constituents, the application of eqn. (6.34) will probably be advantageous. As indicated on p. 113, other methods in addition to ion exchange must be found for determining those constituents that are not separated by a column of reasonable length.

When a set of elution conditions has been found which gives no overlapping of elution curves when each constituent of the sample is eluted separately, the same elution conditions will surely give a quantitative separation when the mixture of these constituents is eluted, provided that the column is not overloaded, i.e. that the quantity of sample is small enough so that the second assumption of p. 94 is valid.

It is necessary to collect and analyze small samples of the eluate only during the development of the chromatographic procedure. Once the elution conditions have been found that give a good separation, it is necessary to collect only as many fractions as there are constituents to be determined, each fraction being chosen so as to contain all of one constituent and none of any of the others. Of course, any change in the elution conditions will require changes in the volumes of eluate fractions containing any one constituent.

Theoretically, the width of the column is unimportant. Two columns of different width but of the same resin and same height will have interstitial volumes, bed volumes, and weights of resin in direct proportion to their cross-sectional areas. They will be equally effective in separations except that the maximum sample weights that can be applied without violating assumption 2 of p. 94 will also be in direct proportion to their cross-sectional areas. The time required for any given separation (with the assumption of equal linear flow rates) will be the same, but the volumes of eluent required will vary as the cross-sectional areas. Thus the use of narrow columns introduces a saving in the quantities of resin and eluent. However, with excessively narrow columns, the volume of eluent containing any one sample constituent or the quantity of the constituent may be too small for accurate and convenient determinations. For ordinary analytical purposes, columns with internal diameters between 1.0 and 2.5 cm ($0.8 < A < 4.9$) are generally used.

D. Mass-transfer or Continuous-flow Plate Theory

Chromatography is a continuous process in which equilibrium is approached but never reached. Molecules of the solute are continually moving back and forth between the fixed and the moving phase. On the leading side of an elution band there are more molecules in the moving phase than there would be at equilibrium, and on the trailing side there are less. The "plate" concept is simply a mental device to allow mathematical treatment of the moving system. One pretends that the solute concentration, instead of varying continuously, changes in a series of steps. One replaces the continuous process by a series of batch processes. The faster is the interchange of molecules between fixed and moving phases, the more "plates" or steps are needed to describe the process.

The theory developed in section C serves well to interpret the chromatographic process and permits useful predictions. It does not, however, account explicitly for the well-known fact that the number of plates p depends on the flow rate, nor for the fact that p depends on the distribution coefficient of the solute between the two phases. It is intellectually clumsy in that it pictures the liquid (or mobile phase) advancing in a series of jerks. It imagines the liquid to remain in one column segment or "plate" for long enough to reach equilibrium with the fixed phase, then instantaneously to move into the next "plate" and repeat the process.

D.I. THE MODEL OF GLUECKAUF

The "plate" concept was modified by Glueckauf[25, 26] to recognize the fact that the resin particles are discontinuous, yet the flow of solution is continuous. The "plate" cannot be much thinner than the diameter of the resin bead. For closely packed uniform spherical beads, the minimum effective thickness is calculated to be 0.82 times the bead diameter. In practice, the "plate thickness" will be several times this value, for reasons which will soon appear.

The model is illustrated in Fig. 6.12. The "theoretical plate" is a segment of volume Δx; the bulk volume of the column from its upper surface down to this "plate" is x. To simplify the discussion we shall consider a column of unit cross-sectional area. Then x will also be the height of the column above the segment that we are considering.

We shall present a summary of Glueckauf's development, and to help the reader who wishes to refer to the original papers, we shall use the symbols in the references cited,[25, 26] with minor changes. These symbols are as follows:

Volume from top of column (length of column of unit area)	$= x$
Volume of liquid that has entered the column	$= v$
Flow rate (presumed uniform) $= v/t$	$= F$
Concentration of interstitial solution (solute in unit volume of solution)	$= c$
Solute in exchanger in unit bulk volume of the column	$= \bar{q}$
Solute in exchanger and solution combined, that is, *total* solute in unit bulk volume of the column	$= q = \bar{q} + ac$
Values of these concentrations when the exchanger *is in equilibrium with* an interstitial solution of concentration c	$= \bar{q}^*, q^*$
Void volume fraction of column	$= a$

Plate height $= H$; number of plates in column $= N$

Distribution relations: $K = \bar{q}^*/c$;

$a = q^*/c$

Barred symbols indicate the resin phase; asterisks indicate equilibrium. The quantity K is called the "column distribution ratio" by Helfferich (ref. 27, p. 452) to emphasize the fact that the solution concentration c is referred to unit volume of interstitial solution, but the exchanger concentration $\bar{q}^*$ is referred to the exchanger in *unit volume of the column*, i.e. to $(1 - a)$ volumes of exchanger.

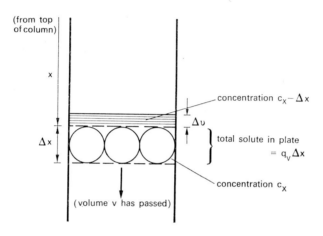

FIG. 6.12. Continuous-flow theoretical-plate model.

At a particular instant in time, let the volume of solvent that has entered the column be v. At this moment let the concentration of the solution in the reference segment or "plate" be c_x, and the concentration in the solution in the segment just above it, $c_{x-\Delta x}$. Let the total concentration in the reference segment at this instant be q_v.

Now let a volume of solution Δv enter the reference "plate". Simultaneously an equal volume Δv must leave the plate through its lower side. The volume Δv may not be larger than the "free" or interstitial volume in one plate, but *it need not equal this volume*; indeed, *it may be as small as one wishes to make it*. This is the essential point of difference between Glueckauf's treatment and the treatment of Mayer and Tompkins, discussed in the last section.

After the volume Δv has passed, let the total concentration of solute in the reference plate be $q_{v+\Delta v}$. We now have the following relationships:

Solute entering reference plate $= \Delta v . c_{x-\Delta x}$

Solute leaving reference plate $= \Delta v . c_x$

Solute in plate before movement $= \Delta x . q_v$

Solute in plate after movement $= \Delta x . q_{v+\Delta v}$

The conservation of mass requires that

$$\Delta v(c_{x-\Delta x} - c_x) = \Delta x(q_{v+\Delta v} - q_v) \qquad (6.47)$$

Since the differences are small, we can use Taylor's theorem to express them in terms of differentials:

$$c_{x-\Delta x} = c_x - \left(\frac{\partial c}{\partial x}\right)_v \Delta x + \left(\frac{\partial^2 c}{\partial x^2}\right)_v \frac{\Delta x^2}{2} - \cdots$$

$$q_{v+\Delta v} = q_v + \left(\frac{\partial q}{\partial v}\right)_x \Delta v + \left(\frac{\partial^2 q}{\partial v^2}\right)_x \frac{\Delta v^2}{2} + \cdots$$

Then eqn. (6.47) becomes

$$-\left(\frac{\partial c}{\partial x}\right)_v + \left(\frac{\partial^2 c}{\partial x^2}\right)_v \frac{\Delta x}{2} = \left(\frac{\partial q}{\partial v}\right)_x + \left(\frac{\partial^2 q}{\partial v^2}\right)_x \frac{\Delta v}{2}. \qquad (6.48)$$

The volume Δx has a certain minimum value, as we noted above. It can be no smaller than the volume of one layer of beads. The volume Δv, on the other hand, can be made as small as we like. The second term on the right in eqn. (6.48) becomes vanishingly small, and the equation may be simplified to read

$$\left(\frac{\partial c}{\partial x}\right)_v + \left(\frac{\partial q}{\partial v}\right)_x - \left(\frac{\partial^2 c}{\partial x^2}\right)_v \frac{\Delta x}{2} = 0. \qquad (6.49)$$

This equation is perfectly general for all types of chromatography, whether or not equilibrium is attained. (We have neglected diffusion along the length of the column.) To solve it we must relate the concentrations q and c. As a first step, let us assume that equilibrium is reached between the solution and the exchanger, i.e., that $q = q^*$. Let us also assume a "linear isotherm", i.e., that the distribution coefficient is independent of the solute concentration. Then $q = ac$, and eqn. (6.49) becomes

$$\left(\frac{\partial c}{\partial x}\right)_v + a\left(\frac{\partial c}{\partial v}\right)_x - \left(\frac{\partial^2 c}{\partial x^2}\right)_v \frac{\Delta x}{2} = 0. \qquad (6.50)$$

D.I.a. *Departure from Equilibrium*

In reality the resin and solution are not in equilibrium with one another. Equation (6.50) describes an ideal situation which never exists in practice. *However, we can take cognizance of the lack of equilibrium without changing the form of eqn. (6.50). All we need to do is to change the coefficient of $(\partial^2 c/\partial x^2)$ in the last term.*

The rate of exchange of ions between the exchanger and the solution is limited by two processes, diffusion within the particles of exchanger and diffusion through a film of liquid surrounding the particles (see Chapter 4). We need not go into details of the kinetics, but we shall assume that the displacements from equilibrium are small enough that *the rate of exchange is proportional to the displacement*. If particle diffusion controls the rate,

$$\left(\frac{\partial \bar{q}}{\partial t}\right)_x = b' \, (\bar{q}^* - \bar{q}), \qquad (6.51a)$$

and if film diffusion controls the rate,

$$\left(\frac{\partial \bar{q}}{\partial t}\right)_x = b'' \, (c - c^*) = \frac{b''}{K} \, (\bar{q}^* - \bar{q}), \qquad (6.51b)$$

where b' and b'' are factors which include diffusion coefficients and geometrical relations. The quantity c^* is the concentration of the solution in immediate contact with the exchanger, in equilibrium with concentration $\bar{q}$ in the exchanger.

If we ignore second differentials we may substitute

$$\left(\frac{\partial \bar{q}}{\partial t}\right)_x \text{ by } \left(\frac{\partial \bar{q}^*}{\partial t}\right)_x = K\left(\frac{\partial c}{\partial t}\right)_x.$$

Then eqns. (6.51a) and (6.51b) reduce to:

$$\bar{q}^* - \bar{q} = q^* - q = b\left(\frac{\partial c}{\partial t}\right)_x, \tag{6.52}$$

where the factor b equals K/b' for particle diffusion, K^2/b'' for film diffusion, or the sum of the two if both diffusion rates are comparable.

Now

$$\left(\frac{\partial c}{\partial t}\right)_x = \frac{\partial c}{\partial v}\frac{\partial v}{\partial t} = F\left(\frac{\partial c}{\partial v}\right)_x; \quad \left(\frac{\partial c}{\partial v}\right)_x = -\left(\frac{\partial c}{\partial x}\right)_v\left(\frac{\partial x}{\partial v}\right)_c.$$

The mass balance of a moving concentration profile which does not change its shape—this condition is equivalent to neglecting second differentials—gives

$$\left(\frac{\partial x}{\partial v}\right)_c = \frac{1}{a}.$$

Substituting in eqn (6.52),

$$q = q^* - b\left(\frac{\partial c}{\partial t}\right)_x = q^* + \frac{bF}{a}\left(\frac{\partial c}{\partial x}\right)_v. \tag{6.53}$$

Now we can evaluate the second term in eqn. (6.49):

$$\left(\frac{\partial q}{\partial v}\right)_x = \left(\frac{\partial q^*}{\partial v}\right)_x + \frac{bF}{a}\frac{\partial^2 c}{\partial v \partial x} = a\left(\frac{\partial c}{\partial v}\right)_x - \frac{bF}{a}\left(\frac{\partial^2 c}{\partial x^2}\right)_v\left(\frac{\partial x}{\partial v}\right)_c$$

$$= a\left(\frac{\partial c}{\partial v}\right)_x - \frac{bF}{a^2}\left(\frac{\partial^2 c}{\partial x^2}\right)_v. \tag{6.54}$$

Substituting into eqn. (6.49),

$$\left(\frac{\partial c}{\partial x}\right)_v + a\left(\frac{\partial c}{\partial v}\right)_x - \left(\frac{bF}{a^2} + \frac{\Delta x}{2}\right)\left(\frac{\partial^2 c}{\partial x^2}\right)_v = 0. \tag{6.55}$$

This is just like eqn. (6.50) except for an extra coefficient of the second differential. Consideration of back-and-forth diffusion along the axis of the column adds one more coefficient of $\partial^2 c/\partial x^2$. The end result is the general equation

$$\left(\frac{\partial c}{\partial x}\right)_v + a\left(\frac{\partial c}{\partial v}\right)_x - \frac{H}{2}\left(\frac{\partial^2 c}{\partial x^2}\right)_v = 0, \tag{6.56}$$

where H is the "height equivalent of the theoretical plate", given by the equation[26, 27]

$$H = 1.64r + \frac{0.14r^2F}{\bar{D}}\frac{K}{a^2} + \frac{0.266r^2F}{D(1 + 70\,rF)}\frac{K^2}{a^2} + \frac{Da\sqrt{2}}{F}, \tag{6.57}$$

$$\text{(I)} \qquad\qquad \text{(II)} \qquad\qquad\qquad \text{(III)} \qquad\qquad\qquad \text{(IV)}$$

where r is the particle radius, D and $\bar{D}$ are diffusion coefficients in solution and exchanger, and we recall that $a = K + \alpha$. Term (I) is the same as Δx in eqn. (6.55), and represents the contribution of finite particle size, assuming close packing of uniform spheres. Terms (II) and (III) show the contributions of particle diffusion and film diffusion, respectively; term (IV) shows the effect of longitudinal diffusion.

Figure 6.13 shows how each of the four terms dominate in certain ranges of distribution coefficient and flow rate. Longitudinal diffusion hardly ever affects column performance in liquid chromatography, for diffusion is very slow compared to the flow rates normally used.

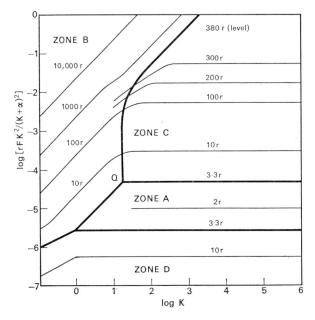

FIG. 6.13. Factors controlling theoretical-plate height. The preponderant factors influencing plate height are particle radius in zone A, particle diffusion in zone B, film diffusion in zone C and diffusion along the length of the column in zone D. (Used by permission of the Chemical Society.)

The distribution coefficient affects the plate height if particle diffusion governs the rate of transfer. The factor $K^2/a^2 = K^2/(K + \alpha)^2$ in the film diffusion term, on the other hand, is very close to unity.

D.I.b. *Form of the Elution Curve*

Equation (6.55) can be solved by Laplace transformations, applying boundary conditions appropriate to the problem at hand. For displacement chromatography in which a solution of concentration c_0 flows continuously through an initially empty column, $c = 0$ if $v = 0$ and $x > 0$, $c = c_0$ if $x = 0$ and $v = 0$. For elution chromatography, the simplest case is that in which *only the first plate* contains solute at the start; then $c = c_0$ if $x = H$ and $v = 0$, and $c = 0$ if $x = H$ and $v > 0$, or $x > H$ and $v = 0$.

The equation is best solved by using the dimensionless parameters $M = v/aH$, $N = x/H$. Under practical conditions of elution chromatography N is large (greater than 10, let us say), and $(M - N)/N$ is small. (This anticipates the result that follows from the general considerations in the next section as well as from the plate model discussed above in section C, namely that $M = N$ at the peak of the band.) The special solution that follows from these simplifications is

$$\frac{c}{c_0} = \frac{1}{(2\pi N)^{\frac{1}{2}}} \exp\left[-\frac{(N - M - 1)^2}{2M}\right] \cdots \tag{6.58}$$

This equation shows that at the peak of the elution band, where c is a maximum, $N = M$ very nearly, or that

$$v_{max} = ax, \tag{6.59}$$

where x is the length of the column and v_{max} is the volume of solvent needed to elute the peak of the band. Substituting for M in eqn. (6.58) we get

$$c = c_{max} \exp\left[-\frac{N}{2}\frac{(v_{max} - v)^2}{vv_{max}}\right], \tag{6.60a}$$

$$c_{max} = \frac{m}{v_{max}}\sqrt{\frac{N}{2\pi}}, \tag{6.60b}$$

where m is the mass of solute eluted and N is the total number of theoretical plates in the column.

We note from eqn. (6.60b) that the "height" of the band c_{max} is directly proportional to the quantity of solute present, given a particular column and constant conditions of elution. Another important quantity is the "width" of the band, which is conveniently measured at the concentration level $c = c_{max}/e$, or $\ln c/c_{max} = -1$. Then

$$\frac{v_{max} - v}{v_{max}} = \sqrt{\frac{2}{N}}. \tag{6.61}$$

These relationships are illustrated in Fig. 6.14.

D.I.c. *Conclusion*

The continuous-flow concept of the theoretical plate has the advantage that it presents the "height equivalent of the theoretical plate" as a quantitative measure of the departure from equilibrium, and allows one to predict how it will be affected by particle diameter, flow rate, and distribution coefficient. It correlates the plate height and the band width. Though it may be difficult to calculate plate heights *a priori* because of ignorance of diffusion coefficients and other factors like unevenness of packing, it is possible to use band widths to calculate plate heights, then use plate heights as a quantitative measure of the resolving power of the column and as a guide to the conditions needed to achieve a desired degree of separation. Illustrations of these calculations are given in the references cited.

E. Tracer-pulse and Concentration-pulse Methods

E.I. RELATION BETWEEN ELUTION VOLUME AND PARTITION RATIO

A simple relationship exists in elution chromatography between the volume of solvent needed to elute the peak of a band and the partition ratio of the solute between solvent

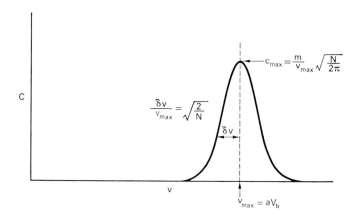

FIG. 6.14. Elution band and plate height.

and absorbent. Essentially, the elution volume is simply the product of the column volume and the partition ratio. We have the relations

$$U^* = V(1 + C),$$ (6.7)

$$v_{max} = ax = aV_b.$$ (6.59)

We recall that U^* and v_{max} are identical except for the fact that we restricted the discussion in section D to columns of unit area. Both symbols denote the volume of solvent passed at the elution of the peak of the band. V and V_b mean the void volume and bulk volume of the column, respectively; C and a are partition ratios defined in sections C.III and D.I of this chapter; $a = a(C + 1)$; $a = V/V_b$. It is understood that the amount of solute is very small and that the column is long enough to include a fair number of theoretical plates, say 20 or more.

We can derive these relations very simply, without reference to a theoretical plate model, as follows. Consider that equilibrium between the fixed and mobile phases (resin and solution) is established instantaneously. Then at any given moment, in any given region of the column, a fraction $C/(C + 1)$ of the solute is in the resin and a fraction $1/(C + 1)$ is in the solution. If the resin were not there, the molecules of solute and eluting solvent would travel at the same rate. Because the resin is there, only the solute molecules *in the solution* are actually moving, and the ratio

$$\frac{\text{mean velocity of solute molecules}}{\text{mean velocity of solvent molecules}} = \frac{1}{C + 1}.$$

To move a solvent molecule from one end of the tube to the other we must pass V volume units; to move a *solute* molecule we must pass $V(C + 1) = U^*$ volume units. This is eqn. (6.7).

The band spreads as it moves down the column because the transfer of solute between the fixed and mobile phases is *not* instantaneous. However, eqn. (6.7) gives the rate of movement of the center of mass of the band, and if the band is symmetrical (as it will be if V and C are large enough), the center of mass coincides with the peak concentration.

So one can calculate elution volumes from experimentally determined values of distribution ratios. These can be found by shaking tests as described in Chapter 3. One needs to know the volume of the column, the quantity of exchanger, and the interstitial volume or void fraction. Determination of interstitial volume is discussed in Chapter 7. A little reflection will show that it need not be known accurately unless C is small.

Clearly one may reverse the process and measure the elution volume experimentally, then use this value to calculate the distribution ratio and hence the selectivity coefficient. The practical advantage of finding selectivity coefficients in this way is that one need not worry about reaching equilibrium; departure from equilibrium broadens the band but does not change the position of the peak. A disadvantage is that the method is only valid for small amounts of the eluted ion. The selectivity coefficient can only be measured for very low equivalent fractions of this ion.

E.II. THE TRACER-PULSE METHOD

A simple way to escape from this difficulty was shown by Helfferich.[28] A column of known volume, containing a known quantity of resin, is prepared, and a solution containing both exchanging ions, in whatever concentrations are desired, is passed continuously until the resin has come to equilibrium with it. Then one introduces at the top of the column a minute amount of a radioactive isotope of one of the exchanging ions and measures the volume of solution that must be passed to elute the peak of radioactivity. From this volume one calculates the distribution ratio and selectivity coefficient. One may then repeat the process with solutions of different compositions until the whole isotherm has been explored.

The tracer-pulse method lends itself to automatic recording and has the advantage that one does not have to verify the attainment of equilibrium while the tracer is flowing through the column. However, one does have to ensure that the column is in equilibrium with the flowing solution before the tracer is introduced. If C is large, enormous volumes of solution must be passed, and this fact makes the method almost useless in such cases.

E.III. THE CONCENTRATION-PULSE METHOD

In this method one adds to the solution a small amount of one of the exchanging ions, raising its concentration slightly. This can be a "pulse", i.e. a brief temporary increase in concentration, or it may be a sustained flow of a more concentrated solution, which causes a "front" to move down the column as illustrated in Fig. 6.15. It is important to realize that the pulse or concentration front does *not* necessarily travel at the same rate as the molecules of the solute which cause it. As Helfferich[28] puts it, the movement of the *disturbance* is quite separate and distinct from the movement of *molecules*. This point will be made clearer from the following derivation, based upon Fig. 6.15.

Consider that in an element of time dt the concentration front, i.e. the boundary between the more concentrated and the less concentrated solution, sweeps through a volume element du of the column. (This is *total* volume of the column, exchanger and solution combined.) During this time let a volume element of solution dv enter the column. Let the concentration of the solution ahead of the disturbance be c, that behind it $c + \Delta c$; let the solute in the exchanger in unit volume of the column ahead of the disturbance be $\bar{q}$, that behind it $\bar{q} + \Delta\bar{q}$. (These definitions of c and $\bar{q}$ are the same as those used in section D above.) Let the

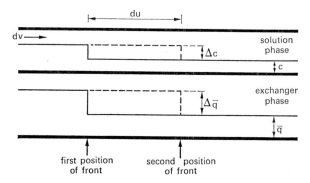

FIG. 6.15. Concentration front and material balance.

interstitial volume fraction be α. Now, considering the column segment of volume du swept out by the concentration front, and taking a mass balance:

Mass of solute entering the segment $= \Delta c (dv/dt)\, dt$.

Increase of solute within the segment $= \Delta c\, \alpha du + \Delta \bar{q}\, du$.

These two quantities are equal. Combining them,

$$\frac{du}{dt} = \frac{\Delta c\, dv/dt}{\Delta c\, \alpha + \bar{q}} \tag{6.62}$$

and

$$\frac{dv}{du} = \frac{\Delta c\, \alpha + \Delta \bar{q}}{\Delta c} = \alpha + \frac{\Delta \bar{q}}{\Delta c}. \tag{6.63}$$

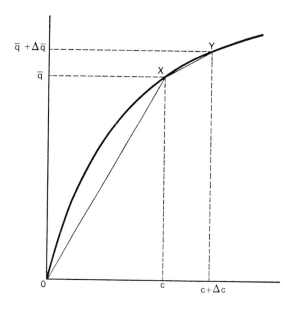

FIG. 6.16. Distribution isotherm, tracer-pulse, and concentration pulse.

The quantity dv/du, the ratio of solution velocity to disturbance velocity, is equal to the ratio of elution volume to bulk column volume for the elution of the "front", and it gives the slope of the distribution isotherm illustrated in Fig. 6.16. By comparison, the ratio of elution volume to bulk column volume for a "tracer-pulse" experiment is simply

$$\left(\frac{dv}{du}\right)_{tracer} = \alpha + \frac{\bar{q}}{c}. \tag{6.64}$$

Thus the two methods complement one another and could be used jointly for the study of exchange or adsorption isotherms.

References

1. R. C. De Geiso, W. Rieman and S. Lindenbaum, *Anal. Chem.*, **26**, 1840 (1954).
2. A. D. Jouy and J. Coursier, *Bull. Soc. chim. France*, **323** (1958).
2a. W. Rieman, *Record Chem. Progr.*, **15**, 85 (1954).
3. J. Beukenkamp, W. Rieman and S. Lindenbaum, *Anal. Chem.*, **26**, 505 (1954).
4. H. L. Rothbart, H. W. Weymouth and W. Rieman, *Talanta*, **11**, 33 (1964).
5. A. J. P. Martin and R. L. M. Synge, *Biochem. J.*, **35**, 1385 (1941).
6. S. W. Mayer and E. R. Tompkins, *J. Am. Chem. Soc.*, **69**, 2866 (1947).
7. A. Breyer and W. Rieman, *Anal. Chim. Acta*, **18**, 204 (1958).
8. H. Schwab, W. Rieman and P. A. Vaughan, *Anal. Chem.*, **29**, 1357 (1957).
8a. W. Rieman and A. Breyer, Chromatography: Columnar liquid–solid ion-exchange processes, in *Treatise on Analytical Chemistry* (I. M. Kolthoff and P. J. Elving, eds.), Part I, vol. 3, p. 1545, Interscience, New York, 1961.
9. W. H. Gerdes and W. Rieman, *Anal. Chim. Acta*, **27**, 113 (1962).
10. K. A. Kraus and F. Nelson, Anion exchange studies of metal complexes, in *The Structure of Electrolytic Solutions* (W. J. Hamer, ed.), Wiley, New York, 1959.
11. F. Helfferich, *Ion Exchange*, pp. 205–222, McGraw-Hill, New York, 1962.
12. Y. Marcus and D. Maydan, *J. Phys. Chem.*, **67**, 979 (1963).
13. D. Maydan and Y. Marcus, *J. Phys. Chem.*, **67**, 987 (1963).
14. R. Sargent and W. Rieman, *Anal. Chim. Acta*, **16**, 144 (1957).
15. R. Sargent and W. Rieman, *J. Phys. Chem.* **60**, 1370 (1956).
16. R. C. Weast, *Handbook of Chemistry and Physics*, 45th edn., p. A 108, Chemical Rubber Publishing Co., Cleveland, 1964.
17. A. J. Goudie and W. Rieman, *Anal. Chim. Acta*, **26**, 419 (1962).
18. A. Varon, F. Jakob, K. D. Park, J. Ciric and W. Rieman, *Talanta*, **9**, 573 (1962).
19. E. C. Freiling, *J. Am. Chem. Soc.*, **77**, 2067 (1955).
20. A. C. Breyer, Ph.D. Thesis, Rutgers, The State University, New Brunswick, N.J., U.S.A., 1958.
21. B. Drake, *Arkiv Kemi*, **8**, 1 (1955).
22. H. Schwab, Ph.D. Thesis, Rutgers, The State University, New Brunswick, N.J., U.S.A., 1956.
23. P. Hamilton, *Anal. Chem.*, **30**, 914 (1958).
24. A. J. Goudie, Ph.D. Thesis, Rutgers, The State University, New Brunswick, N.J., U.S.A., 1961.
25. E. Glueckauf, *Trans. Faraday Soc.*, **51**, 34 (1955).
26. E. Glueckauf, *Ion Exchange and Its Applications*, London, Society of Chemical Industry, 1955, p. 34.
27. F. G. Helfferich, *Ion Exchange*, New York, McGraw-Hill, 1962, chap. 9.
28. F. G. Helfferich and D. L. Peterson, *Science*, **142**, 661 (1963).

CHAPTER 7

TECHNIQUE OF ION-EXCHANGE CHROMATOGRAPHY

THE ion-exchange column is a simple device and the technique of ion-exchange chromatography is basically simple. A certain amount of care must be taken in preparing and operating a column, but the part of the operation which is likely to be most complex is the analysis of the solution flowing out of the column.

A. Preparation of the Column

The preparation and use of an ion-exchange column was discussed in Chapter 5. There is little to add to this discussion except to note that in separating complex mixtures by elution chromatography one must use longer columns than in "nonchromatographic applications" and be much more careful about evenness of flow. "Channelling", or irregular flow caused by uneven packing, will play havoc with any separation scheme.

A.I. SIZE OF RESIN

The width of bands in elution chromatography, which is expressed by the parameter "theoretical-plate height" (Chapter 6), is very sensitive to the particle size of the exchanger, and one may decrease "plate height" more effectively by decreasing the particle radius than by any other parameter. Thus one should use as small a particle size as is practical. The smaller the particles, however, the greater is the hydrodynamic resistance to flow, and one has to strike a compromise. If necessary one can force the liquid through the exchanger bed under pressure, and this is often done. Pumps of various types are available, which give controlled flow rates. With gravity feed and columns of the order of 30–60 cm high, a resin of 200–400 mesh is practical to use. With a head of water about 100 cm from inlet to exit, and a 30-cm column and the mesh size stated, the linear flow velocity will be about 1 cm/min; i.e., 1 ml will flow out of a column 1 cm^2 in cross-sectional area in 1 min. This is a convenient flow rate for exchanges of inorganic ions or organic ions with about four carbons or less.

The observations in Chapter 5 about removing fine particles should be noted; it is essential to get rid of the fine material that is present in most commercial resins. Most of the "fines" can be removed by stirring the resin with water in a large beaker, pausing a moment to let the bulk of the resin settle, then pouring off the fine suspension. This process is repeated a couple of times. The water must of course be free from air bubbles.

Finer particle sizes are necessary for chromatographic separation of large ions and molecules. For many biochemical applications a resin of 325-mesh is used or the so-called "colloidal" resin of particle diameter down to 20 μ and less. Particles as small as this cannot readily be classified for size by screening.

The best way to obtain particles of controlled size is by hydraulic classification[1]. The resin is placed in a large pear-shaped separating funnel (Squibb type, 4–6l) fitted at the top with a rubber stopper bearing a glass tube that bends over and leads into a receiver which can be changed. Water is passed upwards at a controlled rate through the funnel. By gradually increasing the flow rate, fractions of resin of increasing particle size are collected successively in the receivers, and with care, 80 % of the particles will have a diameter within 10 % of the mean.

For the very exacting conditions of amino-acid analysis one must control not only the particle size of the resin but its crosslinking. The better grades of commercial resin are very uniformly crosslinked, but segregation can be performed if necessary by floating the resins in sodium tungstate solutions of known densities.[2] (See also ref. 22, Chapter 2.) Solutions of cane sugar can be used in place of sodium tungstate. It is desirable to have uniform spherical beads with as few broken fragments as possible; this factor is best controlled by the manufacturer.

To see what can be achieved in chromatographic analysis by careful attention to the details one should consult Hamilton[2, 3] who resolved mixtures of 148 amino acids over a 24-hr period on a column 125 cm long and 0.636 cm wide, containing sulphonated polystyrene resin (Dowex 50) of 8.5 % crosslinking and particle diameter 17.5 μ. The flow rate was 0.5 ml/min, the driving pressure 30 atm, and the column temperature 45–60°C. The effluent solution was mixed with ninhydrin reagent and passed continuously through a photometer cell; the light absorbance was continuously recorded. The sensitivity was 10^{-10} mole of amino acid.

A.II. EVENNESS OF PACKING

An ion-exchange column intended for elution chromatography must be packed with care. After the resin is placed in the column it must be backwashed. This can be done by attaching a rubber tube and funnel to the outlet of the column, pouring water into the funnel, and alternately raising and lowering the funnel so as to pass water up and down through the bed, passing it upwards with sufficient speed to loosen the resin. The object of this procedure is partly to remove air bubbles, but also to achieve a size classification within the column, with the largest particles at the bottom and the smallest at the top. In any segment of the column the resin beads should be of about the same size, and their packing should approach the uniform close packing of spheres. For the classification to be effective the column must be vertical and not slanting.

A better way to backwash the column is to drive the water upwards with a variable-speed peristaltic pump. There must be a funnel or reservoir at the top of the column which is big enough to hold the resin. Using a pump for backwashing one can use any solution or solvent he desires.

If tap water is used for backwashing one must be quite sure that it does not carry bubbles of air. On the other hand, Sargent[3a] has found that the disturbance caused by air in the column is much less than is usually believed.

A.III. DETERMINATION OF VOID VOLUME AND EXCHANGE CAPACITY

The "void volume" of an exchange column not only includes the space in between the beads, but it also includes the dead space from the bottom of the exchanger bed to the outlet or detector device. This should be as small as possible. The void volume is best measured by passing through the column a solution of a substance which will not enter the exchanger, and noting the volume that passes before this substance emerges. The choice of substance naturally depends on the type of exchanger. Polyvinyl alcohol has been used, also soluble polysaccharides. Cane sugar is a satisfactory substance for ion-exchange resins of more than 8% crosslinking. Solutions of polyphosphates can be used, and if desired these can be "labeled" with radioactive P-32.

With ion-exchange resins that are highly ionized, and therefore exclude electrolytes through the Donnan equilibrium (Chapter 2), the void column volume can be found very simply by passing a dilute solution of a strong electrolyte which is easily analyzed and has one ion identical with the counter-ion of the exchanger. Thus the void volume of a column of hydrogen-form sulphonic acid exchanger can be found by passing dilute (say 0.001 M) hydrochloric acid. For a column of sodium-form exchanger one would use a solution of sodium chloride and test for chloride ion in the effluent (or, perhaps, use electrical conductivity to mark the break-through). With spherical beads of polystyrene resins, such as the commercially available ion exchangers, the ratio of interstitial volume to bed volume V/V_b, is independent of the mean particle size and increases with increasing divinylbenzene content of the resin.[3b] For almost all purposes, a sufficiently accurate value of V can be found by multiplying the height by the cross-sectional area of the resin bed and multiplying the product by the appropriate value of V/V_b from Table 7.1. These values are averages obtained from four different solutions and are corrected for Donnan equilibrium penetration.

The total exchange capacity can be measured in various ways, depending on the system. Thus the capacity of a column of cation-exchange resin could be found by converting it to the acid or hydrogen form, rinsing out the interstitial acid, and then passing excess sodium or potassium chloride solution and titrating the acid which emerges. If the introduction of hydrogen ions causes more than a small change in swelling, however, the operation will spoil the packing of the column. Generally it is best to find the exchange capacity of a column by setting up a small column of measured volume, say a couple of cubic centimeters, with the same exchanger that is used in the chromatographic column and making the capacity measurement on the small column.

TABLE 7.1. RELATIVE INTERSTITIAL VOLUMES OF RESINS

Resin type	Crosslinking	V/V_b	Standard deviation
Dowex 50	2	0.304	0.010
(polystyrene-sulphonic acid)	4	0.327	0.017
	8	0.379	0.010
	16	0.395	0.016
Dowex 1	2	0.351	0.022
(polystyrene-quaternary	4	0.350	0.012
ammonium chloride)	8	0.390	0.015
	10	0.396	0.024

(Used by permission from *Analytica Chimica Acta*.)

B. Performing the Elution

B.I. APPLYING THE SAMPLE TO THE COLUMN

In elution chromatography the sample to be analyzed is applied to the column as a relatively small volume of relatively concentrated solution. Before elution is begun, the sample constituents must occupy a thin and uniform layer at the top of the column. Thus some attention must be paid to the way in which the sample is introduced. If the sample solution is pipetted on the top of a resin bed, the surface layer of resin must be disturbed as little as possible, and the solution must be allowed to sink into the exchanger before any eluting solution is added. Then one should add a small volume of eluting solution, using it to rinse the inside of the tube above the resin, and let this sink into the exchanger bed before adding any more.

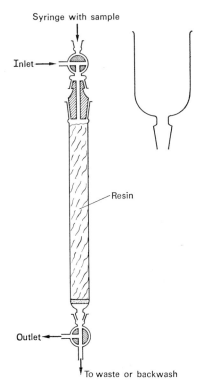

FIG. 7.1. Chromatographic column with stopcocks.

A better way to add the sample, if the operation justifies it, is to use a three-way stopcock at the top of the column (Fig. 7.1) with a socket that will accept a syringe; the sample is introduced from the syringe. In this arrangement the space between the sample inlet and the top of the column should be as small as possible.

Systems can also be arranged with multiport valves to change the eluting solution as desired during the operation. Gradient elution systems can be used (Chapter 6).

B.II. MAINTENANCE OF FLOW RATE

The resolution of the column, or its "number of theoretical plates", depends inversely on the square root of the flow rate if particle diffusion is controlling the exchange, so it is desirable to keep a constant flow rate. If continuous monitoring with automatic recording is used, this is generally based upon the time of flow rather than the volume passed, and a

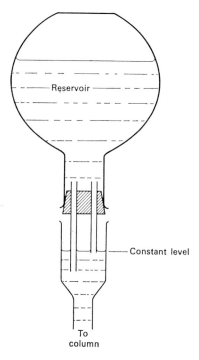

Fig. 7.2. Constant flow device.

constant, reproducible flow rate is essential. Gravity flow is strikingly uniform providing there is no clogging of the column, nor any change in its hydrodynamic resistance during elution. If the flow is to be controlled by gravity a constant-head device is necessary. A very simple device is shown in Fig. 7.2.

If the hydrodynamic resistance of the column is high or variable, mechanical pumping must be used. Peristaltic pumps are convenient for low pressures, and dual-channel peristaltic pumps nearly, but not quite, avoid the pressure fluctuations that characterize this type of pump. Pumps with pistons and valves are used for higher pressures, and multi-cylinder pumps are available that deliver liquids smoothly at flow rates constant to a few tenths of one per cent.

C. Analyzing the Effluent

C.I. FRACTION COLLECTORS

The most common way of analyzing the effluents from chromatographic columns is the discontinuous way. A number of fractions are collected, and these are tested one at a time. Automatic fraction collectors are of various types. All of them carry a series of test-tubes and place them in position, one at a time, below the outlet of the column. The mechanism for shifting the tubes differs according to the equipment. Some instruments count the number of drops of liquid, moving the rack of tubes after a prearranged number of drops have fallen. Others work on a siphon principle; the liquid from the column accumulates in the siphon tube, and when the siphon discharges into the waiting test-tube, it actuates a relay with a slight time delay which moves the receiver rack. The relay may be triggered by an electrical contact or by the refraction of a beam of light passing through the delivery tube of the siphon. Or the movement of the collector rack may take place at preset time intervals. All these systems have their advantages and disadvantages. The siphon and drop systems depend on the surface tension of the flowing liquid, and a system that works well for water may not work for a liquid of low surface tension, such as glacial acetic acid or ethanol.

C.II. CHEMICAL ANALYSIS

Once the samples are collected, they must be tested or analyzed to locate and measure the desired constituents. The simpler and faster the analytical method, the better. Physical methods such as light absorption or gamma-ray counting are ideal. Complicated chemical techniques that have to be performed on many samples should be avoided. The desired peaks can, of course, be located by quick physical means and then an accurate chemical analysis performed on a few selected fractions.

Chemical analysis can be rendered automatic by dispensing reagents with proportional pumps through valves controlled by switches. This is the purpose of the "Auto Analyzer", an extremely versatile instrument that can be used to analyze a succession of individual samples[4] or to monitor a flowing stream.[3] The instrument was invented for the clinical analysis of blood,[5, 6] where many samples must be run one after the other, but it was soon used for analyzing fractions collected in chromatographic analysis. For example, sugars and hydroxy acids are determined automatically by mixing with chromic and concentrated sulphuric acids and then measuring the light absorption of the green chromium(III) produced;[7] mixtures of ortho-, pyro-, and triphosphate are separated by anion exchange and determined by a heteropoly blue reaction.[8] Where time is needed for a slow reaction to take place, the mixed solutions are made to flow through a long coil or spiral of narrow tubing, maintained at whatever temperature is desired. The "finish" or final measurement is usually made by light absorption, but flame emission is also used, and in principle any physical measurement can be made.

C.II.a. *Methods of Continuous Analysis*

Ion-exchange chromatography lacks the sensitive, all-purpose methods of monitoring the effluent stream that have proved so useful in gas chromatography. There is no procedure for liquid chromatography which has the simplicity or universality of the thermal conductivity cell used for gases. Nevertheless, there are ways of continuously following the composition of effluent solutions.

One of these is *light absorption*, either in the visible or the ultraviolet region. Ultraviolet absorption is useful in amino-acid analysis, and even more in the analysis of nucleotides and nucleic acid derivatives. If the substances to be detected do not absorb themselves, they can be combined with a reagent to give a colored substance, and with the aid of solution-metering pumps this can be done continuously as was mentioned in the preceding section.

Another general property specially suited to organic compounds is *refractive index*. Any dissolved substance affects the refractive index of the solution, and one can observe differences in refractive index of the order of 10^{-6} unit and less with a recording differential refractometer.[9, 10] The principle of this instrument is shown in Fig. 7.3. It contains a rectangular glass cell which is bisected diagonally. The flowing liquid first passes through one side of the cell, then through the ion-exchange column, then through the other side of the cell. A beam of light passing through the cell will not be deflected if the liquids in the two halves of the cell have the same refractive index, but if their refractive indices differ the light beam is turned to one side. The effect is doubled by reflecting the beam back through the cell. The returning beam is split by a mirror, and the two halves fall on two photocells. If the currents from these photocells get out of balance, a mechanism is operated which moves a thin glass prism and shifts the beam to restore the balance, at the same time moving the pen of a recording potentiometer. A record of refractive index against time results.

A typical record is that in Fig. 7.4; this shows the separation of propylene imine and ethylene imine by ligand exchange (Chapter 8). The sharp peak that occurs before the two imine peaks is caused by displacement of ammonia from the exchanger.

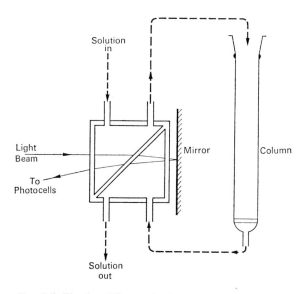

FIG. 7.3. Flowing differential refractometer (schematic).

Refractometric monitoring has the drawback that any change in the composition of the liquid affects the record. Small fluctuations in temperature or pressure cause such changes because they affect the equilibrium between solution and exchanger, and one must shield the column from temperature fluctuations. Refractometric monitoring is not feasible with gradient elution.

Flame photometry has been used for the continuous analysis of effluents containing the alkaline-earth metals.[11] One may ask why, if flame emission is to be used, the elements need to be separated from one another in the first place. The reason is that the presence of one element affects the emission of another, and without separation, tedious calibrations are needed to obtain quantitative results.

Electrical methods of many kinds can be used for special purposes. A continuous record of conductivity is useful for the elution of nitrates, acetates, borates and other ions from a cellulosic exchanger;[12] the eluent is water or very dilute hydrochloric acid. Most solutions in ion-exchange chromatography have such a high conductivity, however, that the small changes accompanying elution would go unobserved. Electrode potential measurements have a limited use; the current at a dropping-mercury electrode placed at the exit from the column is used in the technique of chromato-polarography.[13] This is applicable to organic compounds as well as to the separation of transition-metal ions by cation[14] or anion[15] exchange.

Heat of adsorption is the basis of a new method for monitoring effluents in liquid chromatography, applicable to ionic and nonionic solutes. Great sensitivity and linear response over a wide range are claimed.[16]

Many other ways of column monitoring await the ingenuity of the worker with a problem. The trouble is that most of them will be restricted to use in special circumstances.

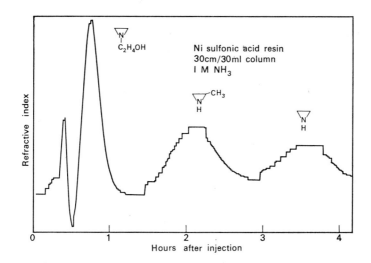

FIG. 7.4. Refractometer record; chromatographic separation of propylene imine and ethylene imine.

References

1. P. B. HAMILTON, *Anal. Chem.*, **30**, 914 (1958).
2. P. B. HAMILTON, *Anal. Chem.*, **35**, 2055 (1963).
3. P. B. HAMILTON and R. A. ANDERSON, *Anal. Chem.*, **31**, 1504 (1959).
3a. R. N. SARGENT, *Ind. Eng. Chem., Process Design Develop.*, **2**, 89 (1963).
3b. G. D. MANALO, R. TURSE and W. RIEMAN, *Anal. Chim. Acta*, **21**, 383 (1959).
4. L. T. SKEGGS, *Am. J. Clin. Pathol.*, **28**, 311 (1957).
5. L. T. SKEGGS, *Anal. Chem.*, **38**, 31A (May 1966); review entitled "New dimensions in medical diagnoses".
6. L. T. SKEGGS, *Clin. Chem.*, **10**, 918 (1964).
7. B. ALFREDSSON, S. BERGDAHL and O. SAMUELSON, *Anal. Chim. Acta*, **28**, 371 (1963).
8. D. P. LUNDGREN, *Anal. Chem.*, **32**, 824 (1960); D. P. LUNDGREN and N. P. LOEB, *ibid.*, **33**, 366 (1961).
9. R. D. COULSON, *Rev. Sci. Instr.*, **34**, 1418 (1963).
10. K. SHIMOMURA and H. F. WALTON, *Anal. Chem.*, **37**, 1012 (1965).
11. F. H. POLLARD, G. NICKLESS and D. SPINCER, *J. Chromatog.*, **11**, 542 (1963); *ibid.*, **13**, 224 (1964).
12. C. DUHNE and O. S. DE ITA, *Anal. Chem.*, **34**, 1074 (1962).
13. W. KEMULA in *Progress in Polarography* (ed. by P. Zuman and I. M. Kolthoff), vol. 2, p. 397, Wiley-Interscience Publishers, New York, 1962
14. W. J. BLAEDEL and J. W. TODD, *Anal. Chem.*, **30**, 1821 (1958).
15. R. L. REBERTUS, R. J. CAPPEL and G. W. BOND, *Anal. Chem.*, **30**, 1825 (1958).
16. K. P. HUPE and E. BAYER, *J. Gas Chromatog*, **5**, 197 (1967).

CHAPTER 8

APPLICATIONS OF ION-EXCHANGE CHROMATOGRAPHY

"CHROMATOGRAPHIC" separations, for the purpose of this chapter, are those in which two or more components are eluted successively from an ion-exchange column. A vast number of such separations are reported and continue to be reported year by year. Tables 8.1 and 8.2 summarize the more important ones. In this chapter we shall restrict ourselves to a few representative separations that will illustrate the various possibilities.

A. Separation of Inorganic Ions
A.I. METALS

Separations of metallic ions depend, as a rule, on two cooperating effects—the selectivity of the ion exchanger and the selectivity of complex-ion equilibria in solution. Most metallic ions associate with the common anions in aqueous solution to form complexes, and the association is strengthened by admixture with nonaqueous solvents that lower the dielectric constant. This point will be discussed again later on. Association of cations with complexing anions decreases the positive charge or changes it to a negative charge, thus modifying the attraction to the exchanger. In the presence of high concentrations of complexing anions, such as chloride, metal ions are displaced from cation exchangers and absorbed by anion exchangers. Some very important chromatographic separations depend on this fact.

The only metal ions that do not commonly form complexes—apart, that is, from aquo-complexes or hydrated ions—are the ions of the alkali and alkaline-earth metals. We shall consider these first.

A.I.a. *Separation of Noncomplexed Ions*

The alkali and alkaline-earth ions are separated most effectively by chromatography on inorganic exchangers such as zirconium phosphate and tungstate. The distribution ratios between exchanger and solution vary so greatly from ion to ion that progressively increasing eluent concentrations are needed to displace successive ions. The early separations by Kraus and his colleagues are well known. From a column of zirconium tungstate, lithium was eluted by 0.05 M ammonium chloride, sodium by 0.1 M, potassium by 0.5 M, rubidium by 1.0 M, and cesium by 3 M ammonium chloride;[1] from a column of zirconium molybdate, calcium, strontium, barium, and radium were separated by slightly acidified solutions of ammonium chloride which were 0.2 M, 0.5 M, 1 M, and 4 M respectively[2] (see Fig. 8.1). A similar separation on zirconium molybdate was reported by Campbell;[3] magnesium ions were eluted by ammonium sulphate, then calcium, strontium, and barium ions successively with ammonium nitrate.

TABLE 8.1. SEPARATIONS OF INORGANIC IONS

Abbreviations: A = anion, C = cation, Cell. = cellulosic, Chel. = chelating,
I = inorganic, L = liquid ion exchanger held on solid support.
The order of presentation follows the periodic system, with the actinides last.
The references are listed at the end of the table.

Elements	Separated from	Exchanger	Eluent	Notes	Ref.
Alkali metals	Each other	C	Alc. HCl		A22
		CI	NH₄Cl	Zr tungstate, molybdate	A53
Li	Na, K, etc.	C	Methanol-HCl	In sea water	A82, 100
Cs	Rb, etc.	CI	HNO₃, HCl	Ferrocyanide exchangers	A11, 32, 54, 74
			NH₄NO₃	Mo-P-tungstate	A55
Cu	Waste waters	Chel.	HNO₃	Pb, Zn abs.	A8, 97
	Sea water	C	—	Carboxylate res.	A12
	Na, Ca, Mn	A	HCl	1 N HCl elutes Cu	A84
	Ni, Zn, Sr, Fe	A	HCl	In brass	A30
Ag	Many elts.	CL	HNO₃	8 M HNO₃ elutes others, 1 M elutes Ag	A67
	Many elts.	C	HNO₃-H₂SO₄	—	A80
Au	Many elts.	Chel.	—	Pt metals abs.	A52
	Cu	A	HCl-HNO₃	—	A62
Alkaline earths	Each other	C	Lactate	Gradient	A74
		CI	NH₄Cl	—	A53
Be	Al, Mg	A	HCl	Aq. isopropanol	A48
	Other elts.	A	—	—	A37, 96
Mg	Ca, Sr	C	(NH₄)₂SO₄	In saline water	A16
Ca	NaOH	Chel.	HCl	Trace det.	A98
	Sr, Ba	C	HNO₃	—	A60, 79
Sr	Environmental samples	C	EDTA	Radiochemical	A39
Zn	Cu, Cd, Ni	A	HCl	Solders; 0.01 N HCl elutes Zn	A41
	Bronze	A	Tartrate	Zn eluted last	A17
Cd	Zn, much Hg	Cell.	Ether–HCl	—	A63
Hg	Many elts.	Chel.	—	—	A9
	Meteorites	A	Thiourea	Trace sep.	A25
	Sn, Sb	A	HNO₃	Trace sep.	A42
B	Alk. salts	A	HCl	Mannitol complex	A101
	U	A, C	Acids	Mannitol used	A64,102
Al	Cu, Pb, Sn	A	Alc. HCl	Al passes	A71
	Fe, Ti	C	HCl	Fe, Ti complexed	A3
	Other elts.	Chel.	—	Al held	A59, 94

TABLE 8.1—*continued*

Elements	Separated from	Exchanger	Eluent	Notes	Ref.
Sc	Al, lanthanides	C	$(NH_4)_2SO_4$	Sc passes	A57
	Rocks	C, A	HCl	Sc passes	A85
Y	Sr	C	Citrate	Y eluted	A77
Sc, La	Al, Fe, Ga	C	HCl	—	A91
Lanthanides	Each other	C	Hydroxyiso-butyric acid	Gradient	A26, 61, 105
Ga, In	Al, Fe, U	A	HBr–methanol	Ga retained	A47
In	Zn, Sn	A	HCl, $(NH_4)_2SO_4$	HCl elutes In	A78, 90
Tl	Al, Mn, Cu, Fe	C	H_2SO_4	4N elutes Tl	A81
Si	Phosphate	A	$HF–H_2SO_4$	—	A95
Ti	Fe, Zn, Pb, Mn	C	Sulphosalicylate	Fe, Ti eluted first	A40
Ti, Zr	Fe, Mo, W, Nb	A	HF–HCl	Elution order Fe, Ti, W, Nb	A21
Zr, Hf	Each other	A	H_2SO_4	Sep. from other elts.	A34, 93
Ge	As, Fe	A	HCl–HOAc	Ge passes	A24, 49
Sn	Mn, Fe, Co	C	HCl	Sn passes	A14
	Many elts.	A	HF	Sn absorbed	A15
Pb	Many elts.	A	HCl	—	A58, 92
	Bi, Hg, Cd	C	HBr, 0.5 M	Order Bi, Hg, Cd, Pb	A28
P	Oxy-anions	A	$NaNO_3$, etc.	—	A75, 103
	Thio-anions	A	NH_4Cl	—	A76
V	Ti	C	NaOH	V eluted	A86
	Many elts.	Chel.	—	V, Ti, Fe held	A72
Nb, Ta	Zr, Hf	A	$NH_4Cl–HF$	Nb eluted before Ta	A33, 37
As	Sb, Sn, Hg	A	HCl	As passes	A42
	Sb, Various	A	HCl–HI	—	A104
Sb	As, Sn, Hg	A	$HCl–NH_4F$	Sb passes	A42
	Many elts.	A	$HCl–H_2O_2$	—	A6
Bi	Many elts.	A	HCl–HBr	—	A28, 29
	Pb, Sn, Sb	A	$HCl–HNO_3$	—	A70
	U, many elts.	C	HNO_3–tetra-hydrofuran	Bi, U pass	A46
S	Thionates	A	NaCl	—	A83
	Polysulphides	A	—	—	A1
Se, Te	As	A	HCl	Order Se, As, Te	A88
	Each other	C	HBr	Order Se, Te	A68

TABLE 8.1—*continued*

Elements	Separated from	Exchanger	Eluent	Notes	Ref.
Te	As, Sn, Fe, Zn	A	HF	Te retained	A2
Cr	Other elts.	Chel.	—	Cr retained	A56
	Limestone	C	Acetate	CrO_4 passes	A18
Mo	W, other elts.	C	HBr	Order Mo, W	A68
W	Mo, Ti	C, A	$HCl-H_2O_2$	—	A87
	Steels	C	NaOH	W passes	A10
F	Fe, P	A	HCl	F passes	A31
Halides	Each other	A	$NaNO_3$	Order F, Cl, Br, I	A13, 38
Halates	Each other	A	$NaNO_3$	Order IO_3, BrO_3, ClO_3	A89
I	IO_3	A	$NaNO_3$	—	A4
Mn	U, other elts.	A	HCl	Mn passes	A66
Re	Mo, other elts.	A	HNO_3–tetra-hydrofuran	ReO abs.	A23, 46
Fe, Co, Ni	Each other	A	HCl, nonaq.	Order Fe, Ni, Co	A48
Fe	Other elts.	Chel.	—	Fe abs.	A20, 36
Co	Other elts.	A	HCl–acetone	Co bound	A50
Ni	Fe, Cd, Zn	A	HCl–acetone	Ni bound	A27
Platinum metals	Other elts.	Chel.	—	Au retained too	A52
	Each other and other elts.	A, C	—	—	A5, 7, 99
Th	Many elts., Zr	C	$HClO_4-H_2SO_4$	Th retained	A65
	Many elts., La		HNO_3-acetone	Th retained	A43, 51
U	Many elts., Bi	C	HNO_3–tetra-hydrofuran	U passes	A44
	Water	A	—	U absorbed	A35
	Pu, etc.	A	HCl–HF	U eluted last	A69
Pu	Environmental	C	EDTA	Pu retained	A19, 96

Table 8.1: References.

A1. P. ALGREN and N. HARTLER, *Svensk. Kem. Tidscr.*, **78**, 404 (1966).
A2. P. I. ARTYUKHIN, E. N. GIL'BERT and V. A. PRONIN, *Zh. Analit. Khim.*, **22**, 111 (1967).
A3. G. N. BABATCHEV, *Chim. Anal.*, **48**, 258 (1966).
A4. W. T. BAKER and G. OWEN, *Nature*, **211**, 641 (1966).
A5. L. M. BANBURY and F. E. BEAMISH, *Z. Anal. Chem.*, **211**, 178 (1965).
A6. F. BAUMGÄRTNER, H. STÄRK and A. SCHÖNTAG, *Ibid.*, **197**, 424 (1963).
A7. F. E. BEAMISH, *Talanta*, **14**, 991, 1133 (1967).
A8. D. G. BIECHLER, *Anal. Chem.*, **37**, 1054 (1965).
A9. E. BLASIUS and M. LASER, *J. Chromatog.*, **11**, 84 (1963).

A10. R. S. Bottei and A. Trusk, *Anal. Chim. Acta*, **37**, 409 (1967).
A11. A. L. Boni, *Anal. Chem.*, **38**, 89 (1966).
A12. R. R. Brooks, *Analyst*, **85**, 745 (1960).
A13. F. Burriel-Marti and C. A. Herrero, *Inform. Quim. Anal.*, **17**, 77 (1963).
A14. F. Burriel-Marti and C. A. Herrero, *Chim. Anal.*, **48**, 602 (1966).
A15. B. R. Chamberlain and R. J. Leech, *Talanta*, **14**, 597 (1967).
A16. R. Christova and A. Kruschevska, *Anal. Chim. Acta*, **36**, 392 (1966).
A17. A. K. De and A. K. Sen, *Z. Anal. Chim.*, **211**, 243 (1965).
A18. A. K. De and A. K. Sen, *Talanta*, **13**, 1313 (1966).
A19. M. C. De Bortoli, *Anal. Chem.*, **39**, 375 (1967).
A20. R. C. De Geiso, L. G. Donaruma, and E. A. Tomic, *Ibid.*, **34**, 845 (1962).
A21. E. J. Dixon and J. B. Headridge, *Analyst*, **89**, 185 (1964).
A22. H. G. Doege and H. Gross-Ruyken, *Talanta*, **12**, 73 (1965).
A23. H. G. Doege and H. Gross-Ruyken, *Mikrochim. Acta*, **98** (1967).
A24. R. M. Dranitskaya and C. C. Liu, *Zh. Analit. Khim.*, **19**, 769 (1964).
A25. W. D. Ehmann and J. F. Lovering, *Geochim. Cosmochim. Acta*, **31**, 357 (1967).
A26. S. C. Foti and L. Wish, *J. Chromatog.*, **29**, 203 (1967).
A27. J. S. Fritz and J. E. Abbink, *Anal. Chem.*, **37**, 1274 (1965).
A28. J. S. Fritz and B. B. Garralda, *Anal. Chem.*, **34**, 102 (1962).
A29. J. S. Fritz and B. B. Garralda, *Ibid.*, p. 1387.
A30. W. H. Gerdes and W. Rieman, *Anal. Chim. Acta*, **27**, 113 (1962).
A31. O. S. Glasoe, *Ibid.*, **28**, 543 (1963).
A32. B. Gorenc and L. Kosta, *Z. Anal. Chem.*, **223**, 410 (1966).
A33. J. F. Hague and L. A. Machlan, *J. Res. Nat. Bur. Standards*, **62**, 11, 53 (1959).
A34. J. L. Hague and L. A. Machlan, *Ibid.*, **A 65**, 75 (1961).
A35. I. Hazan, J. Korkisch and G. Arrhenius, *Z. Anal. Chem.*, **213**, 182 (1965).
A36. R. Hering, *Z. Chem.*, **5**, 402 (1965).
A37. J. O. Hibbits, H. Oberthein, R. Liu and S. Kallmann, *Talanta*, **8**, 209 (1961).
A38. H. Holzappel and O. Guertter, *J. Prakt. Chem.*, **35**, 113 (1967).
A39. R. D. Ibbett, *Analyst*, **92**, 417 (1967).
A40. J. Inczedy, P. Gabor-Klatsmanyi and L. Erdey, *Acta Chim. Acad. Sci. Hung.*, **50**, 105 (1966).
A41. S. L. Jones, *Anal. Chim. Acta*, **21**, 532 (1959).
A42. W. Kiesel, *Z. Anal. Chem.*, **227**, 13 (1967).
A43. J. Korkish and S. S. Ahluwalia, *J. Inorg. Nucl. Chem.*, **28**, 264 (1966).
A44. J. Korkisch and S. S. Ahluwalia, *Anal. Chem.*, **38**, 497 (1966).
A45. J. Korkisch and S. S. Ahluwalia, *Anal. Chim. Acta*, **34**, 308 (1966).
A46. J. Korkisch and F. Feik, *Anal. Chim. Acta*, **37**, 364 (1967).
A47. J. Korkisch and I. Hazan, *Anal. Chem.*, **37**, 707 (1965).
A48. J. Korkisch and F. Feik, *Ibid.*, p. 757.
A49. J. Korkisch and F. Feik, *Separation Sci.*, **2**, 1 (1967).
A50. J. Korkisch and F. Feik, *Ibid.*, 169 (1967).
A51. J. Korkisch and I. Hazan, *Talanta*, **11**, 523, 721, 1157 (1964).
A52. G. Koster and G. Schmuckler, *Anal. Chim. Acta*, **38**, 179 (1967).
A53. K. A. Kraus, H. O. Phillips, T. A. Carlson and J. S. Johnson, *Proc. 2nd Intern. Conf. on Peaceful Uses of Atomic Energy*, **28**, 3 (1958).
A54. J. Krtil, *J. Inorg. Nucl. Chem.*, **27**, 233 (1965).
A55. J. Krtil and I. Krivy, *J. Inorg. Nucl. Chem.*, **25**, 1191 (1963).
A56. G. Kuehn and E. Hoyer, *J. Prakt. Chem.*, **35**, 197 (1967).
A57. R. Kuroda, Y. Nakagoni and K. Ishida, *J. Chromatog.*, **22**, 143 (1966).
A58. M. Leclerq and G. Duyckaerts, *Anal. Chim. Acta*, **29**, 139 (1963).
A59. M. Lederer and L. Ossicini, *J. Chromatog.*, **22**, 200 (1966).
A60. K. H. Lieser and H. Bernhard, *Z. Anal. Chem.*, **219**, 401 (1966).
A61. S. F. Marsh, *Anal. Chem.*, **39**, 641 (1967).
A62. A. Mizuika, Y. Iida, K. Yamada and S. Hirano, *Anal. Chim. Acta*, **32**, 428 (1965).
A63. R. A. A. Muzzarelli, *Talanta*, **13**, 809 (1966).
A64. R. A. A. Muzzarelli, *Anal. Chem.*, **39**, 365 (1967).
A65. B. N. Nabivanets and L. N. Kudritskaya, *Zh. Analit. Khim.*, **21**, 40 (1966).
A66. F. Nakashima, *Anal. Chim. Acta*, **30**, 167, 255 (1964).
A67. F. Nelson, *J. Chromatog.*, **20**, 378 (1965).
A68. F. Nelson and D. A. Michelson, *J. Chromatog.*, **25**, 414 (1966).
A69. F. Nelson, D. A. Michelson and J. H. Holloway, *Ibid.*, **14**, 258 (1964).
A70. S. Onuki, *Bunseki Kagaku*, **12**, 844 (1963).
A71. S. Onuki, K. Watanuki and Y. Yoshino, *Ibid.*, **15**, 928 (1966).

A72. G. Petrie, D. Locke and C. Meloan, *Anal. Chem.*, **37**, 919 (1965).
A73. H. G. Petrow and H. Levine, *Ibid.*, **39**, 360 (1967).
A74. F. H. Pollard, G. Nickless and D. Spincer, *J. Chromatog.*, **11**, 542 (1963); *ibid.*, **13**, 224 (1964).
A75. F. H. Pollard, G. Nickless, D. E. Rogers and M. T. Rothwell, *Ibid.*, **17**, 157 (1965).
A76. F. H. Pollard, G. Nickless and J. D. Murray, *Nature*, **209**, 396 (1966).
A77. C. R. Porter and B. Kahn, *Anal. Chem.*, **36**, 676 (1964).
A78. B. A. Raby and C. V. Banks, *Anal. Chim. Acta*, **29**, 532 (1963).
A79. A. T. Rane and K. S. Bhatki, *Anal. Chem.*, **38**, 1598 (1966).
A80. A. V. Rangnekar and S. M. Khopkar, *Mikrochim. Ichnoanal. Acta*, **1965**, 642.
A81. A. V. Rangnekar and S. M. Khopkar, *Indian J. Chem.*, **4**, 318 (1966).
A82. R. Ratner and Z. Ludmer, *Israel J. Chem.*, **2**, 21 (1964).
A83. M. Schmidt and T. Sand, *Z. Anal. Chem.*, **330**, 188 (1964).
A84. W. G. Schrenk, K. Graber and R. Johnson, *Anal. Chem.*, **33**, 106 (1961).
A85. T. Shimizu, *Anal. Chim. Acta*, **37**, 75 (1967).
A86. D. A. Shishkov and L. Shishkova, *Compt. Rend. Acad. Bulgare Sci.*, **16**, 833 (1963).
A87. D. A. Shishkov, L. Shishkova and E. G. Koleva, *Talanta*, **12**, 857, 865 (1965).
A88. M. Simek, *Chem. Listy*, **60**, 817 (1966).
A89. J. K. Skloss, J. A. Hudson and C. J. Cummiskey, *Anal. Chem.*, **37**, 1240 (1965).
A90. E. Stanchevska, I. P. Alimarin and E. P. Tsintsevich, *Zavodsk. Lab.*, **28**, 156 (1962).
A91. F. W. E. Strelow, *Anal. Chim. Acta*, **34**, 387 (1966).
A92. F. W. E. Strelow, *Anal. Chem.*, **39**, 1454 (1967).
A93. F. W. E. Strelow and C. J. C. Bothma, *Ibid.*, p. 595.
A94. V. Sykora and F. Dubsky, *Coll. Czech. Chem. Commun.*, **32**, 3342 (1967).
A95. T. Tagaki, T. Hashimoto and M. Sasaki, *Bunseki Kagaku*, **12**, 618 (1963).
A96. T. Y. Toribara, *Separation Sci.*, **2**, 283 (1967).
A97. R. Turse and W. Rieman, *Anal. Chim. Acta*, **24**, 202 (1961).
A98. A. J. Vander Reyden and R. L. M. Van Lingen, *Z. Anal. Chem.*, **187**, 241 (1962).
A99. J. C. Van Loon and F. E. Beamish, *Anal. Chem.*, **36**, 1771 (1964).
A100. K. Vesugi and T. Murakami, *Bunseki Kagaku*, **15**, 482 (1966).
A101. S. Y. Vinkovetskaya and V. A. Nazarenko, *Zavod. Lab.*, **32**, 1202 (1966).
A102. A. W. Wenzel and C. E. Pietri, *Anal. Chem.*, **36**, 2083 (1964).
A103. J. H. Wiersma and A. A. Sandoval, *J. Chromatog.*, **20**, 374 (1964).
A104. W. M. Wise and J. P. Williams, *Anal. Chem.*, **36**, 19 (1964).
A105. L. Wish and S. Foti, *J. Chromatog.*, **20**, 585 (1965).

Table 8.2. Organic Chromatographic Applications

Abbreviations: As in Table 8.1, with Lig = ligand exchange.

Compounds	Exchanger	Eluent, method, notes	Ref.
Acetophenone	C	Frontal analysis; sep. from β-naphthol	B46
Acids			
Aldonic	A	Acetate buffer	B29
Aliphatic:			
carboxylic	A	Nitrate–borate; esp. hydroxy acids	B21, 42
		NaOAc (gradient, automatic)	B57
		Formic acid	B15
		$Mg(OAc)_2$; $Zn(OAc)_2$; hydroxy acids	B28, 30
	C	Acetone–CH_2Cl_2–H_2O eluent	B40
Aromatic:			
carboxylic	A	HCl; elution follows acid strengths	B22
		NaCl–CH_3OH; subst benzoic acids;	
		strongest acid eluted last	B20
sulphonic	A	NaCl, $CaCl_2$, KBr	B17, 19, 47
Chloroacetic	A	In herbicides; HCl, NaCl eluent	B48

TABLE 8.2—*continued*

Compounds	Exchanger	Eluent, method, notes	Ref.
Acids—*cont.*			
Chlorobenzoic	C	Aq. NaCl; salting-out chrom.	B18
Naphthenic: phenolic	Lig.	In petroleum products; CH_3OH eluent	B52
Unsaturated	Lig.	$CHCl_3$–CH_3OH eluent	B16
Uronic	A	Acetate buffer	B26, 38
Alcohols			
Glycols, Polyols	A	Borate eluent (or water)	B45, 55
Aldehydes and ketones	A A, C	Bisulphite eluent Salting-out with $(NH_4)_2SO_4$	B9 B7
Alkaloids	AL	Citrate eluent; paper chrom.	B43
Amines			
Aliphatic	CI Lig.	Zr phosphate; HCl eluent Aq. ammonia	B36 B24, 41
Aromatic	C	HCl eluent	B51
Ethanolamines	C	Borax eluent; trieth. eluted first	B56
Phenolic	C	Aq. ammonia	B53
Amino acids	C Lig.	Citrate buffers Cd resin, NH_3 eluent	B3, 23, 44 B2
Sulphur-containing	C	Citrate buffers	B32
Amino sugars	C	Citrate buffers	B6, 49
Anthocyanines	Lig.	On aluminosilicate gel	B4
Antibiotics	C	Water eluent	B31
Carbohydrates	A	Aq. ethanol (up to 88% EtOH); sulphate form resin	B14, 37
Monosaccharides	A	Borate–NaCl–glycerol	B50
Coumarins	C	10% methanol	B35
Esters, unsat.	Lig.	On Ag resin; aq. methanol eluent	B39, 54
Hydrazines	Lig.	Aq. NH_3 eluent; dimethylhydrazine first	B41
Nitro compounds	C	Frontal analysis	B46
Nucleosides, nucleotides	A C	Sodium acetate gradient Formate buffer	B1, 12 B5
Peptides	Lig.	Sep. from amino acids	B8
Phenols	A	On paper; butanol–H_2O HOAc	B10

TABLE 8.2—*continued*

Compounds	Exchange	Eluent, method, notes	Ref.
Purines,	A	Acetate gradient	B1
pyrimidines	C	HCl eluent; pyrimidines first	B11
	C	Citrate gradient	B13
	Lig.	NH_3 eluent; pyrimidines first	B41
Sulphonamides	C	Conc. aq. NH_3	B33
Sulphoxides	C	Water and methanol eluents	B27, 34
Vitamin B	C	Components sep. on thin layer	B25

Table 8.2: References.

B1. N. G. ANDERSON, J. G. GREEN, M. L. BARBER, and F. C. LADD, *Anal. Biochem.*, **6**, 153 (1963).
B2. Y. ARIKAWA and I. MAKIMO, *Proc. Federated Biological Societies* (Abstract), **25**, 786 (1966).
B3. J. V. BENSON, Jr. and J. A. PATTERSON, *Anal. Chem.*, **37**, 1108 (1965).
B4. L. BIRKOFER, C. KAISER, and M. DONIKE, *J. Chromatog.*, **22**, 303 (1966).
B5. F. R. BLATTNER and H. P. ERICKSON, *Anal. Biochem.*, **18**, 220 (1967).
B6. K. BRENDEL, N. O. ROSZEL, R. W. WHEAT, and E. A. DAVIDSON, *Anal. Biochem.*, **18**, 147 (1967).
B7. A. BREYER and W. RIEMAN, *Anal. Chim. Acta*, **18**, 204 (1958).
B8. N. R. M. BUIST and DONOUGH O'BRIEN, *J. Chromatog.*, **29**, 398 (1967).
B9. K. CHRISTOFFERSON, *Anal. Chim. Acta*, **33**, 303 (1965).
B10. I. T. CLARA, *J. Chromatog.*, **15**, 65 (1964).
B11. W. E. COHN, *Science*, **109**, 377 (1949).
B12. W. E. COHN, *J. Am. Chem. Soc.*, **72**, 1471 (1950).
B13. C. F. CRAMPTON, F. R. FRANKEL, A. M. BENSON, and A. WADE, *Anal. Biochem.*, **1**, 249 (1960).
B14. J. DAHLBERG and O. SAMUELSON, *Acta Chem. Scand.*, **17**, 2136 (1963).
B15. C. DAVIES, R. D. HARTLEY, and G. J. LAWSON, *J. Chromatog.*, **18**, 47 (1965).
B16. S. P. DUTTA and A. K. BARUTA, *J. Chromatog.*, **29**, 263 (1967).
B17. W. FUNASAKA, T. KOJIMA, K. FUJIMURA, and S. KUSHIDA, *Bunseki Kagaku*, **12**, 1170 (1963).
B18. W. FUNASAKA, T. KOJIMA, K. FUJIMURA, and S. KUSHIDA, *Ibid.*, **15**, 835 (1966).
B19. W. FUNASAKA, T. KOJIMA, K. FUJIMURA, and T. MINAMI, *Ibid.*, **12**, 466 (1963).
B20. W. FUNASAKA, T. KOJIMA, G. TAKESHIMA, and J. KIMOTO, *Ibid.*, **11**, 434 (1962).
B21. A. J. GOUDIE and W. RIEMAN III, *Anal. Chim. Acta*, **26**, 419 (1962).
B22. D. K. HALE, A. R. HAWDON, J. I. JONES, and D. I. PACKHAM, *J. Chem. Soc.*, **1952**, 3503.
B23. P. B. HAMILTON, *Anal. Chem.*, **35**, 2055 (1963).
B24. Sister A. G. HILL, R. SEDGELEY, and H. F. WALTON, *Anal. Chim. Acta*, **33**, 84 (1965).
B25. R. HUTTENRAUCH, L. KLOTZ, and W. MULLER, *Z. Chem.*, **3**, 193 (1963).
B26. S. JOHNSON and O. SAMUELSON, *Anal. Chim. Acta*, **36**, 1 (1966).
B27. L. H. KRULL and M. FRIEDMAN, *J. Chromatog.*, **26**, 336 (1967).
B28. U. B. LARSSON, T. ISAKSSON, and O. SAMUELSON, *Acta Chem. Scand.*, **20**, 1966 (1965).
B29. U. B. LARSSON, I. NORSTEDT, and O. SAMUELSON, *J. Chromatog.*, **22**, 102 (1966).
B30. K. S. LEE and O. SAMUELSON, *Anal. Chim. Acta*, **37**, 359 (1967).
B31. H. MAEHR and C. P. SCHAFFNER, *Anal. Chem.*, **36**, 104 (1964).
B32. C. DE MARCO, R. MOSTI, and D. CAVALLINI, *J. Chromatog.*, **18**, 492 (1965).
B33. R. P. MOONEY and N. R. PASARELA, *J. Agricul. Food Chem.*, **12**, 123 (1964).
B34. I. OKUNO, D. R. LATHAM, and W. E. HAINES, *Anal. Chem.*, **39**, 1830 (1967).
B35. D. J. PATEL and S. L. BAFNA, *Nature*, **211**, 963 (1966).
B36. R. L. REBERTUS, *Anal. Chem.*, **38**, 1089 (1966).
B37. O. SAMUELSON and B. SWENSON, *Acta Chem. Scand.*, **16**, 2056 (1962).
B38. O. SAMUELSON and L. THEDE, *J. Chromatog.*, **30**, 556 (1967).
B39. C. R. SCHOLFIELD and E. A. EMKEN, *Lipids*, **1**, 235 (1966).
B40. T. SEKI, *J. Chromatog.*, **22**, 498 (1966).
B41. K. SHIMOMURA, L. DICKSON, and H. F. WALTON, *Anal. Chim. Acta*, **37**, 102 (1967).
B42. K. SHIMOMURA and H. F. WALTON, *Anal. Chem.*, **37**, 1012 (1965).
B43. E. SOCZEWINSKI and M. ROJOWSKA, *J. Chromatog.*, **27**, 206 (1967).
B44. D. H. SPACKMAN, W. H. STEIN, and S. MOORE, *Anal. Chem.*, **30**, 1190 (1958).

B45. N. Spencer, *J. Chromatog.*, **30**, 566 (1967).
B46. H. D. Spitz, H. L. Rothbart, and W. Rieman III, *Talanta*, **12**, 395 (1965).
B47. H. Stenerle, *Z. Anal. Chem.*, **220**, 413 (1966).
B48. I. K. Tsitovich and E. A. Kuzmenko, *Zh. Anal. Khim.*, **22**, 603 (1967).
B49. E. F. Walborg, B. F. Colb, M. Adams-Mayne, and D. N. Ward, *Anal. Biochem.*, **6**, 367 (1963).
B50. E. F. Walborg, L. Christensson, and S. Gardell, *Anal. Biochem.*, **13**, 177 (1965).
B51. S. R. Watkins and H. F. Walton, *Anal. Chim. Acta*, **24**, 334 (1961).
B52. P. V. Webster, J. N. Wilson, and M. C. Franks, *Anal. Chim. Acta*, **38**, 193 (1967).
B53. T. A. Wheaton and I. Stewart, *Anal. Biochem.*, **12**, 585 (1965).
B54. C. F. Wurster, J. H. Copenhaver, and P. R. Shafer, *J. Am. Oil Chemists Soc.*, **40**, 513 (1963).
B55. F. Yaku and Y. Matsushima, *Nippon Kagaku Zasshi*, **87**, 969 (1966).
B56. Y. Yoshino, H. Kinoshita, and H. Sugiyama, *Nippon Kagaku Zasshi*, **86**, 405 (1965).
B57. R. C. Zerfing and H. Veening, *Anal. Chem.*, **38**, 1312 (1966).

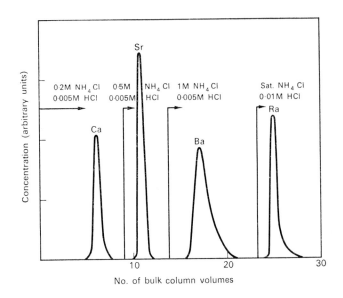

FIG. 8.1. Separation of alkaline-earth elements on a column of zirconium molybdate. Eluents: mixtures of ammonium chloride and hydrochloric acid of the concentrations indicated. (Reproduced by permission from K. A. Kraus, J. O. Phillips, T. A. Carlson, and J. S. Johnson, *Proc. 2nd Intern. Conf. Peaceful Uses Atomic Energy*, **28**, 10 (1958).)

Studies of separations in these groups have been stimulated by the need to recover radioactive cesium and strontium from fission-product wastes produced in atomic reactors. A number of very effective separations of rubidium and cesium have been described, using such exchangers as ammonium phosphomolybdate,[4] phosphomolybdotungstate,[5] and hexacyanocobalt(II) ferrate(II).[6] From the phosphomolybdotungstate, for example, rubidium was eluted with 1 M ammonium nitrate and cesium by 10 M ammonium nitrate. Strontium is separated from calcium, barium, and other elements by elution with ammonium nitrate from a stannic phosphate exchanger.[7] It should be noted that these exchangers bind protons strongly and their selectivity depends greatly on pH, being greatest at low pH, i.e., when their capacity for holding ions other than hydrogen ions is at a minimum. At high pH the selectivity order may even be reversed.

Sulphonated polystyrene resins show little selectivity between cesium and rubidium, but serve well to separate the other alkali metals. Lithium, sodium, and potassium are eluted successively by hydrochloric acid.[8] It is easy to separate the alkali metals as a group from the alkaline earths[9] on the basis of their charge; in dilute (0.2–0.5 M) hydrochloric acid the divalent cations are held much more strongly. As the acid concentration is raised the differences in affinity diminish in accordance with the mass law and the equilibrium constants (see Chapter 3).

Separation of cations according to their charge was performed by Fritz and Karraker[10] using a sulphonated polystyrene resin and solutions of ethylenediammonium perchlorate, $C_2H_4(NH_3)_2(ClO_4)_2$ as eluents. The cation of this salt, being doubly charged, is held strongly by the resin, yet the excess of salt that accompanies the eluted fractions is easily removed by evaporation and heating. Singly charged cations are displaced immediately, then most divalent ions are displaced by 4–20 bulk column volumes of 0.1 M solution. The concentration is then raised to 0.5 M to displace the trivalent ions of aluminum, scandium, and yttrium. Cerium and lead are eluted with the trivalent ions.

Strontium is separated from yttrium, and calcium from scandium, in radiochemical work by elution with dilute nitric acid from a sulphonated polystyrene resin; the divalent ions emerge first, the trivalent ions later with 2 M acid.[11]

A.I.b. *Cation Exchange with Complexing Agents*

It is usual in the ion-exchange chromatography of metals to add to the solution something that will form complex ions of differing stabilities with the metals to be separated. Ions having the "inert-gas configuration", i.e., ions of the alkaline earths and the scandium–yttrium–lanthanum group, form complexes preferentially with hydroxy compounds and oxy anions in contrast to the post-transition ions like those of zinc and cadmium, which prefer ammonia, iodide, or sulphide ions.[12] Complex formation by the alkaline-earth metal ions is governed primarily by electrostatic forces and therefore becomes weaker as the ionic radius increases. Ion-exchange distribution coefficients *increase* as the ionic radius increases (with exchangers of the sulphonic acid type; see Chapter 3), so the two effects reinforce each other; the resultant attraction for the resin increases rapidly with increasing ionic radius.

The simplest organic hydroxy acids are glycollic acid, $CH_2OH.COOH$, and lactic acid, $CH_3.CHOH.COOH$. Others frequently used in chromatographic separations are 2-methyllactic or α-hydroxy*iso*butyric acid, $(CH_3)_2COH.COOH$, tartaric, and citric acids. These are all weak acids, and the concentrations of their anions, which are the actual complex-formers, can be varied within wide limits by simply changing the pH. Pollard *et al.*[13] used buffers of lactic and 2-methyllactic acid to separate the alkaline earth elements on a column of sulphonated polystyrene resin. The best results were obtained with 2-methyllactic acid-ammonia buffers, as follows: 0.8 M, pH 4.15 eluted magnesium; 1 M, pH 5.0 eluted calcium and then strontium; 2 M, pH 6.20 eluted barium. The solutions were analyzed by flame photometry.

Hydroxy-acid buffers are used to separate the rare-earth or lanthanide ions. One of the earliest triumphs of ion-exchange elution chromatography was the separation of these elements[14] (obtained as fission products from uranium) on sulphonic acid resins by citrate buffers at pH values ranging from 3.2 to 4.5. Lutecium is eluted first, lanthanum last. The strength of binding by the resin *decreases* with increasing atomic number instead of increasing because of the "lanthanide contraction"; the ionic radius decreases with increasing atomic number. Between two consecutive elements the difference in elution volumes is small, but from one end of the series to the other the difference is great. Gradient elution, therefore, is advantageous.[15]

Buffers of lactate,[15, 16] 2-methyllactate, or α-hydroxy*iso*butyrate[17] and EDTA[18] have been studied intensively for rare-earth separations; additional references are cited in Table 8.1. Both lactate and α-hydroxy*iso*butyrate give better separations of the "yttrium earths", yttrium, samarium, europium, gadolinium, and terbium than does citrate. (It must not be forgotten that the element yttrium has an ionic radius falling midway in the lanthanide series, and therefore is associated with the lanthanides in separation processes.) Ethylene-diaminetetraacetic acid (EDTA) gives good discrimination in distribution coefficients and elution volumes, but the peaks are broader and overlap more than those observed with hydroxy acids, undoubtedly because of slow formation and dissociation of the complexes.[18]

Hydroxy-acid buffers have the drawback that the excess of eluent must sometimes be removed from the fractions after elution, and this is a tedious operation. An excess of EDTA is somewhat easier to remove, as the acid itself is not very soluble in water and it can be crystallized out of the solutions after acidification. The low solubility is a drawback, how-ever, if solutions of pH below 3 have to be used in the elution.

These reagents form more stable complexes with trivalent ions than with divalent ions, and are used to separate ions of different charges. In the analytical determination of traces of strontium-90, for example, the strontium is absorbed on a cation-exchange resin column, left for a day or two to produce some yttrium-90 by beta-decay, and then the yttrium is eluted, leaving the strontium behind. It is counted by its gamma-ray activity. Elution may be with citrate, lactate, or EDTA.[19]

Ethylenediaminetetraacetic acid (EDTA or H_4Y) is often used to elute metals from cation-exchange columns. The complex-forming ability depends on the activity of the fully deprotonated ion Y^{-4}, and this can be controlled over some fifteen powers of ten by adjusting the pH. Fritz and Umbreit[20] studied the distribution of certain metals between 0.015 M EDTA solutions and a cation-exchange resin in the low pH range, 0.6–4.0. Comparing cations of different charge they noted that increasing charge favored binding by the resin more than it favored complexing in the solution. Nevertheless the ions of bismuth(III), zirconium(IV), and copper(II) were more than 99% extracted from the resin above pH 1. Cations that are hard to remove, such as thorium(IV), can nearly always be removed from columns by passing EDTA.

The stability of metal-EDTA complexes generally decreases with increasing (unhydrated) ionic radius, favoring displacement of smaller ions from the resin, but there are exceptions. In the pairs Mg–Ca and Al–Sc the larger ion (Ca or Sc) forms the more stable chelate, probably because of a strain effect in the chelate ring.

Other eluting agents which form complexes include oxalate, which removes iron(III) and aluminum and leaves beryllium;[21] tartrate, which removes aluminum and leaves copper;[22] sulphosalicylate, which does the same;[22] and hydrogen peroxide, which removes those transition-metal ions that form peroxy-complexes, such as titanium(IV),[23] vanadium(V), and molybdenum(VI).[24] Many others could be mentioned.

Hydrochloric acid solutions are of special interest because of their use in anion-exchange separations. If hydrochloric acid puts metals on to anion-exchange resins by forming anionic chloride complexes, it should take metals off cation-exchange resins for the same reason. The effect on cation exchange is not nearly as great, however. The resin environment generally favors the absorption of highly charged ions; anion-exchange resins absorb and stabilize the fully coordinated chloride complexes whereas cation-exchange resins favor the uncomplexed cations, since these have the maximum positive charge. This point is discussed at more length in Chapter 11. Nevertheless, hydrochloric acid is more efficient than nitric or sulphuric acid in eluting metals which form strong chloride complexes, such as mercury(II), zinc(II), cadmium(II), iron(III),[25] zirconium (IV),[26] beryllium,[27] and palladium.[28] Sulphuric acid, which fixes uranium(VI) on an anion exchanger, takes it off a cation exchanger; nitric acid acts similarly towards thorium(IV), eluting it preferentially from cation exchangers.[29]

All these effects increase markedly as the acid concentration increases, which is only to be expected, since the proportion of highly coordinated complexes rises rapidly with the concentration of the complexing ions. Strelow[30] studied the distribution of some fifty metals between a sulphonated polystyrene resin and solutions of nitric, sulphuric, and hydrochloric acids up to 4 N. By changing from one acid to another during an elution and gradually increasing the concentrations, a number of excellent chromatographic separations can be made (see Table 8.3 and Fig. 8.2).

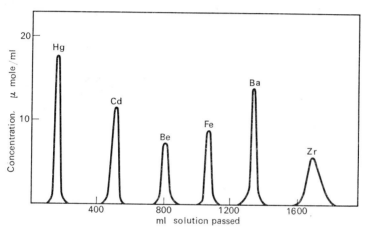

FIG. 8.2. Cation-exchange separation of metal ions. The resin was sulphonated polystyrene, 8% crosslinking, in a column containing 20 g of resin on a dry basis. The eluent was changed between elution peaks as follows: to elute Hg, 0.20 N HCl; to elute Cd, 0.50 N HCl; to elute Be, 1.20 N HNO$_3$; to elute Fe(III), 1.75 N HCl; to elute Ba, 2.50 N HNO$_3$; to elute Zr, 5.00 N HCl. (Reproduced by permission from F. W. E. Strelow, R. Rethemeyer, and C. J. C. Bothma, *Anal. Chem.*, **37**, 106 (1965).)

At still higher concentrations of acid many selectivity coefficients, instead of falling with rising hydrogen-ion concentration as the mass law leads one to expect, start to rise, and selectivity orders are drastically changed. Iron(III) is strongly absorbed by a sulphonated polystyrene resin from 6 N hydrochloric acid; the minimum absorption falls at about 2–4 N. In 12 N perchloric acid the calcium–strontium–barium selectivity order is reversed; calcium is held most strongly and barium the least, the distribution ratios being 900, 70, and 5 ml/g for these three ions.[31] Probably the reversal is due to dehydration of the metal ions. The invasion of the resin by the co-ions is another effect which is important at high concentrations.

A.I.c. *Effect of Organic Solvents*

Cation-exchange selectivities change markedly if another solvent is substituted for water, or if mixed organic-aqueous solvents are used. This was noted many years ago with alumino-

TABLE 8.3. CATION-EXCHANGE DISTRIBUTIONS IN NITRIC AND
SULPHURIC ACIDS
(Selected data of Strelow *et al.*, *Anal. Chem.* **37**, 106, 1965)

Cation	*D* in HNO$_3$ of concentration				
	0.2 N	0.5 N	1 N	2 N	4 N
Ag(I)	86	36	18	7.9	4.0
Al(III)	3900	392	79	16	5.4
Ba(II)	1560	271	68	13	3.6
Be(II)	183	52	15	6.6	3.1
Bi(III)	305	79	25	7.9	3.0
Ca(II)	480	113	35	9.7	1.8
Cd(II)	382	91	33	11	3.4
Co(II)	392	91	29	10	4.7
Cr(III)	1620	418	112	28	11
Cu(II)	356	84	27	8.6	3.1
Fe(III)	4100	362	74	14	3.1
Ga(III)	4200	445	94	20	5.8
Hg(II)	1090	121	17	5.9	2.8
La(III)	10^4	1870	267	47	9.1
Mg(II)	295	71	23	9.1	4.1
Mn(II)	389	89	28	11.4	3.0
Ni(II)	384	91	28	10	7.3
Pb(II)	1420	180	36	8.5	4.5
Th(IV)	10^4	10^4	1180	120	25
Ti(IV)	460	70	15	6.5	3.4
U(VI)	260	69	24	11	6.6
Zn(II)	350	80	25	7.5	3.6
Zr(IV)	10^4	10^4	6500	650	31

Cation	*D* in H$_2$SO$_4$ of concentration				
	0.2 N	0.5 N	1 N	2 N	4 N
Al(III)	8300	540	126	28	4.7
Be(II)	300	80	27	8	2.6
Bi(III)	10^4	6800	235	32	6.4
Cd(II)	540	140	46	15	4.3
Co(II)	433	126	43	14	5.4
Cr(III)	176	126	55	19	0.2
Cu(II)	505	130	41	13	3.7
Fe(III)	2050	255	58	13.5	1.8
Ga(III)	3500	620	137	27	4.9
Hg(II)	1790	320	100	35	12
La(III)	10^4	1860	330	68	12
Mg(II)	480	120	41	13	3.4
Mn(II)	610	165	59	17	5.5
Ni(II)	590	140	46	16.5	2.8
Th(IV)	3900	260	52	9.0	1.8
Ti(IV)	225	46	9	2.5	0.4
U(VI)	118	29	10	3.2	1.8
Zn(II)	550	135	43	12	4.0
Zr(IV)	474	98	5	1.4	1.0

Notes: (a) The resin was the polystyrene sulphonate, Dowex 50-X8.
(b) The table gives *D* (Chapter 3, section A.II) in units of
ml/g, at a uniform resin loading of 40% of the exchange
capacity.
(c) Values have been rounded to two significant figures in
many cases; the reference cited gives additional data.
(Used by permission from *Analytical Chemistry*.)

silicate exchangers; adding alcohol lessened the discrimination between the alkali-metal ions, making sodium nearly as strongly bound as potassium, whereas in water it was much less strongly bound than potassium. This effect was attributed to the decreased hydration, and this explanation is still a good one. According to current theories of cation-exchange selectivity (Chapter 3) the equilibrium depends on the balance between the forces of hydration in the solution and the attraction of the fixed ions in the exchanger for the counter-ions. In the resins, where the hydrogen ion can be used as a reference ion, the attraction for the alkali and alkaline-earth ions increases considerably as the organic solvent is added.[32] However, Nelson et al.[33] found that lithium, sodium, and potassium were more efficiently separated by hydrochloric acid in 80% methanol than in water.

As an organic solvent is substituted for water the dielectric constant is lowered from a value of 80 in water to 33 in methanol, 24 in ethanol, 21 in acetone, and 2.5 in dioxane. Lowering the dielectric constant increases the forces between ions and therefore favors association or complex-forming. The well-known Bjerrum theory of ion pairs indicates that the association between ions of unlike charge increases very rapidly below a certain critical value of the dielectric constant which depends on the charges and the radii of the ions. The prediction is fulfilled by experiment; in water-acetone and water-alcohol mixtures having about 60–80% of the organic liquid, metals that form chloride complexes are much more easily eluted from cation-exchange resins than in water, and the selectivity of elution increases. This was noted by Fritz and Rettig[34, 35] and confirmed by other workers.[36] Thus cobalt is selectively eluted by hydrochloric acid from a cation-exchange column, leaving nickel behind, and cadmium and zinc are eluted in that order and ahead of ions that do not form such stable chloride complexes. Thiocyanate, too, acts as a complexing eluent in aqueous acetone solutions to strip metal ions from a cation exchanger or put them on to an anion exchanger.[37]

The effects of organic solvents (acetone and eight alcohols, including glycerol and ethylene glycol) on the cation-exchange separation of rare earths was intensively studied by Alexa.[38] Some experiments were made with no complex-former, others with α-hydroxyisobutyric acid. Adding the organic solvent decreased the preference for lanthanides over sodium, and first increased, then decreased the separation between two lanthanides. The effect depended partly on dielectric constant, partly on the mole fraction of the organic solvent.

Adding organic solvent stabilizes complex formation in anion-exchange resins, too, and permits the absorption of metal salts by anion exchangers to occur at much lower concentrations of the complexing agent than in water solutions. This is a distinct practical advantage. Selectivity orders are modified, and improved separations are possible, especially among the heavier elements. This is an active field of research at present, and the publications of J. Korkisch and coworkers[39] deserve special mention. Two points must be noted; first, that the formation of uncharged ion pairs, such as $H^+FeCl_4^-$, in the solution will impede absorption on anion-exchange as well as cation-exchange resins; thus, iron(III) is eluted from an anion exchanger ahead of nickel(II) by hydrochloric acid in aqueous acetone;[40] and second, that the less polar solvents do not swell the resin as much as water, and that it may be an advantage to use macroreticular or porous resins.

A.I.d. *Anion Exchange with Complexing Agents*

A.I.d.1. *Hydrochloric acid.* Around 1953 two groups, that of Kraus[41] in the United States and Jentzsch[42] in Germany, began to experiment with the chromatography of metal ions on anion-exchange resins in hydrochloric acid solutions. They found that many metals were absorbed from hydrochloric acid by anion-exchange resins of the polystyrene-quaternary ammonium type, and that in some cases the distribution ratios changed with the hydrochloric acid concentration over several powers of ten. This made possible some striking chromatographic separations.

For example[41] a solution of six metal salts in 12 M hydrochloric acid was placed at the top of a resin column and more 12 M hydrochloric acid was passed. Nickel chloride came through immediately, as it was not absorbed by the resin. The acid concentration was then reduced by stages, and the metals were eluted as follows: 6 M, Mn(II); 4 M, Co(II); 2.5 M, Cu(II); 0.5 M, Fe(III); 0.005 M, Zn(II). The connection with complex-ion equilibria was nowhere more evident than with cobalt, which was absorbed by the resin as the blue anion $CoCl_4^{2-}$ and eluted as the pink cation $Co(OH_2)_6^{2+}$.

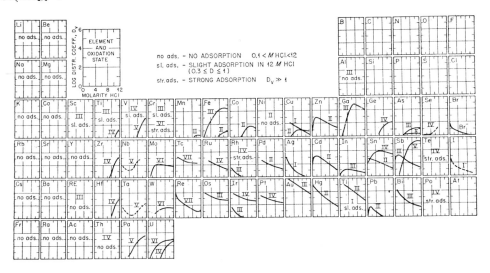

FIG. 8.3. Distribution of elements between a strong-base anion-exchange resin with 10% crosslinking and hydrochloric acid. The ordinates are logarithms of K, the "column distribution coefficient" as defined in Chapter 6, section D1. One liter of Kraus's column contained 450 g of dry chloride-form resin. (Reproduced by permission from K. A. Kraus and F. Nelson, *Proc. 1st Intern. Conf. Peaceful Uses At. Energy*, **7**, 118 (1955).)

To obtain a comprehensive view of the absorption of metal ions from hydrochloric acid, Kraus and his colleagues[43] measured the equilibrium distribution of a great many elements over a range of hydrochloric acid concentrations from zero to 12 M, using radioactive tracers and small metal-ion loadings. They presented the data in a series of semilogarithmic graphs (Fig. 8.3). The typical distribution coefficient first rises with increasing acid concentration, then reaches a maximum and falls. The early part of the curve is related to the charge of the ionic species which is absorbed (see Chapter 11). The fall in absorption at high acid concentrations is due to displacement of the complexes from the resin by the chloride ions of the acid, as well as to electrolyte invasion.

The distribution coefficients depend partly on the formation constants in the solution, partly on the selectivity of the exchanger toward the particular complexes that are formed. The second factor is quite as important as the first. The high distribution coefficient for iron(III), for example, is not due to the stability of $FeCl_4^-$ in the solution, for this is actually

...haracteristic curve, and an examination of these
...hic separations.

...vards quaternary-base anion exchangers in hydro-
...l thus:

...all. These are the alkali and alkaline-earth metals,
...d the lanthanides, thorium, and nickel.

...absorbed. Those whose distribution coefficients
..., Au(III), Hg(II), Tl(III), Bi(III), and Sb(V).

...monstration purposes because of the colors of the
...el.

...onvenient size) is washed with 8 M hydrochloric acid,
...salt (enough to load about a tenth of the column) in
...n and allowed to soak slowly in. A dark-colored zone
...part being visibly blue. Then 8 M hydrochloric acid is
...size of column suggested). Soon green nickel chloride
...llected. As more 8 M acid is passed the cobalt is slowly
...a broad blue zone, completely separated from the
...gth of clear white resin. At this point the eluant may be
...is now forced rapidly down the column, the top of the
...de complex is broken down. It flows out of the column
...y pink solution. After the cobalt is out of the column
...drochloric acid (0.1 N) or by 1 M nitric acid. (This dis-
...esin absorbs nitrate ions more strongly than chloride.
...th hydrochloric acid before re-use.)

...d some preliminary experimentation is needed to
...eparation. The separation of cadmium and zinc is
...cients are shown in Fig. 8.4. For good separations,
...2 M; at 0.1 M, $D = 10$ for zinc, 150 for cadmium.
...od separation, for one has to strike a compromise
...oefficients and small elution volumes. The influent
...acid concentration should be reduced as soon as one is sure that the zinc is out of the
column. The separation is improved by nonaqueous solvents; zinc and cadmium are eluted
consecutively with 0.01 N hydrochloric acid, with good separation between the peaks, if
the solvent is 10% methanol.[44]

A typical anion-exchange procedure used for the analysis of silver solder[45] is as follows.

Dissolve alloy in nitric and hydrochloric acids, evaporate, dilute, filter silver chloride, adjust to
6 M hydrochloric acid, pass through anion-exchange resin column. 6 M acid elutes nickel; 1 M acid
elutes copper; 0.01 M acid elutes zinc; water elutes cadmium.

The analytical determination may be accomplished in many ways. A good one is titration with
ethylenediaminetetraacetic acid. One should remember that EDTA titrations with "metalochromic
indicators" depend for their success on the proper balance of several equilibria; for titrating nickel,
for instance, one needs an ammonia–ammonium chloride buffer solution with a high pH but only a
moderate concentration of free ammonia. It is a mistake to neutralize a 6 M hydrochloric-acid
solution directly with ammonia. To approach the desired pH of 10, one must add a large excess of
ammonia, and this complexes the nickel so strongly that it cannot be titrated. One should neutralize
nearly all the acid with sodium hydroxide and then add ammonia, or, better, evaporate the solution
to remove most of the hydrochloric acid then add ammonia to bring to pH 10.

A.I.d.2. *Hydrofluoric acid.* Anion exchange in hydrofluoric acid is used for certain metals
whose salts are easily hydrolyzed. These are metals of high ionic charge, such as titanium,

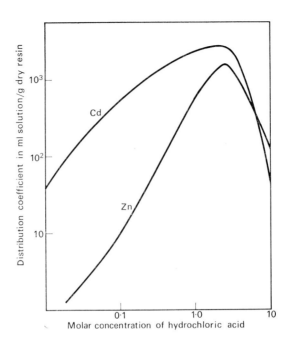

FIG. 8.4. Distribution of tracer cadmium and zinc between a strong-base anion-exchange resin (10% crosslinked) and hydrochloric acid. (Reproduced by permission from K. A. Kraus and F. Nelson, *Proc. 1st Intern. Conf. Peaceful Uses Atomic Energy*, **7**, 121 (1955).)

zirconium and hafnium, niobium, tantalum, protactinium, and tin (IV). Even in 1 M hydrochloric acid solutions these metals form hydrolytic products of high molecular weight which precipitate on glassware yet pass through ion-exchange resin columns without being absorbed. Because these products are polymeric, their formation is not so pronounced at the tracer level, and serious errors may be made by applying data obtained with tracers to operations at higher concentrations. Hydrolysis and polymerization can be avoided by adding hydrofluoric acid, for the fluoride complexes are more stable than the hydroxy complexes. The negatively charged fluoride complexes can be absorbed on anion-exchange resins and desorbed again without loss.

Mixed solutions of hydrofluoric and hydrochloric acid are used as a rule. Anion-exchange behavior of metals in these solutions has been comprehensively studied by Nelson *et al.*[46] Faris[47] has surveyed distributions in hydrofluoric acid solutions up to 24 M and found, as would be expected, that aluminum and boron are absorbed from hydrofluoric acid solutions, and so are many metals that form fluoride complexes (see Fig. 8.5).

One of the earliest chromatographic separations of metals in hydrofluoric acid solutions was that of zirconium and hafnium by Huffman and Lilly.[48] The anions ZrF_6^{2-} and HfF_6^{2-} were eluted in that order from a strong-base anion-exchange resin by a solution 0.2 M in hydrochloric acid and 0.01 M in hydrofluoric acid. The bands overlapped somewhat. Solvent extraction with the reagent thenoyltrifluoroacetone gives better separations. So does anion exchange in sulphuric acid solutions (see below).

Niobium and tantalum are more effectively separated.[49] A separation scheme for a

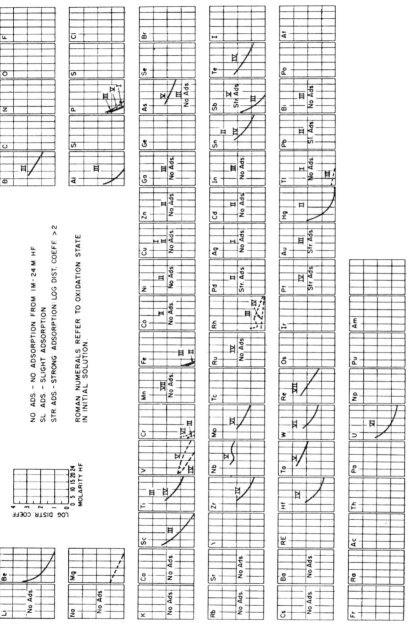

FIG. 8.5. Distribution of elements between a strong-base anion-exchange resin (10% crosslinked) and hydrofluoric acid solutions. The ordinates are logarithms of the distribution coefficient in milliliters of solution per gram of dry fluoride-form resin. (Reproduced by permission from J. P. Faris, *Anal. Chem.*, **32**, 521 (1960).)

high-temperature alloy containing niobium, tantalum, molybdenum, and tungsten is the following, due to Wilkins.[50]

The alloy is dissolved in hydrofluoric acid and the solution placed on quaternary ammonium-type resin in the fluoride form: 1.25 M HF elutes Al, Fe, Co, Ni; 8 M HCl elutes Ti; (7 M HCl, 2.5 M HF) elutes W; (3 M HCl, 5 M HF) elutes Mo; (1 M HF, 2.5 M NH$_4$Cl) elutes Nb; (1 M NH$_4$F, 2.5 M NH$_4$Cl) elutes Ta. Note that because of the weak ionization of hydrofluoric acid, the fluoride ion concentration is the highest in the last-named solution.

Frequently it is unnecessary to isolate each individual element in a separate fraction. A partial separation or a separation into groups of a few elements each may permit the determination of the individual elements by physical methods. A good example is the separation of trace impurities from aluminum by Girardi and Pietra.[51] The aluminum was irradiated by neutrons for several days, and the radioactive gamma-emitting nuclides were separated into groups and determined by gamma-ray spectroscopy. The method was as follows.

A sample of 200 mg was dissolved in hydrochloric acid; the solution was cooled to 0° and saturated with hydrogen chloride gas; most of the aluminum chloride precipitated and was filtered. The solution was placed on an anion-exchange resin column of bulk volume 2 ml, which was then washed with solutions as follows:
12 M hydrochloric acid removes alkali metals, chromium, scandium and rare earths; (12 M HCl; 0.5 M HF) removes hafnium and protoactinium with traces of cadmium; 4 M HCl removes zinc; water removes cadmium. Silver remains with the crystallized aluminum chloride and is separated later by filtering the silver chloride; chromium is oxidized to chromium(VI) and retained on another resin column.

This is one example of many that could be cited to show the use of ion-exchange separations in activation analysis. Activation by neutrons, followed by separation, is a common technique for determining trace constituents in very pure materials. The need for elaborate separations in activation analysis is diminishing as gamma-ray spectrometers with high resolution become available. On the other hand, the need for analyzing trace constituents in high-purity materials is increasing, and special ion-exchange techniques have been devised to that end.[51a] The bulk of the major constituent is precipitated within the resin bed, then all the trace elements are eluted.

A.I.d.3. *Sulphuric acid.* Anion exchange in dilute sulphuric acid has been used for some time to recover uranium(VI) from low-grade ores by hydrometallurgy; it is also used to absorb and concentrate uranium in the analysis of these ores.[52] The species absorbed[43] is probably $UO_2(SO_4)_2^{2-}$. Thorium is absorbed, too, but its distribution coefficient is much smaller.

Recently several systematic studies have been made of anion exchange in sulphuric acid solutions.[53a, 53b, 53c] The most complete study is that by Strelow and Bothma[53c] who investigated some fifty elements in sulphuric acid concentrations ranging from 0.01 N to 4.0 N. A selection of their data appears in Table 8.4. The elements most strongly absorbed are chromium, scandium, and the transition metals of the fourth, fifth, and sixth groups of the periodic table. Good chromatographic separations can be made, for example:

(1) *Mixture of Y(III), Th(IV), U(VI), and Mo(VI)* in 0.7 N sulphuric acid, passed into anion-exchange resin column; yttrium passes immediately; 0.7 N H$_2$SO$_4$ elutes thorium; 2.0 N H$_2$SO$_4$ elutes uranium; (2 N NH$_4$NO$_3$; 0.5 N NH$_3$) elutes molybdenum.

(2) *Mixture of Th(IV), Hf(IV), Zr(IV), Mo(VI)* in 0.7 N H₂SO₄; 0.7 N H₂SO₄ elutes thorium; 1.25 N H₂SO₄ elutes hafnium; 2.0 N H₂SO₄; elutes zirconium; (2 N NH₄NO₃; 0.5 N NH₃) elutes molybdenum.

Hafnium and zirconium were separated very effectively, but the authors point out that it is essential to start with a solution in concentrated sulphuric acid and to begin the separation immediately after diluting before there is time to form polymeric hydrolysis products. They were not working at the tracer level, but with amounts of the order of 1 mmole, so that the formation of polynuclear hydrolytic species was easily possible.

Scandium is easily separated from thorium and the rare earths, and by eluting with dilute potassium sulphate solutions a partial separation of the rare earth metals is possible.[53a]

TABLE 8.4. ANION-EXCHANGE DISTRIBUTIONS IN SULPHURIC ACID
(Selected data of Strelow and Bothma, *Anal. Chem.*, **39**, 595, 1967)

Ion	D in H₂SO₄ of concentration					
	0.10 N	0.20 N	0.50 N	1.0 N	2.0 N	4.0 N
As(III)	0.9	0.6	—	—	—	—
Bi(III)	18	4.7	2.1	0.9	0.5	—
Cr(III)	2.1	0.7	0.5	—	—	—
Cr(VI)	12000	7800	4400	2100	800	300
Fe(III)	16	9	3.6	1.4	0.9	—
Ga(III)	0.6	—	—	—	—	—
Hf(IV)	4700	700	57	12	3.2	1.2
In(III)	2.4	0.8	—	—	—	—
Mo(VI)	530	670	480	230	50	4.6
Th(IV)	35	21	8.3	3.7	2.0	0.6
U(VI)	520	250	90	27	9.3	2.9
V(V)	6.5	3.3	1.6	0.7	—	—
Zr(IV)	1350	700	210	47	11	2.9

Notes: (a) The resin was the quaternary ammonium type, Dowex 1-X8.
      (b) The table gives D (Chapter 3, Section A.II) in units of ml/g.
      (c) More extensive data are given in the reference cited.

(Used by permission from *Analytical Chemistry*.)

A.I.d.4. *Nitric acid.* Extensive studies have been made of anion exchange in nitrate solutions; these have been reviewed by Faris.[54] Graphs of distribution coefficients vs. nitric acid concentration[55] show that a number of metals are strongly absorbed, including Mo(VI), Tc(VII), Re(VII), Nb(V), Ta(V), La, Ce, Th, Ni, and others. Thorium is absorbed some ten times more strongly than uranium(VI); therefore anion exchange in nitric acid nicely supplements the exchange in sulphuric acid for the analysis of mixtures of uranium and thorium. The trans-uranium elements, especially neptunium and plutonium, are strongly absorbed.[56] A summary of distribution coefficients is given in Fig. 8.6.

As happens with chloride solutions, the addition of nonaqueous solvents greatly increases the association between metal ions and nitrate ions and facilitates their binding on anion-exchange resins. An extreme example is barium, which in 90% dioxane with 0.002 M nitric acid is bound by a quaternary-base anion exchanger with a distribution ratio of 500. The alkaline-earth metals, magnesium, calcium, strontium, and barium are effectively separated by chromatography in 60–90% dioxane with decreasing concentrations of nitric

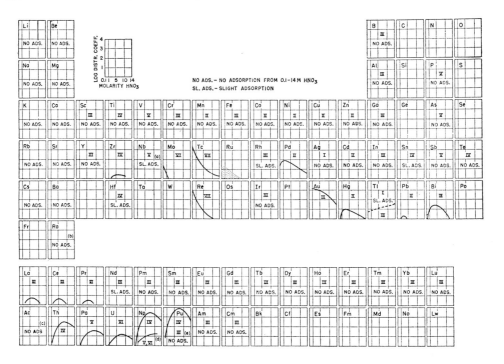

FIG. 8.6. Distribution of elements between a strong-base anion-exchange resin and nitric acid solutions. The ordinates are logarithms of the distribution coefficient in milliliters of solution per gram of dry resin. (Reproduced by permission from J. P. Faris and R. F. Buchanan, *Anal. Chem.*, **36**, 1158 (1964).)

acid.[57] Separations of uranium, thorium, and the lanthanides are improved by adding acetone and methanol;[58a] copper, lead and bismuth are eluted in that order by 5 N nitric acid, with gradual decrease in the proportion of ethyl alcohol in the solvent;[58b] beryllium is retained while aluminum and magnesium pass, with a solvent 1 part of 5 N nitric acid to 9 parts of ethyl alcohol by volume.[59]

A.I.d.5. *Other complexing agents.* Any anion which forms negatively charged complex ions with metals can be used to put metals on to anion-exchange resins, and a number have been used besides chloride, fluoride, sulphate, and nitrate. The thiocyanate ion is an obvious choice. Chromium,[60] molybdenum,[61] cobalt,[60] and iron(III) form thiocyanate complexes and can be absorbed on an anion-exchange resin while other ions pass. Again, nonaqueous solvents increase the absorption and permit many chromatographic separations, including thorium and the lanthanides.[62] Limited use has been made of EDTA,[63] oxalate[64, 124], and other complex-forming anions. All these ions have the disadvantage that it is difficult to get rid of them after the separations have been performed.

A.I.e. *Combinations of Anion and Cation Exchange*

In analyzing complex mixtures it is advantageous to separate some of the constituents by cation exchange and others by anion exchange. Comprehensive separation schemes include additional techniques such as solvent extraction and vaporization. To illustrate what can be done we cite the work of Ahrens *et al.*[65] on the separation and spectroscopic determination of some thirty minor elements in silicate rocks. Since the final determinations were done by emission spectroscopy, complete separation of the individual elements was not required. On the other hand, many of the trace elements in rocks commonly occur in concentrations below the limit of spectroscopic detection; examples are silver, bismuth, molybdenum, tin, and zinc. These must be concentrated to be detected at all, and for quantitative determination, additional concentration and separation is necessary.

We may note that with the development of selective analytical techniques such as atomic-absorption spectroscopy, the role of ion-exchange chromatography in the future will be to accomplish the rapid separation of groups of metals, rather than of individual metals in the pure state.

In the procedure of Ahrens *et al.* the silicate rock is dissolved in a mixture of hydrofluoric, hydrochloric, and nitric acids. Perchloric or sulphuric acid is added, and the mixture is evaporated to fumes. In spite of this evaporation, all the zirconium, most of the titanium, and some of the aluminum remain as fluoride complexes. The residue from the evaporation is dissolved in 2 M hydrochloric acid. Aliquot portions are then taken, equivalent to 1–5 g of rock, and treated as follows:

No. 1. Passed through quaternary-ammonium type anion-exchange resin (20 × 1 cm, 100/200 mesh, elution rate 0.5 ml/min) and eluted with 0.25 M nitric acid:

Immediately eluted: alkali and alkaline-earth metals, rare earths, Al, Ni, V, Fe, Ti, Mn, Co.

Then follow In; Zn and Cd (combined); Sn, Bi, Tl(III).

No. 2. Hydriodic acid is added to a concentration 1 M; two portions are extracted, one with *diethyl ether*, which quantitatively separates indium and thallium, the other with *tributyl phosphate*, which dissolves molybdenum; this is back-extracted with water.

No. 3. This portion is evaporated and dissolved in 11.3 M hydrochloric acid. (A fresh sample of rock may be used if desired.) The solution is passed into a quaternary-ammonium type anion-exchange resin (22 × 2 cm, 100/200 mesh, elution rate 1 ml/min) and eluted with successive concentrations of hydrochloric acid, as follows:

11.3 M elutes alkali metals, etc.; this effluent is used for the cation-exchange step, No. 4.

8 M elutes Ti, Zr.

4 M elutes Co.

1 M elutes Fe, Ga.

(The great excess of iron over gallium would require further separation were it not for the easy spectrographic determination of gallium.)

No. 4. This portion is analyzed by cation exchange. It may be the first effluent from portion 3, as indicated, or the first effluent from portion 1, or a solution from a fresh sample of rock. It is made 2 M in hydrochloric acid, and passed through a sulphonic-acid cation-exchange resin (38 × 1.7 cm, 200/400 mesh, elution rate 0.25 ml/min); elution is with 2 M hydrochloric acid except at the end, when 3–6 M hydrochloric acid may be used. Some 1200 ml of effluent is collected, equivalent to 1 g of rock.

The first solution to appear contains Zr, Sn, Ti (presumably as their fluoride complexes) plus some Al; then follow Li, Na, Mg, Mn with Fe, Co, Ni; then K, Rb, Cs; then Al and Ca; then, preferably with 6 M HCl, come Y, Sr, Nd, La, and finally Ba.

An improved ion-exchange analysis of silicate rocks has been described by Strelow[124].

A.II. NONMETALS

Anion-exchange chromatography is used to separate mixtures of inorganic anions, some of which are very hard to separate in any other way. The complex anions of phosphorus and sulphur, for example, can be analyzed by anion-exchange chromatography.

The *halide ions*, chloride, bromide, and iodide, are separated on a quaternary-base resin column by elution with sodium nitrate (0.5 M for chloride, 2.0 M for bromide and iodide);[66] this separation was discussed in Chapter 6. As was noted in Chapter 3, the differences in equilibrium constants are much greater for the anion-exchange of halides than for the cation-exchange of alkali metals. Indeed, iodide ions are held so strongly that they are difficult to remove from the resin. Mild oxidation to iodine may be used to hasten their removal. At the other extreme, fluoride ions are held very weakly, permitting the retention of other anions while fluoride is eluted by hydrochloric acid or sodium hydroxide. This is a great advantage in the photometric determination of fluoride by the zirconium–alizarin or alizarin–complexone method, where many ions cause interference (see Chapter 5). Thus fluoride can be separated from iron, aluminum, and phosphate by anion exchange and elution with concentrated hydrochloric acid[67] or an ammonia-ammonium chloride buffer.[68] Eluting with sodium hydroxide separates fluoride (eluted first) from silicate, aluminate,[69] and phosphate.

The *halate ions*, iodate, bromate, and chlorate, are bound in the reverse order from the corresponding halides, iodate being held most weakly.[70, 71] Elution is with sodium hydroxide or, better, sodium nitrate. Perchlorate is held much more strongly than chlorate.

Separation of the various anions of *phosphorus(V)* and of phosphorus in lower oxidation states is a difficult matter. Besides orthophosphate, PO_4^{3-}, P(V) forms the anions $P_2O_7^{4-}$, $P_3O_{10}^{5-}$, $P_2O_6^{2-}$, and other polymers of (non-existent) PO_3^-. In general, the larger the ion and the higher its charge the more strongly it is absorbed. Beyond a certain size, however, the ion cannot get into the resin bead, and highly polymerized phosphates are excluded by a "screening effect".

Quaternary-base resins are used, with various eluents; one of the first to be used was potassium chloride[72] buffered with acetate to pH 5. Mixtures of linear polymers up to and including tridecaphosphate, plus three cyclic polymers, have been separated.[73] Nitrate ions are more strongly bound by the resin than chloride ions, and nitrate solutions are better eluents for the phosphates of higher molecular weight. Because of the importance of these analyses in the detergent industry and elsewhere, methods for automatic analysis have been developed by several groups of workers.[74, 75] The distribution coefficients are affected by temperature,[76] and a combination of temperature, pH, and salt-concentration gradients may be used. As a rule the effluents are analyzed by hydrolyzing the complex phosphates to orthophosphate and determining the latter by molybdenum blue. Refractometric monitoring can be used, but it is not sensitive and not adapted to gradient elution.

The anions of phosphorus in lower oxidation states (phosphites, hypophosphites, etc.) as well as complex phosphorus anions containing nitrogen and sulphur can be separated by anion-exchange chromatography.[75]

Sulphur forms many kinds of anions also; besides sulphate, sulphite, and thisulphate there are the thionic acids from $H_2S_2O_6$. These can be separated by anion-exchange chromatography. Sodium chloride eluent separates the thionates.[77] For more complex mixtures a succession of eluents may be used; potassium hydrogen phthalate elutes sulphite and thiosulphate, then an acetate buffer, followed by hydrochloric acid solutions up to 9 M, elutes trithionate, tetrathionate and pentathionate ions.[78] The effluents are analyzed by iodate titration.

B. Separation of Organic Compounds

Ion-exchange chromatography of organic compounds is well developed. Probably the most important analytical application of ion-exchange chromatography, if the number of analyses performed per day is any criterion, is the analysis of amino-acid mixtures by cation exchange. Organic acids and bases can be separated by ion-exchange chromatography, and so can many nonionic species which can be made to condense or coordinate with simple ions. Examples of these compounds are aldehydes, which form addition compounds with bisulphite ions, and conjugated diolefines, which form coordination compounds with silver ions. Furthermore, many substances are absorbed by ion-exchange resins without ionic charges being involved. An important example of this type of absorption is the mono-saccharides, which can be separated by elution chromatography on quaternary-base anion-exchange resins (see Chapter 9).

In comparing ion exchange of organic compounds with that of simple inorganic ions, certain contrasts should be kept in mind. It is very difficult to predict selectivity orders. The higher the molecular weight of an organic acid or base the more strongly it is held by ion-exchange resins, but beyond a certain ionic size there enters a "sieve effect"; the ions become too big to enter a crosslinked resin network, and such penetration as occurs is very slow. Large ions may, however, be held on exchangers with an open physical structure, such as the porous or "macroreticular" resins, the cellulose-based exchangers and the micro-crystalline inorganic exchangers. Nonionic "solvent effects" become important; a polystyrene-based exchanger binds aromatic ions, such as the benzylammonium cation, the cation of 1,10-phenanthroline or the anion of *p*-toluene-sulphonic acid very strongly. An inorganic exchanger, on the other hand, will hold such ions very weakly. When it comes to the absorption of nonionic materials the chemical nature of the exchanger matrix or "backbone" is much more important than the type of ionic groups.

The use of nonaqueous solvents in ion-exchange chromatography of organic compounds is very common, of course, and the nature of the solvent has a great deal to do with the distribution of the solutes between the exchanger and the solution.

Differences in acid or base strengths influence selectivity, and by controlling the pH of the eluent it is possible to make separations on this basis. However, the ionization of the acid or base is only one factor, and quantitative correlations with selectivity orders are hard to find.

In assessing the utility of ion-exchange chromatography in organic analysis one must recognize that most organic compounds are most conveniently analyzed by gas chromatography. Ion exchange has the advantage for compounds of low volatility and for the analysis of aqueous solutions. It has a very wide application in pharmaceutical and biochemical analysis, and we can do no more than touch on these applications here. Ion exchangers are not only used in columns, but in thin-layer chromatography of biological products.

B.I. CATIONS

B.I.a. *Amines and Quaternary Ammonium Salts*

Amines can be separated by cation exchange in acid solution. Aromatic amines, aniline, benzylamine, and pyridine, are much more strongly absorbed from aqueous solutions than aliphatic amines, and there seems to be a correlation with the basic strength, even in 1 M hydrochloric acid; aniline and pyridine, which are very weak bases, are eluted before benzylamine, which is stronger. The elution volumes are much less in alcohol than they are in water, the order of elution is changed, and better separations result; aniline, *n*-butylamine, benzylamine, and pyridine are eluted in that order by 1 M hydrochloric acid in ethanol.[80]

Amine cations and quaternary ammonium ions are separated on a column of sulphonated polystyrene resin (4% crosslinking) by elution with hydrochloric acid in methanol; this is part of a comprehensive scheme of detergent analysis in which anionic components are caught on a column of anion-exchange resin and selectively eluted.[81] The large elution volumes that are common with amines and quaternary ammonium salts can be reduced by using an inorganic exchanger which has less "solvent action" or nonionic absorption for the organic compounds; thus Rebertus[82] used zirconium phosphate for the cation-exchange chromatography of amines. To minimize "tailing" he operated the column at 80–90°. The cations $n\text{-}C_8H_{17}NH_3^+$, $C_2H_5NH_3^+$, $CH_3NH_3^+$, and $(CH_2NH_3)_2^{2+}$ were eluted successively in that order by 0.1 N hydrochloric acid. The selectivity order for the first three of these cations is the reverse of that found with the resinous exchangers. The ethanolamines, too, were separated on zirconium phosphate; triethanolamine was eluted first by 0.1 N hydrochloric acid, then monoethanolamine, then ammonium ions.

Mono-, di-, and triethanolamine have also been separated on a column of sulphonated polystyrene exchanger by elution in alkaline solution, by 0.05 N sodium borate buffered at pH 9.2.[82a] This pH value is suggested by the ionization constants of the three bases; the pK values for their conjugate acids BH^+ are 9.60, 9.00, and 7.90 respectively. Triethanolamine (the weakest base) is eluted first, then di-, then mono-. The distribution coefficients of their cations between resin and solution can be correlated only qualitatively with the pK values, however, the pH values inside the resin beads are uncertain, and probably the hydrogen bonding of the hydroxyl groups in the aqueous solution is a decisive factor. This would favor the preferential elution of triethanolamine.

Whatever the reasons for the selectivity, the analysis of alkanolamines by ion exchange is of real practical value, for it is hard to analyze these mixtures by gas chromatography.

Finally, we note the possibility of using the exchanger in its ammonium form, letting the uncharged amines become absorbed on the exchanger as cations with displacement of ammonia, and then eluting with an aqueous ammonia solution. This is really a kind of "ligand exchange" (section B.III.b). Some interesting selectivity orders are found which roughly parallel the base strengths of the amines; the stronger the amine is as a base, the more strongly it is absorbed.[83] This kind of elution chromatography has been used to separate and determine phenolic amines in fruit juices.[84]

B.I.b. *Amino acids*

The separation of amino acids by elution chromatography on cation-exchange resins has been exhaustively studied and brought to a state approaching perfection. Automatic equipment is available and several arrangements of columns, eluting agents, and temperatures are used, depending on the particular needs of the analysis. For full details, recent reviews[85] and the original literature should be consulted. The work of Stein and Moore[86] and of Hamilton[87] is especially noteworthy.

Columns of sulphonated polystyrene resin are used, normally with 8% crosslinking, though this can be varied, and small particle diameter, about 50 μ or less. It is very important for the particles to be spherical[88] and uniform in size and crosslinking;[87, 88] the resolution and speed of operation are increased greatly if these conditions are met. Pulverized resins of fine particle size have been used, but it is much better to use resins made by polymerizing very fine droplets of liquid monomer, then sulphonating and classifying the beads according to particle size by an upflow hydraulic technique.

The amino acids are eluted with sodium citrate solutions at selected sodium-ion concentrations, pH values, and temperatures. The first acids to be eluted are the "acidic" amino acids, those with two carboxyl groups in the molecule, the commonest of which are aspartic acid and glutamic acid (Table 8.5). Then follows the large group of "neutral" amino acids with one carboxyl and one amino group in the molecule. Of these the ones containing aromatic rings, tyrosine, and phenylalanine are retained the longest; the "solvent" action of the resin is clearly seen. Last to be eluted are the "basic" amino acids including lysine, tryptophane, and arginine. Because this group is so much more strongly held than the others it is customary to divide the sample into two and run two columns, one a long one for the acidic and neutral amino acids, the other a short one for the basic amino acids.

In the scheme of Spackman, Stein, and Moore[86] the first is 150 × 0.9 cm (resin dimensions) and the second 15 × 0.9 cm; the particle diameter is about 40 μ; elution in the first column is by 0.2 N sodium citrate at pH 3.24–4.25; in the second column, by 0.4 N sodium citrate at pH 5.26. The temperature is 50°C. Using finer and more perfectly spherical resin particles (diameter 22 ± 6 μ for the long column, 15 ± 6 μ for the short one) the lengths can be reduced to 57 cm and 5 cm respectively; with the same eluents as before, and flow rates exceeding 1 ml/min, a complete analysis of 5 mg of a protein hydrolysate can be performed in 4–5 hr.[88]

Methods employing a single column and a series of concentration and pH gradients have also been devised.[89]

TABLE 8.5. AMINO ACID ELUTION SEQUENCE

Name of acid	Formula	Elution volume (ml)
Hydroxyproline	HO— ... —COOH (ring with N–H)	111
Aspartic acid	$HOOC . CH_2 . CHNH_2 . COOH$	115
Threonine	$CH_3 . CHOH . CHNH_2 . COOH$	134
Serine	$CH_2OH . CHNH_2 . COOH$	143
Sarcosine	$CH_3NH . CH_2 . COOH$	160
Glutamic acid	$HOOC . (CH_2)_2 . CHNH_2 . COOH$	178
Proline	(ring with N–H) —COOH	186
Glycine	$CH_2NH_2 . COOH$	219
Alanine	$CH_3 . CHNH_2 . COOH$	230
Cystine	$(—SCH_2 . CHNH_2 . COOH)_2$	239
Valine	$(CH_3)_2CH . CHNH_2 . COOH$	271
Norvaline	$CH_3(CH_2)_2 . CHNH_2 . COOH$	337
Isoleucine	$CH_3 . CH_2 . CH . CHNH_2 . COOH$ $\quad\quad\quad CH_3$	368

TABLE 8.5—*continued*

Name of acid	Formula	Elution volume (ml)
Leucine	$(CH_3)_2CHCH_2 . CHNH_2 . COOH$	400
Tyrosine	$HO-\!\!\bigcirc\!\!-CH_2.CHNH_2.COOH$	490
Phenylalanine	$\bigcirc\!\!-CH_2.CHNH_2.COOH$	500
β-alanine	$H_2N . (CH_2)_2 . COOH$	517
Lysine	$H_2N . (CH_2)_4 . CHNH_2 . COOH$	805
Histidine	(imidazole ring)—$CH_2.CHNH_2.COOH$	848
Tryptophane	(indole ring)—$CH_2.CHNH_2.COOH$	970
Arginine	$H_2N . C . NH . (CH_2)_3 . CHNH_2 . COOH$, with $\|\,NH$	1167

Notes: (a) Column was 10 mm diameter by 160 cm length.
(b) Gradient elution was used, starting with a sodium citrate–citric acid solution of pH 3.30, 0.20 M; the pH was raised by adding 2.5 M sodium acetate.
(c) Data taken from H. Stegemann, *Naturwiss.*, **46**, 110 (1959).

(Used by permission from *Naturwissenschaften*.)

To show the sensitivity and precision of amino-acid analysis by ion-exchange chromatography we cite a recent study of the amino acids in human fingerprints. The tips of fingers were placed in 0.5 ml of citrate solution for 15 secs and the resulting solution passed through an amino-acid analyzer. A characteristic and essentially constant amino-acid composition was found with different subjects, indicating a source of contamination in experimental studies, and it was suggested that reports of the occurrence of amino-acids in meteorites were the result of contamination in handling the samples.[90]

The analysis of the column effluents is done by mixing them with ninhydrin,

in an acetate buffer at 90–100°. Red, purple, or yellow compounds are formed with molar absorptivities of the order of 2500 (this is the value for the lysine compound at 570 mμ).

The solutions are mixed automatically and continuously and the absorbance is recorded automatically, at two different wavelengths if desired.[89]

B.I.c. *Purines and Pyrimidines: Nucleic Acid Derivatives*

One of the earliest applications of ion-exchange resins to biochemical analysis was the separation of the bases derived from the complete hydrolysis of nucleic acid. This was done by Cohn.[91] On a small column of sulphonated polystyrene cation exchanger, 300 mesh, by elution with 2 N hydrochloric acid, he separated uracil, cytosine, guanine, and adenine, eluting them in that order. The single-ring pyrimidine bases, uracil and cytosine (and thymine, which was eluted with uracil) were less strongly bound to the resin than the double-ring, purine bases, guanine and adenine:

| uracil | cytosine | guanine | adenine |

The selectivity order can be correlated with the basic strengths; the stronger the bases, the more ionized they are and the more strongly they are bound to the cation exchanger; but this is not the only factor, and once again the "solvent" effect of the resin is evident. Almost the same order of elution (cytosine first, then uracil, thymine, guanine, and adenine) is observed when these substances, which are amphiprotic in nature, are eluted from a polystyrene-based quaternary ammonium type anion exchanger by an ammonia-ammonium chloride buffer.[91]

The nucleosides, whose molecules consist of ribose condensed with the purine and pyrimidine bases, can also be separated by cation exchange in acid solution,[92] and so can the nucleotides, which are more complex as they contain phosphate groups in addition. Resins of low crosslinking (2% divinylbenzene) are used. They can also be separated by anion exchange, and resins of 8% crosslinking are used for this purpose. It is surprising that ions of such large size can diffuse in and out of the resin beads.

B.II. ANIONS

B.II.a. *Aliphatic Carboxylic Acids*

These acids can be separated by chromatography on a quaternary ammonium-type anion-exchange resin, and various eluents have been used. Some workers have used nitrate and acetate solutions of pH 6 or higher, at which pH value most of the acids are almost completely ionized; others have used acidic solutions, principally formic acid. The practical interest in separating these acids has been in the analysis of fruit juices and also in the study of the acids of the metabolic cycles.

One of the earliest studies was that of Schenker and Rieman[93] who separated the "fruit acids" with an eluent of sodium nitrate, sodium borate, and boric acid of pH 6.15. The purpose of the borate was to buffer the solution and also to form anionic complexes with the hydroxy acids. The order of elution was: succinic, lactic, malic, tartaric, oxalic, citric. Separation was slow, needing up to 8 hr, and elution with an acetate solution[94] gave shorter elution times. Fractions were collected and analyzed by permanganate titration or else by reaction with chromic acid. Shimomura and Walton[95] modified the technique by continuous recording of the refractive index of the effluent (Chapter 7); nitrate solutions of pH 7–8 buffered with a small proportion of borate, were used. With a typical eluent (0.08 M sodium nitrate, pH 8) the following elution volumes were recorded (multiples of the bulk column volume): acetic, lactic and 2-methyllactic acids, 1.3; succinic, 4.0; malic, 4.2; malonic, 4.5; tartaric, 5.3; maleic, 6.3; oxalic, 7.9; fumaric, 8.3; citric, 10.0. The time needed for elution was of the order of 4 hr. The refractometric method is not suited to gradient elution, and Zerfing and Veening[96] used chromic acid oxidation and an auto analyzer to analyze the effluent continuously and spectrophotometrically. They used a column heated to 67°, and a sodium acetate solution of concentration gradually increasing from zero to 1.2 M. They eluted pyruvic, glutaric, citric, 2-ketoglutaric, and aconitic acids in that order.

The effect of borate on the elution of hydroxy acids is hard to predict. It forms anion complexes with cis-1,2-diols, and these will be attracted to the resin, but the borate ion itself will tend to displace these complex ions. Borate solutions have proved very useful as eluents to separate acids derived from sugars.[97]

Acid solutions can also be used to displace the organic acids from anion-exchange resins, and although the equilibria are different, involving the ionization of the acids as well as the binding of their anions by the resin, the elution sequences are substantially the same. The simple acids of low molecular weight, like acetic and lactic, are eluted first, the dicarboxylic acids later and the tricarboxylic acids still later. Acetic acid has been used as an eluent,[98] and so have acidified calcium chloride solutions,[99] but the most thorough work has been done with formic acid, whose concentration can be increased progressively up to 100% (25 M); ninety-four acids, aliphatic and aromatic, were eluted successively by formic acid.[100] An early application of formic acid elution was to the acids of the Krebs cycle.[101]

B.II.b. *Aromatic Acids and Phenols*

In general these are more strongly held than aliphatic acids because of the "solvent" action of the polystyrene framework. A series of studies of aromatic sulphonic acids has been made by Funasaka et al.[102] Among the carboxylic acids there are some interesting correlations with acid strengths; thus, γ-resorcylic acid (2,6-dihydroxybenzoic acid) is a much stronger acid than its isomer β-resorcylic acid (3,5-dihydroxybenzoic acid) because of hydrogen-bonding and chelation in the anion of the 2,6-acid; thus when the two are absorbed together on an anion-exchange resin column, the weaker acid is eluted by 0.01 M hydrochloric acid while the stronger acid has to be eluted by 0.1 M hydrochloric acid.[103]

Phenols are bound more strongly than their weak ionization would suggest, again because of the "solvent" action of the resins, whether they be of the polystyrene or condensation type. In fact, phenols are absorbed from dilute aqueous solutions to an amount considerably exceeding the ion-exchange capacity; they are absorbed as neutral molecules as well as ions. This fact is very useful in recovering phenols from waste waters and in other industrial applications, but has not found much application in analysis by ion-exchange chromatography.

B.III. UNCHARGED MOLECULES

Uncharged molecular species are absorbed by ion-exchange resins because of the "solvent" effect we have repeatedly noted. The nature and charge of the functional groups of the resin are not as important as the kind of polymer network. However, in the absorption of the sugars and saccharides, the anion-exchange resins of the quaternary-base type in the hydroxyl form are the most effective. Many papers describe separations of saccharides on these resins, including methods of automatic analysis.[104] Absorption is greater from aqueous alcohol than from water, and mixed water–alcohol solvents are preferred.[105] It is not essential to use the hydroxyl form of the resin; the sulphate form has been used in acidic solutions.

The methods of solubilization and salting-out chromatography use nonionic absorption by resins and will be discussed in Chapter 9. In the rest of this chapter we shall describe methods of chromatographic analysis that depend on the combination of neutral molecules with ions to form ionic derivatives.

B.III.a. *Complex Anions*

B.III.a.1. *Borate complexes.* The separation of polyhydroxy compounds, and especially sugars, by absorption of their borate complexes on a strong-base resin in its borate form has been known for years, following the pioneering work of Khym and Zill.[106] *Cis*-1,2-,diols form ring complexes of the types:

These are formed by compounds as simple as ethylene glycol and glycerol, and a method for recovering glycerol from dilute (1 %) solutions uses absorption on a resin in its borate form.[107] Mixtures of 1,2-glycols and glycerol were analyzed chromatographically by absorbing them on a column of borate-form resin and eluting with sodium borate solutions;[108] sugar phosphates and sugar alcohols[109] have been separated in a similar way. Of the sugars, disaccharides are absorbed more weakly than monosaccharides and are eluted first by borax (sodium borate) solutions of 0.005 M concentration; of the monosaccharides, fructose is eluted before glucose.[106] A procedure for chromatographic analysis of neutral monosaccharides[110] uses a temperature of 50° and an eluant 0.4 M in boric acid, 1 M in glycerol (which helps to displace the sugars), and 0.05 M in sodium chloride, adjusted to pH 6.8.

Polysaccharides can be separated by anion exchange of their borate complexes if resins of high porosity or exchangers based on cellulose are used.[111]

B.III.a.2. *Bisulphite complexes.* Bisulphite ions form addition complexes with aldehydes and certain ketones:

$$RCH(OH)SO_3^-$$

These differ in stabilities, the complexes with ketones (acetone, cyclohexanone, methyl ethyl ketone) being considerably less stable than the aldehyde complexes, so that ketones, after absorption, are simply eluted with water. Aldehydes are eluted successively by sodium bisulphite solutions or solutions of sodium chloride or bicarbonate.[112] Gradient elution with bisulphite solutions can be used for complex mixtures containing furfural and vanillin.[113]

The monosaccharides, because of their aldehyde and ketone character, form bisulphite addition compounds in aqueous-alcoholic solution and can be separated by chromatography on bisulphite-form resins,[114] but this is not nearly as effective as the separation through their borate complexes.

B.III.b. *Complex Cations: "Ligand Exchange"*

Cations of transition and post-transition metals form complexes with ammonia and amines, and some form complexes with other uncharged molecules such as olefines. Ammonia and amines can be absorbed by cation-exchange resins which contain these metal ions, and can be desorbed again or exchanged for one another in the course of chromatographic analysis. For this type of exchange the name "ligand exchange" was proposed by Helfferich.[115] Helfferich used cation-exchange resins loaded with the ions Cu^{2+} and Ni^{2+} in the form of their ammonia complexes to absorb the diamine 1,3-diaminopropane-2-ol from a dilute aqueous solution which also contained ammonia.[116] One diamine molecule replaced two ammonia molecules; thus the equilibrium was shifted by concentration, and the diamine could be displaced again by passing concentrated aqueous ammonia.

Amines are bases and can be absorbed on cation-exchange resins as protonated cations from acid solutions; see section B.I.a above. Ligand exchange has certain advantages over conventional cation exchange, however. First, the capacity of the resin is substantially increased; two fixed sulphonate ions bind one Ni^{2+} ion, which in its turn can bind up to six amine molecules, giving a threefold increase in capacity. Second, absorption and desorption take place in an alkaline environment, permitting the chromatography of substances sensitive to acid hydrolysis, such as ethyleneimine and its derivatives. Finally, many variations of selectivity can be made by changing the nature of the metal ion and the exchanger.

Metal-ammonia and metal-amine complexes are just as stable in sulphonic-acid resins as they are in water, and the maximum coordination capacity of the metal ions is unchanged.[117] Resins with fixed carboxylate and phosphonate ions show decreased stability of the metal-ammonia complex ions, though it seems that copper and nickel ions in carboxylate resins can still bind four and six ammonia molecules respectively if the ammonia concentration in the surrounding solution is high enough.[83, 117]

Ligand exchange elution chromatography is performed by absorbing the mixture of amines on a column containing the metal-ammonia complex, then eluting them with aqueous or aqueous-alcoholic solutions of ammonia. A problem that arises is that the ammonium ions which these solutions contain displace some of the metal ions from the exchanger by cation exchange. It is desirable to keep the loss of metal ions as small as possible, and exchangers with carboxylate, phosphonate, and iminodiacetate groups hold the metal ions more firmly than exchangers with sulphonate ions.[116, 118, 119] However, the disadvantage of the reduced capacity for holding ammonia molecules outweighs the advantage of lower metal leakage. For most purposes, the combination of nickel ions with a sulphonic acid resin is satisfactory. The leakage of nickel ions with 1 M ammonia is of the order of 10^{-4} moles/l. Copper ions, however, are much less tightly held, and if copper is to be used as the coordinating ion, exchangers with functional groups other than sulphonate must be used.

The selectivity orders found in ligand exchange chromatography depend less on the nature of the metal ion than on the nature of the exchanger matrix. On polystyrene-based resins with nickel as the coordinating cation, the four amines diethanolamine, ethanolamine, dimethylamine and *n*-butylamine were eluted in that order, diethanolamine being held most weakly. On a zirconium phosphate exchanger, also with nickel ions, the order of elution was exactly reversed.[118] It seems that the "solvent action" of the polystyrene matrix favors the hydrocarbon chain of the butylamine, while the hydrophilic absorptive property of the zirconium phosphate favors diethanolamine—and it favors triethanolamine even more. Good separations of mono, di, and triethanolamine have been achieved at the tracer level (using C-14 compounds), but higher concentrations show considerable overlapping of bands, probably because of nonlinear absorption isotherms. Caution is always necessary in extrapolating chromatographic data obtained at the tracer level.

Most ligand exchange studies of amines have been made with sulphonated polystyrene resins containing nickel ions, and in the examples to follow, some selectivity orders obtained with this combination will be discussed. We shall also note some applications of ligand exchange to compounds other than amines.

B.III.b.1. *Aliphatic amines.*[119] Elution orders are the following, the most weakly bound amine being listed first: $(C_2H_5)_3N$, $C_2H_5NH_2$, CH_3NH_2, $(C_2H_5)_2NH$, $n\text{-}C_4H_9NH_2$; $(CH_3)_2CHNH_2$, $n\text{-}C_3H_7NH_2$; $(CH_3)_3C.NH_2$, $C_2H_5CH(CH_3)NH_2$, $(CH_3)_2CHCH_2NH_2$, $n\text{-}C_4H_9NH_2$. There is no correlation with the strengths of the amines as bases, but in the propylamines and butylamines we see that the more highly branched is the carbon chain, the more weakly it is held. No correlation with the stabilities of the nickel–amine complexes in solution can be made, since these stabilities are too low to be measured.

B.III.b.2. *Diamines.*[83] Ethylenediamine and 1,2-propanediamine are very strongly held indeed, so strongly, in fact, that they cannot be displaced by passing ammonia, and the only way to get them off the resin seems to be to elute amine and metal together by passing an acid. Next strongly held is 1,3-propanediamine, followed by 1,6-hexanediamine, then 1,4-butanediamine.

B.III.b.3. *Hydrazines.*[119] Hydrazine itself is much more strongly held than its derivatives monomethylhydrazine and 1,1'-dimethylhydrazine; all three can be separated cleanly by elution with ammonia, the dimethylhydrazine being eluted first. It is important to note that these compounds suffer little or no decomposition in 24 hr of residence in a resin containing nickel ions.

B.III.b.4. *Ethyleneimines.*[119a] Ethyleneimine, too, can be absorbed on a sulphonic-acid resin containing nickel-ammonia complex ions and eluted again several hours later with little or no decomposition. Ethyleneimine itself is more strongly held than C-methylethyleneimine, and N-hydroxyethylethyleneimine is hardly bound at all.

B.III.b.5. *Purine and pyrimidine bases.*[119] The bases thymine, cytosine, adenine, and guanine are eluted by ammonia in that order, but the last two show considerable tailing and the separation is poorer than that obtained by conventional cation exchange.

B.III.b.6. *Amino acids.* Amino acids have been successfully separated by Arikawa and Makimo[120] on columns of sulphonic-acid resins containing zinc, cadmium, and other cations which form amino-acid complexes. The eluent was an acetate buffer containing the metal ion in a concentration of 10^{-3} M; this made up for the elution of metal ions as their amino-acid complexes. For reproducible results to be obtained it was necessary to keep the metal content of the resin constant. Mixtures of 20 amino acids were analyzed in this way, and an automatic analyzer was operated on this principle.

There is no reason for not using metal salt solutions as eluents in ligand exchange, provided that the metal ions in the effluent do not interfere with the subsequent analyses. The amino acids were determined by the ninhydrin method, and no interference was noted.

B.III.b.7. *Other amino compounds.* Nicotinic acid hydrazide has been absorbed from aqueous solutions by cation-exchange resins loaded with copper and nickel ions, and eluted by water.[121]

B.III.b.8. *Olefines and unsaturated compounds.* Silver ions form coordination complexes with olefines, and more stable ones with 1,2-dienes. Powdered silver nitrate has been used as the stationary phase in the gas chromatography of olefines, and the *cis-* and *trans-*isomers of dienes have been separated in methanolic solution on a macroreticular cation-exchange resin containing silver ions.[122] Methanol was used as the eluent. Esters of the unsaturated acids, oleic and linoleic, have likewise been separated by chromatography on a silver-loaded cation-exchange resin, with a methanol–water mixture as eluent or a solution of butene in methanol.[123]

The principle of ligand exchange has also been applied to gas chromatography, but this topic is beyond the scope of this book.

References

1. K. A. Kraus, T. A. Carlson and J. S. Johnson, *Nature*, **177**, 1128 (1956).
2. K. A. Kraus, H. O. Phillips, T. A. Carlson and J. S. Johnson, *Proc. 2nd Intern. Conf. on Peaceful Uses of Atomic Energy*, **28**, 3 (1958).
3. M. H. Campbell, *Anal. Chem.*, **37**, 252 (1965).
4. J. Van R. Smit, W. Robb and J. J. Jacobs, *J. Inorg. Nucl. Chem.*, **12**, 104 (1959).
5. J. Krtil and I. Krivy, *J. Inorg. Nucl. Chem.*, **25**, 1191 (1963).
6. H. G. Petrow and H. Levine, *Anal. Chem.*, **39**, 360 (1967).
7. Y. Inoue, *Bull. Chem. Soc., Japan*, **36**, 1316, 1324 (1963).
8. R. C. Sweet, W. Rieman and J. Beukenkamp, *Anal. Chem.*, **24**, 952 (1952).
9. L. E. Reichen, *Anal. Chem.*, **30**, 1948 (1958).
10. J. S. Fritz and S. K. Karraker, *Anal. Chem.*, **32**, 957 (1960).
11. A. T. Rane and K. S. Bhatki, *Anal. Chem.*, **38**, 1598 (1966).
12. G. Schwarzenbach, *Angew. Chem.*, **70**, 451 (1958).
13. F. H. Pollard et al., *J. Chromatog.*, **10**, 212, 215 (1963); *ibid.*, **11**, 542 (1963); *ibid.*, **13**, 224 (1964).
14. E. R. Tompkins, J. X. Khym and W. E. Cohn, *J. Am. Chem. Soc.*, **69**, 2769 (1947).
15. W. E. Nervik, *J. Phys. Chem.*, **59**, 690 (1955).
16. S. W. Mayer and E. C. Freiling, *J. Am. Chem. Soc.*, **75**, 5647 (1953).
17. G. R. Choppin and R. J. Silva, *J. Inorg. Nucl. Chem.*, **3**, 153 (1956).
18. F. W. Cornish, G. Phillips and A. Thomas, *Can. J. Chem.*, **34**, 1471 (1956).
19. W. E. Blake, G. Oldham and D. Sumpter, *Nature*, **203**, 862 (1964).
20. J. S. Fritz and G. R. Umbreit, *Anal. Chim. Acta*, **19**, 509 (1958).
21. D. I. Ryabchikov and V. E. Buhtiarov, *Zhur. Analit. Khim.*, **9**, 196 (1954).
22. D. I. Ryabchikov and V. E. Buhtiarov, *Zhur. Analit. Khim.*, **7**, 377 (1952).
23. I. P. Alimarin et al., *Zhur. Analit. Khim.*, **18**, 468 (1963).
24. J. S. Fritz and L. H. Dahmer, *Anal. Chem*, **37**, 1272 (1965).
25. F. W. E. Strelow, *Anal. Chem.*, **32**, 1185 (1960).
26. F. W. E. Strelow, *Anal. Chem.*, **31**, 1201 (1959).
27. F. W. E. Strelow, *Anal. Chem.*, **33**, 542 (1961).
28. W. M. McNevin and W. B. Crummett, *Anal. Chim. Acta*, **10**, 323 (1954).
29. E. A. Huff, *Anal. Chem.*, **37**, 533 (1965).
30. F. W. E. Strelow, R. Rethemeyer and C. J. C. Bothma, *Anal. Chem.*, **37**, 106 (1965).
31. F. Nelson, T. Murase and K. A. Kraus, *J. Chromatog.*, **13**, 503 (1965).
32. V. Nevoral, *Z. Anal. Chem.*, **195**, 332 (1963).
33. F. Nelson, D. A. Michelson, H. O. Phillips and K. A. Kraus, *J. Chromatog.*, **20**, 107 (1965).
34. J. S. Fritz and T. A. Rettig, *Anal. Chem.*, **34**, 1562 (1962).

35. J. S. FRITZ and J. E. ABBINK, *Anal. Chem.*, **37**, 1274 (1965).
36. J. C. GIDDINGS, *J. Chromatog*, **13**, 301 (1964).
37. D. J. PIETRZYK and D. L. KISER, *Anal. Chem.*, **37**, 233, 1578 (1965).
38. J. ALEXA, *Coll. Czech. Chem. Commun.*, **30**, 2344, 2351, 2361, 2368 (1965).
39. I. HAZAN, J. KORKISCH and G. ARRHENIUS, *Z. Anal. Chem.*, **213**, 182 (1965).
40. I. HAZAN and J. KORKISCH, *Anal. Chim. Acta*, **32**, 46 (1965).
41. K. A. KRAUS and G. E. MOORE, *J. Am. Chem. Soc.*, **75**, 1460 (1953).
42. D. JENTZSCH *et al.*, *Z. Anal. Chem.*, **144**, 8, 17 (1955); **146**, 88 (1955) and subsequent papers.
43. K. A. KRAUS and F. NELSON, *Proc. 1st Intern. Conf. on Peaceful Uses of Atomic Energy*, **7**, 113 (1955); *The Structure of Electrolytic Solutions* (ed. W. H. Hamer), New York, Wiley, 1959, chap. 23.
44. E. W. BERG and J. T. TRUEMPNER, *Anal. Chem.*, **30**, 1827 (1958).
45. S. L. JONES, *Anal. Chim. Acta*, **21**, 532 (1959).
46. F. NELSON, R. M. RUSH and K. A. KRAUS, *J. Am. Chem. Soc.*, **82**, 339 (1960).
47. J. P. FARIS, *Anal. Chem.*, **32**, 520 (1960).
48. E. H. HUFFMAN and R. C. LILLY, *J. Am. Chem. Soc.*, **71**, 4147 (1949).
49. J. L. HAGUE and L. A. MACHLAN, *J. Res. Nat. Bureau Standards*, **62**, 53 (1959).
50. D. H. WILKINS, *Talanta*, **2**, 355 (1959).
51. F. GIRARDI and R. PIETRA, *Anal. Chem.*, **35**, 173 (1963).
51a. F. TERA, R. R. RUCH and G. H. MORRISON, *Anal. Chem.*, **37**, 358, 1565 (1965).
52. S. FISHER and R. KUNIN, *Anal. Chem.*, **29**, 400 (1957).
53a. H. HAMAGUCHI, A. CHUCHI, N. OMURA and R. KURODA, *J. Chromatog.*, **16**, 396 (1964); H. HAMAGUCHI *et al.*, *Anal. Chem.*, **36**, 2304 (1964).
53b. L. DANIELSSON, *Acta Chem. Scand.*, **19**, 670 (1965).
53c. F. W. E. STRELOW and C. J. C. BOTHMA, *Anal. Chem.*, **39**, 595 (1967).
54. J. P. FARIS and R. F. BUCHANAN, in D. C. STEWART and H. A. ELION (eds.), Progress in Nuclear Energy, Series IX, *Analytical Chemistry*, vol. 6, Pergamon Press, 1966.
55. F. ICHIKAIWA, S. URUNO and H. IMAI, *Bull. Chem. Soc. Japan*, **34**, 952 (1961).
56. I. K. KRESSIN and G. R. WATERBURY, *Anal. Chem.*, **34**, 1598 (1962).
57. R. R. RUCH, F. TERA and G. H. MORRISON, *Anal. Chem.*, **36**, 2311 (1964).
58a. I. HAZAN, S. S. AHLUWALIA and J. KORKISCH, *Z. Anal. Chem.*, **206**, 324 (1964).
58b. F. FEIK and J. KORKISCH, *Talanta*, **11**, 1585 (1964).
59. J. KORKISCH and F. FEIK, *Anal. Chem.*, **37**, 757 (1965).
60. J. B. TURNER, R. H. PHILP and R. A. DAY, *Anal. Chim. Acta*, **26**, 94 (1962).
61. H. GROSSE-RUYKEN and H. G. DOGE, *Talanta*, **12**, 73 (1965).
62. D. J. PIETRZYK and D. L. KISER, *Anal. Chem.*, **37**, 1578 (1965).
63. J. MINCZEWSKI and R. DYBEZYNSKI, *J. Chromatog.*, **7**, 98 (1962).
64. W. H. GERDES and W. RIEMAN, *Anal. Chim. Acta*, **27**, 113 (1962).
65. L. H. AHRENS, R. A. EDGE and R. R. BROOKS, *Anal. Chim. Acta*, **28**, 551 (1963).
66. D. C. DeGEISO, W. RIEMAN and S. LINDENBAUM, *Anal. Chem.*, **26**, 1840 (1954).
67. O. S. GLASO, *Anal. Chim. Acta*, **28**, 543 (1963).
68. A. C. D. NEWMAN, *Anal. Chim. Acta*, **19**, 471, 580 (1958).
69. J. COURSIER and J. SAULNIER, *Anal. Chim. Acta*, **14**, 62 (1956).
70. M. KIKUDAI, *Compt. rend.*, **240**, 1110 (1955).
71. J. K. SKLOSS, J. A. HUDSON and C. J. CUMMISKEY, *Anal. Chem.*, **37**, 1240 (1965).
72. S. LINDENBAUM, T. V. PETERS and W. RIEMAN, *Anal. Chim. Acta*, **11**, 530 (1954); T. V. PETERS and W. RIEMAN, *ibid.*, **14**, 131 (1956).
73. H. L. ROTHBART, H. W. WEYMOUTH and W. RIEMAN, *Talanta*, **11**, 33, 43 (1964).
74. D. P. LUNDGREN and N. P. LOEB, *Anal. Chem.*, **33**, 366 (1961).
75. F. H. POLLARD, G. NICKLESS, D. E. ROGERS and M. T. ROTHWELL, *J. Chromatog.*, **17**, 157 (1965).
76. N. SHIRAKISI and T. IBA, *Bunseki Kagaku*, **13**, 883 (1964).
77. M. SCHMIDT and T. SAND, *Z. anorg. Chem.*, **330**, 188 (1964).
78. F. H. POLLARD, G. NICKLESS and R. B. GLOVER, *J. Chromatog.*, **15**, 533 (1964).
79. R. SARGENT and W. RIEMAN, *Anal. Chim. Acta*, **17**, 408 (1957); *ibid.*, **18**, 196 (1958).
80. S. R. WATKINS and H. F. WALTON, *Anal. Chim. Acta*, **24**, 334 (1961).
81. K. BEY, *Fette Seifen Anstrichmittel*, **67**, 25 (1965).
82. R. L. REBERTUS, *Anal. Chem.*, **38**, 1089 (1966).
82a. Y. YOSHINO, H. KINOSHITA and H. SUGIYAMA, *Nippon Kagaku Zasshi*, **86**, 405 (1965).
83. J. J. LATTERELL and H. F. WALTON, *Anal. Chim. Acta*, **32**, 101 (1965).
84. T. A. WHEATON and I. STEWART, *Analyt. Biochem.*, **12**, 585 (1965).
85. G. PATAKI, E. BAUMANN, U. P. GEIGER, P. JENKINS and W. KUPPER, chap. 15 in E. HEFTMANN (ed.), *Chromatography*, 2nd ed., Reinhold, 1967.
86. S. MOORE and W. H. STEIN, *J. Biol. Chem.*, **192**, 663 (1951); D. H. SPACKMAN, W. H. STEIN and S. MOORE, *Anal. Chem.*, **30**, 1190 (1958).

87. P. B. Hamilton, *Anal. Chem.*, **30**, 914 (1958); *ibid.*, **32**, 1779 (1960); P. B. Hamilton, D. C. Bogue and R. A. Anderson, *ibid.*, **32**, 1782 (1960).
88. J. V. Benson and J. A. Patterson, *Analyt. Biochem.*, **13**, 265 (1965); *Anal. Chem.*, **37**, 1108 (1965).
89. K. A. Piez and L. Morris, *Analyt. Biochem.*, **1**, 187 (1960).
90. J. Oro and H. B Skewes, *Nature*, **207**, 1042 (1965).
91. W. E. Cohn, *Science*, **109**, 377 (1949).
92. W. E. Cohn, chap. 22 in E. Heftmann (ed.), *Chromatography*, 2nd edn., Reinhold, 1967.
93. H. H. Schenker and W. Rieman, *Anal. Chem.*, **25**, 1637 (1953).
94. A. J. Goudie and W. Rieman, *Anal. Chim. Acta*, **46**, 419 (1962)
95. K. Shimomura and H. F. Walton, *Anal. Chem.*, **37**, 1012 (1965).
96. R. C. Zerfing and H. Veening, *Anal. Chem.*, **38**, 1312 (1966).
97. U. B. Larsson and O. Samuelson, *J. Chromatog.*, **19**, 404 (1965).
98. A. J. Courtoisier and J. Ribereau-Gayon, *Bull. Soc. Chim.*, 1963, p. 350.
99. W. Funasaka, T. Kojima and K. Fujimura, *Bunseki Kagaku*, **13**, 42 (1964).
100. C. Davies, R. D. Hartley and G. J. Lawson, *J. Chromatog.*, **18**, 47 (1965).
101. H. Busch, R. B. Hurlbut and Van R. Potter, *J. Biol. Chem.*, **196**, 717 (1952).
102. W. Funasaka *et al.*, *Bunseki Kagaku*, **11**, 434, 936 (1962); *ibid.*, **12**, 466, 1170 (1963).
103. D. K. Hale, A. R. Hawdon, J. I. Jones and D. I. Packham, *J. Chem. Soc.*, 3503 (1952).
104. L. I. Larsson and O. Samuelson, *Acta Chem. Scand.*, **19**, 1357 (1965).
105. O. Samuelson and B. Swenson, *Acta Chem. Scand.*, **16**, 2056 (1962); *Anal. Chim. Acta*, **28**, 426 (1963).
106. J. X. Khym and L. P. Zill, *J. Am. Chem. Soc.*, **74**, 2090 (1952).
107. S. E. Zager and T. C. Doody, *Ind. Eng. Chem.*, **43**, 1070 (1951).
108. R. Sargent and W. Rieman, *Anal. Chim. Acta*, **16**, 144 (1957).
109. J. X. Khym *et al.*, *J. Am. Chem. Soc.*, **75**, 1153, 1339 (1953).
110. E. F. Walborg, L. Christensson and S. Gardell, *Analyt. Biochem.*, **13**, 177 (1956).
111. G. Neukom, H. Deuel, W. J. Heri and W. J. Kundig, *Helv Chim. Acta*, **43**, 64 (1960).
112. G. Gabrielson and O. Samuelson, *Acta Chem. Scand.*, **6**, 729 (1952).
113. K. Christofferson, *Anal. Chim. Acta* **33**, 303 (1965).
114. O. Samuelson and E. Sjöstrom, *Svensk. Kem. Tidskr.*, **64**, 305 (1952).
115. F. G. Helfferich, *Nature*, **189**, 1001 (1961).
116. F. G. Helfferich, *J. Am. Chem. Soc.*, **84**, 3237, 3242 (1962).
117. R. H. Stokes and H. F. Walton, *J. Am. Chem. Soc.*, **76**, 3327 (1954).
118. A. G. Hill, R. Sedgley and H. F. Walton, *Anal. Chim. Acta*, **33**, 84 (1965).
119. K. Shimomura, L. Dickson and H. F. Walton, *Anal. Chim. Acta*, **37**, 102 (1967).
119a. K. Shimomura, unpublished work.
120. Y. Arikawa and I. Makimo, *Proc. Federated Biological Societies*, 1967 (abstract).
121. A. Tsuji, *Nippon Kagaku Zasski*, **81**, 847, 1090 (1960).
122. E. A. Emken, C. R. Scholfield and H. F. Dutton, *J. Am. Oil Chemists' Soc.*, **41**, 388 (1964).
123. C. F. Wurster, J. H. Copenhaver and P. R. Shafer, *J. Am. Oil Chemists' Soc.*, **40**, 513 (1963).
124. F. W. E. Strelow, C. J. Leibenberg and F. von S. Toerien, *Anal. Chim. Acta*, **47**, 251 (1969).

CHAPTER 9

SALTING-OUT CHROMATOGRAPHY
AND RELATED METHODS

ALTHOUGH ion-exchange resins serve as the stationary phase for the chromatographic procedures discussed in this chapter, the separations are accomplished without any ion-exchange reaction and depend on van der Waals or Donnan forces.

A. Ion Exclusion

If a mixture of sodium chloride and ethandiol is eluted through a column of strong-acid ion-exchange resin in the sodium form or strong-base resin in the chloride form, a separation occurs,[1] the sodium chloride appearing in the eluate before the glycol (Fig. 9.1). The same conditions that yield nearly Gaussian elution curves in ion-exchange chromatography (small sample, fine resin particles, and slow flow) also give symmetrical curves for the sodium chloride and glycol.

This phenomenon, which was discovered by Wheaton and Bauman,[1a] may be explained as follows: Donnan's principle prevents, in large measure, the electrolyte (subscript e) from diffusing into the resin. Therefore its distribution ratio is essentially zero

$$C_e \cong 0. \tag{9.1}$$

On the other hand, the nonelectrolyte (subscript n) diffuses freely into the resin. Its C value is

$$C_n = K_a V_{wr}/V, \tag{9.2}$$

where V_{wr} denotes the total volume of water inside the resin of the column. Although K_a varies over a large range (see Table 2.7, p. 31), its value is always appreciably greater than zero. Therefore $C_n > C_e$ and a chromatographic separation is practicable.

The method is applicable to mixtures of other electrolytes and other nonelectrolytes. In order to avoid ion-exchange reactions, the counter ion of the resin should be the same as the similarly charged ion of the electrolyte. The hydrogen form of weak-acid resins and the hydroxide form of weak-base resins should not be used in ion exclusion because these resins are only slightly swollen and hence give small values of V_{wr}/V and C_n; also the failure of the Donnan principle with nonionized resins makes eqn. (9.1) inapplicable.

The degree of crosslinking is important. Since resins of low crosslinkage have relatively large Donnan invasion and hence large values of C_e, it would seem advantageous to use resins of high crosslinkage. On the other hand, these resins have smaller values of V_{wr}/V and C_n than the resins of lower crosslinkage. In general, resins of moderate crosslinkage, about 8%, are preferred.

175

Strong acids can be separated from weak acids by ion exclusion. The latter behave more or less like nonelectrolytes, especially since their ionization is partly repressed by the strong acid in the interstitial solution and even more by the sulphonic acid of the internal solution. Even weak acids of different strengths can often be separated[1a] by ion exclusion as indicated by Table 9.1. The strength of the stronger acid seems to be more important that the difference between the pK values.

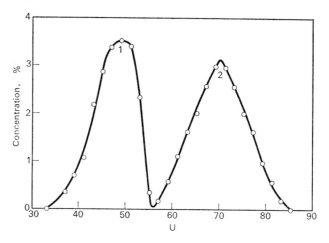

FIG. 9.1. Separation of sodium chloride from ethandiol by ion exclusion. Sample = 400 mg of each in 10 ml of water. Column = 55 cm × 1.8 cm², Dowex 50-X8, 50–100 mesh. Curve 1 = sodium chloride. Curve 2 = ethandiol. (Reproduced by permission from *Chemical Engineering Progress*.)

TABLE 9.1. SEPARATION OF ACIDS BY ION EXCLUSION

Stronger acid	pK	Weaker acid	pK	ΔpK	Separation
Hydrochloric		Acetic	4.8		Excellent
,,		Chloroacetic	2.8		Excellent
,,		Dichloroacetic	1.3		Excellent
,,		Trichloroacetic	0.7		Fair
Trichloroacetic	0.7	Acetic	4.8	4.1	Good
,,	0.7	Chloroacetic	2.8	2.1	Good
,,	0.7	Dichloroacetic	1.3	0.6	Good
Dichloroacetic	1.3	Acetic	4.8	3.5	Fair
,,	1.3	Chloroacetic	2.8	0.5	Fair
Chloroacetic	2.8	Acetic	4.8	2.0	None

(Data used by permission from *Industrial and Engineering Chemistry*.)

Ion exclusion is not very successful in the separation of electrolytes from nonelectrolytes of large molecular dimensions. Sucrose, for example, is not readily separated from sodium chloride because it diffuses too slowly into the resin beads. The use of resins of low cross-linkage facilitates the diffusion of sucrose into the resin but also diminishes the exclusion of the sodium chloride.

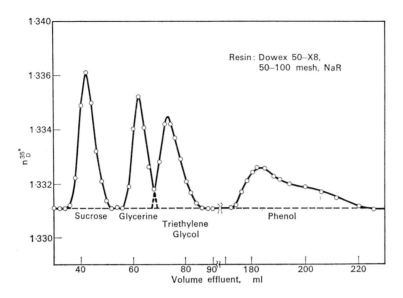

FIG. 9.2. Separation of nonelectrolytes by elution with water. Column = 55 cm × 1.8 cm², Dowex 50-X8, NaR, 50–100 mesh. (Reproduced by permission of New York Academy of Sciences.)

A typical laboratory application of ion exclusion is the separation of hydrogen sulphide from thiocyanic acid with a strong-acid resin.[2] Ion exclusion has not found extensive use in analytical chemistry.

B. Separations by Elution with
Water or Dilute Aqueous Buffer Solutions

For any given column of ion-exchange resin, V_{wr}/V is a constant. Thus according to eqn. (9.2) the distribution ratios of nonelectrolytes will be proportional to their K_a values. The wide spread of K_a values indicated in Table 2.7 suggests that simple elution with water through an ion-exchange resin should serve to separate some mixtures of nonelectrolytes. This is confirmed by Fig. 9.2, which shows the nearly complete separation of four compounds.[3] Maehr and Schaffner[4] have separated the three neomycins (A, B, and C) from each other by elution with water through a column of Dowex 1-X2 in the hydroxide form. In a similar elution of the four catenulins, C and D emerged as the first and second compounds, each quantitatively separated from the others; but A and B overlapped each other.

In the elution of weak electrolytes it is sometimes advantageous to use a dilute aqueous buffer solution instead of pure water as the eluent. For example, Seki[5] achieved a nearly quantitative separation of phenol and the three cresols by elution through a 15-cm column with a buffer consisting of 0.11 M sodium bicarbonate, 0.080 M sodium carbonate, and 0.0015 M disodium ethylenediaminetetraacetate. The pH was 9.70. The pK values[6] of the sample constituents increase in the order phenol (10.00), m-cresol (10.09), p-cresol (10.26), and o-cresol (10.29). In water, these compounds would be almost entirely nonionized; but in the alkaline buffer, there is appreciable ionization. The degree of ionization decreases with increasing pK. Since the resin absorbs the nonionized solutes more readily than their ions, we should expect these compounds to be eluted in the order of their increasing pK values. The actual order of emergence from the column was phenol, m-cresol, o-cresol, and p-cresol. This is the same as the order of increasing pK values except for the reversal of o- and p-cresol. These compounds have pK values so nearly identical that the small inherent preference of the resin for the ortho isomer overbalances its slightly greater ionization.

C. Salting-out Elution Chromatography

The chromatographic separation of nonelectrolytes with ion-exchange resins is greatly facilitated by the use of an aqueous solution of an electrolyte instead of pure water as the eluent. Of course, the counter ion of the resin must be the same as one ion of the eluent in order to avoid ion exchange.

To illustrate the advantage of using a salt solution instead of water as the eluent, let us consider the separation of ethanol and methanol with Dowex 50-X8, ammonium form, as stationary phase. With water as eluent, the C values are 0.826 and 0.803 respectively.[7] If P is assumed to be 9.0, it can readily be calculated by eqn. (6.32) that a column of 40,000 cm is required for a quantitative separation. On the other hand, with 3.0 M ammonium sulphate as eluent, the respective C values are 4.57 and 2.29. Then the required length of column is only 14 cm.

The separation of nonelectrolytes (or weak electrolytes) by elution with an aqueous salt solution, usually in moderate or large concentration, is called salting-out chromatography. Seki's[5] separation of phenol and the three cresols with a dilute carbonate buffer is not an example of salting-out chromatography because the concentration of the carbonate was too small to exert an important salting-out effect and because the separation depended on the buffering action of the mixed carbonates.

C.I. THEORY OF SALTING-OUT ELUTION CHROMATOGRAPHY

The same conditions that favor Gaussian elution graphs in ion-exchange chromatography (small sample, small particle size, slow flow) also favor Gaussian graphs in salting-out chromatography. The strong-acid and strong-base polystyrene resins of moderate cross-linkage are generally also preferred for salting-out chromatography unless the nonelectrolyte molecules are rather large and require 4% DVB for ready diffusion through the resin.

Most of the equations of the plate theory of ion-exchange elution chromatography are also applicable to salting-out elution chromatography. These include eqns. (6.7) and (6.29) through (6.46) but not (6.11) or (6.14).

C.I.a. *Relationship between the Distribution Ratio and the Concentration of Eluent in Salting-out Chromatography*

Sargent[7] eluted eleven alcohols separately four or five times, each time using a different concentration of ammonium sulphate as eluent. From each elution he evaluated U^* and C. The results with the ammonium form of Dowex 50-X8 are presented graphically in Fig. 9.3. Plots of log C vs. the concentration of ammonium sulphate give good straight lines. This leads to the empirical equation

$$\log C = \log C_0 + kM_s, \tag{9.3}$$

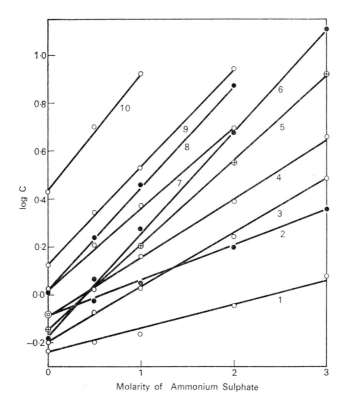

FIG. 9.3. Plots of log C vs. concentration of ammonium sulphate for several alcohols with Dowex 50-X8. 1, glycerol; 2, methanol; 3, propylene glycol; 4, ethanol; 5, propanol-2; 6, t-butyl alcohol; 7, propanol-1; 8, butanol-2; 9, isobutyl alcohol (the graph for butanol-1 coincides with this within experimental error); 10, pentanol-1.

where M_s represents the concentration of the eluent and C_0 is the value of C when $M_s = 0$, i.e. when the eluent is pure water; k is a constant characteristic of the salt and nonelectrolyte. A very similar figure for the same alcohols and electrolyte with the sulphate form of Dowex 1-X8 has been published elsewhere.[7] These graphs also follow eqn. (9.3).

Actually the divergence of the points from the graphs is slightly greater for Dowex 1 than for Dowex 50 because the authors used an erroneous value for V/V_b, 0.44 instead of 0.39 as given on p. 134. This error does not seriously affect the values of log C_0 or k as read from the graph.

The values of log C_0 and k as read from the two figures are given in Table 9.2. For any given nonelectrolyte, the values of log C_0 with the two resins differ from each other appreciably.

This is to be expected. Since C_0 is the distribution ratio of the alcohol in pure water, its value depends both on the alcohol and on the resin. Furthermore, the value of log C_0 generally increases with increasing hydrophobic nature of the nonelectrolyte. This is also to be expected in the light of the discussion on p. 32.

On the other hand, for any one alcohol, the two values of k check very well (Table 9.2). This means that k expresses some function of the interaction in aqueous solution between the nonelectrolyte and the electrolyte. The values of k also show a tendency to increase as the hydrophobic nature of the nonelectrolyte increases.

TABLE 9.2. COMPARISON OF VALUES OF LOG C_0 AND k, AMMONIUM SULPHATE AS ELUENT SALT

Nonelectrolyte	Dowex 1-X8		Dowex 50-X8	
	log C_0	k	log C_0	k
Glycerol	−0.465	0.129	−0.282	0.120
Propandiol-1,2	−0.362	0.210	−0.188	0.224
Methanol	−0.252	0.158	−0.095	0.151
Ethanol	−0.208	0.256	−0.083	0.243
Propanol-2	−0.170	0.353	−0.135	0.350
t-Butyl alcohol	−0.125	0.414	−0.130	0.408
Propanol-1	+0.070	0.336	+0.043	0.330
Butanol-2	0.150	0.422	0.039	0.419
Isobutyl alcohol	0.33	0.38	0.130	0.407
Butanol-1	0.35	0.40	0.125	0.405
Pentanol-1	0.82	0.45	0.44	0.50

(Used by permission from *Journal of Physical Chemistry*.)

C.I.b. *Significance of the Salting-out Constant*

Many years ago Setschenow[8] studied the effect of salts on the solubilities of nonelectrolytes in water. His work led to the empirical equation

$$\log S = \log S_0 - k M_s, \qquad (9.4)$$

where S_0 is the solubility of the nonelectrolyte in pure water, S is its solubility in M_s molar aqueous salt solution, and k is the salting-out constant dependent on both the nonelectrolyte and the salt.

It is interesting to note that eqn. (9.3) can be derived from eqn. (9.4) if it is assumed that the addition of salt to the system water + resin + nonelectrolyte causes neither shrinkage of the resin nor Donnan invasion by the salt. In the absence of salt

$$C_0 = \frac{v_{wr}}{v} \frac{M_{io}}{M_{eo}},$$

where v_{wr} denotes the volume of water inside the resin of one plate, v denotes the interstitial volume of one plate, and M denotes the molarity of the nonelectrolyte. The subscripts e and i pertain respectively to the internal and external (or interstitial) solution, and the subscript o indicates the absence of salt. In the presence of salt

$$C = \frac{v_{wr}}{v} \frac{M_i}{M_e}.$$

If v_{wr} does not change on addition of the salt,

$$\frac{C}{C_0} = \frac{M_i M_{eo}}{M_e M_{io}}.$$

Since the internal and external solution are in equilibrium with each other both in the absence and presence of salt,

$$M_{io} \gamma_{io} = M_{eo} \gamma_{eo},$$
$$M_i \gamma_i = M_e \gamma_e,$$

where γ denotes the activity coefficient. By combination of the last three equations

$$\frac{C}{C_0} = \frac{\gamma_e \, \gamma_{io}}{\gamma_{eo} \, \gamma_i}.$$

From the foregoing assumption it follows that the ionic strength of the internal solution is independent of the concentration of salt in the external solution. Furthermore, the activity coefficient of a dilute nonelectrolyte depends largely on the ionic strength, not on its own concentration. Therefore $\gamma_{io} = \gamma_i$ and

$$C/C_0 = \gamma_e/\gamma_{eo}. \qquad (9.5)$$

When the solubility of a nonelectrolyte is determined, i.e. when the solution is in equilibrium with the pure nonelectrolyte, the activity of the nonelectrolyte in solution is the same as that of the pure compound. This is true whether salt is present or not. Therefore

$$S \, \gamma_e = S_o \, \gamma_{eo}.$$

Combination of this with Setschenow's eqn. (9.4) yields

$$\log \frac{S_o}{S} = \log \frac{\gamma_e}{\gamma_{eo}}.$$

Combination of this with eqns. (9.4) and (9.5) yields eqn. (9.3).

It follows that k of eqn. (9.3) should have the same value for any given combination of salt and nonelectrolyte as the k of eqn. (9.4). The values of k of propanol-1 and butanol-2 as determined by solubility measurements in the presence of ammonium sulphate are respectively 0.341 and 0.420, in very good agreement with the values of Table 9.2. This good agreement is rather surprising in view of Sargent's data[9] showing that the ammonium form of Dowex 50-X8 absorbs 0.80 g of water per g of dry resin from pure water and only 0.63 from 3.0 M ammonium sulphate. This shrinkage, of course, violates the assumptions made in the derivation of eqn. (9.3).

In the case of the ammonium form of Dowex 50-X4, the shrinkage is greater, the absorption of water being 1.53 g per g of dry resin from pure water and 0.99 from 3 M ammonium sulphate. Although this resin and Dowex 1-X4 give linear plots of log C vs. the concentration of salt and k values that agree with each other, these k values do not agree with those of resins with 8% DVB. On the other hand, resins with 12% and 16% DVB have the same k values as resins with 8% DVB. These facts are illustrated[10] in Table 9.3.

TABLE 9.3. VALUES OF LOG C_0 AND k WITH RESINS OF VARIOUS CROSSLINKAGE, AMMONIUM SULPHATE AS ELUENT SALT

Resin	Propanone		Butanone	
	log C_0	k	log C_0	k
Dowex 1-X4	−0.175	0.283	0.000	0.360
Dowex 50-X4	+0.183	0.272	0.282	0.358
Dowex 1-X8	−0.170	0.318	0.105	0.397
Dowex 50-X8	+0.010	0.312	0.151	0.398
Dowex 50-X12	−0.030	0.330	0.125	0.397
Dowex 50-X16	+0.045	0.322	0.200	0.385

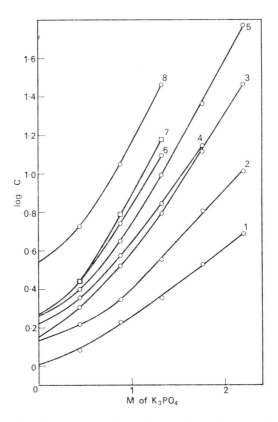

FIG. 9.4. Plots of log C vs. concentration of tripotassium phosphate for several amines with Dowex 50-X4. 1, ethanolamine; 2, methylamine; 3, ethylamine; 4, propanol-1; 5, iso-propylamine; 6, n-propylamine; 7, diethylamine; 8, n-butylamine. (Reproduced by permission from *Analytica Chimica Acta*.)

Linear plots of log C vs. concentration of ammonium sulphate have been observed when alcohols,[7] ethers,[11] aldehydes,[12] and ketones[12] were eluted with solutions of ammonium sulphate. However, the curves for the elution of amines[13] with solutions of tripotassium phosphate are linear only at high concentrations (Fig. 9.4). The explanation is found in the hydrolysis of the tertiary phosphate ion.

$$PO_4^{3-} + H_2O \rightleftharpoons HPO_4^{2-} + OH^-$$

The fraction hydrolyzed is insignificant at high concentrations but becomes greater as the concentration decreases. Therefore the resin exists almost entirely as R_3PO_4 at large concentrations, but contains appreciable and increasing mole fractions of R_2HPO_4 as the concentration of salt is decreased. A linear graph is not to be expected under these conditions. The graph for propanol-1 is included in Fig. 9.4 to indicate that the curvature is caused by the eluent, not by the organic compound.

C.I.c. *Choice of Eluent*

The electrolyte that is selected as eluent in salting-out chromatography should meet the following requirements.

(1) It should not interfere in the determination of the sample constituents in the fractions of eluate. Most water-soluble organic compounds are conveniently determined by treating the solution with an equal volume of unstandardized solution of sodium dichromate (about 0.1 N as an oxidizing agent) in concentrated sulphuric acid, heating for a brief period, and finally measuring the concentration of green chromium(III). Even compounds that react nonstoichiometrically with dichromate have a constant ratio of chromium(III) produced to compound taken if the heating conditions are standardized.[14] The eluent salt that is used in this method must not be oxidized or reduced under the conditions of the determination. This rules out halides and nitrate as anions, also cations such as cerium(IV) and iron(II). Aromatic compounds are conveniently determined by ultraviolet spectrophotometry and permit the use of any salt that does not absorb too greatly the radiation of the desired wavelength.

(2) The chosen electrolyte should have a large salting-out effect, i.e. the value of k in eqn. (9.3) should be great with respect to the constituents of the sample. Unless this requirement is fulfilled, there is little to be gained by using a salt solution as eluent.

(3) The electrolyte should be highly soluble so that great changes in C values can be caused by large changes in the concentration of electrolyte. This is rather unimportant since almost all salts that have large k values also have large solubilities.

(4) Specificity of the salting-out is also desirable. If a given salt has equal values of k toward all the compounds of a given mixture, very little is gained by using a solution of this salt as eluent.

Ammonium sulphate meets the foregoing requirements very well and has been used in salting-out chromatography more than any other salt. Its superiority over ammonium nitrate in regard to points (2) and (3) above may be seen from a comparison[10] of the k values of the two salts toward propanone, butanone, and pentanone-2. The respective values for ammonium sulphate are 0.312, 0.398, and 0.478, whereas those for ammonium nitrate are 0.034, 0.047, and 0.058.

Another consideration is the hydrolysis of the eluent salt. Ammonium sulphate does not undergo serious hydrolysis except at very low concentrations where its salting-out action would be negligible. Therefore it gives straight lines in the graphs of log C vs. M_s. Tripotassium phosphate, on the other hand, undergoes considerable hydrolysis and gives curved plots. Ammonium sulphate is therefore generally to be preferred over tripotassium phosphate as an eluent. Nevertheless, the latter salt was selected for the separation of the amines because of its alkaline nature. At the concentrations used in salting-out chromatography, the pH of its solutions is high enough to suppress almost completely the ionization of most amines. With a neutral salt as eluent, the amine would yield sufficient hydroxide ion and substituted ammonium ion to undergo appreciable ion exchange with the resins.

The monoalkyl esters of the alkanephosphonic acids can be separated more easily by salting-out chromatography than by ion-exchange.[15] The eluent consisted of hydrochloric acid to repress the ionization of the organic acids and lithium chloride to supply a large salting-out effect.

C.I.d. Low-capacity Resins

Salting-out and ion-exclusion chromatography may properly be considered as special cases of partition chromatography in which the resin or its internal water serves as the stationary phase. When nonelectrolytes are eluted with water through a column of crosslinked polystyrene without any ionogenic groups, their C values are essentially zero. This indicates that the internal water rather than the resin itself is the stationary solvent. Manalo *et al.*[16] eluted a number of nonelectrolytes with water or aqueous ammonium sulphate through a series of partly sulphonated polystyrene resins. A plot of log C_0, as calculated from the elution graphs, against the specific exchange capacity of the dry hydrogen-form resins is shown in Fig. 9.5.

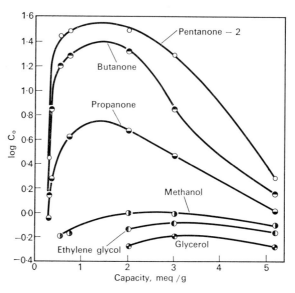

FIG. 9.5. Log C_0 values with partly sulphonated resins with 8% DVB. (Reproduced by permission from *Journal of Physical Chemistry*.)

The following points should be noted in the graphs. (1) There is a maximum in each graph, occurring at a specific capacity of about 3 meq/g for the more hydrophylic compounds but at about 1.5 meq/g for the less hydrophylic solutes. (2) In general, the values of $\Delta \log C_0$, i.e. the vertical distances between any two graphs, are larger with partly sulphonated resins than with fully sulphonated exchangers; this suggests that the partly sulphonated resins may be useful in separations.

Although attempts[17] to derive an equation for log C_0 as a function of the specific capacity have been unsuccessful, a qualitative explanation of the shape of the graphs can be given. An unsulphonated benzene ring exerts greater van der Waals forces on an organic solute than does a sulphonated ring. This explains why resins with specific capacities slightly less than the normal value of 5.2 have larger C_0 values than the fully sulphonated resins. It also explains why the slope of the graphs near the right end are steeper for the more hydrophobic compounds. On the other hand, decreases in the specific capacity of the resins cause decreases in the quantity of internal water (Table 9.4) and hence tend to decrease the sorption of the nonelectrolytes. This second consideration takes precedence over the first to the left of the maxima of the graphs.

It should be reiterated that the C_0 values of Fig. 9.5 were calculated from elutions. Since equilibrium is reached slowly in slightly swollen resins, the C_0 values of the figures may be subject to a large error in the cases of the resins of very low specific capacity.

TABLE 9.4. ABSORPTION OF WATER BY PARTLY
SULPHONATED POLYSTYRENES WITH 8% DVB

Specific capacity (meq per g of dry hydrogen-form resin)	Absorption of water (g per g of dry resin)
5.2	1.09
3.02	0.85
2.03	0.56
0.76	0.32
0.56	0.30
0.38	0.21
0.32	0.19
0.03	0.20
0.00	0.00

(Data used by permission from *Journal of Physical Chemistry*.)

As is to be expected, the values of k in eqn. (9.4) are independent of the capacity of the resin.

From Fig. 9.5, it might be concluded that a sulphonated polystyrene with a specific capacity of about 3.0 meq/g would be best for the separation of the three ketones, and presumably also for other mixtures of rather hydrophobic nonelectrolytes. This is not the case. The rate of approach to equilibrium decreases so sharply with decreasing capacity that sulphonated resins with capacities much less than 4 meq/g fail to give satisfactory separations.

C.II. APPLICATIONS OF SALTING-OUT ELUTION CHROMATOGRAPHY

A graph for the separation of nine alcohols[7] with stepwise decreases in the concentration of eluent is shown in Fig. 9.6. Similar separations have been performed with ethers,[11] carbonyl compounds,[12] and esters.[18]

Weak electrolytes can also be separated from each other by salting-out chromatography. In the case of weak bases, the eluent should be basic so as to repress the ionization of the bases of the sample. An almost quantitative separation of eleven amines[13] has been performed by elution with tripotassium phosphate through Dowex 50-X4. In the case of weak acids, an acidic eluent is used to repress the ionization. Organic acids of phosphorus have been separated by elution through a cation-exchange resin with a mixture of lithium chloride and hydrochloric acid as eluent.[15, 19] Maleic and fumaric acids[20] have been separated by elution through a weak-acid cation-exchange resin (Amberlite CG-50) with 4 M calcium chloride acidified to a pH of 1.5. Since the pK_1 values of these acids are respectively about 1.8 and 3.0, they were mostly nonionized. At higher pH values, the acids are more ionized and are partly excluded from the resin by the Donnan principle; hence their C values decrease with increasing pH.

Even electrolytes that are ordinarily regarded as strong can be separated from each other by salting-out chromatography. Examples are the separation of 2-naphthol-8-sulphonic acid from the 2-6 isomer[21] with a 27-cm column of Amberlite CG-50 and 5 M sodium chloride at pH 2.70, the separation of 1-naphthol-2-sulphonic acid from the 1-4 isomer[22] with a 24-cm column of Amberlite CG-50 and 5 M sodium chloride at pH 2.8, and the separation of naphthalene-1,5-disulphonic acid from the 1-6 isomer[23] with a 22-cm column of the same resin and 2 M calcium chloride at pH 2.0. The resin probably absorbed these sulphonic acids as ion pairs (or as triplets in the cases of the disulphonic acids). Keily *et al.*[24] separated 1-toluenesulphonate from 2,4-dimethylbenzenesulphonate in a 4-hr elution with a 3.5 M ammonium sulphate through Dowex 50-X2. They also separated dodecyl-benzenesulphonate from mixtures of mono- and dimethylbenzenesulphonates, using 1.0 M ammonium sulphate to elute the mono- and dimethyl derivatives and water to remove the long-chain compound.

Even proteins have been fractionated by salting-out chromatography with Sephadex G-50 as stationary phase. This is a crosslinked dextran with ionogenic groups. Sargent and Graham[25] eluted pooled human blood plasma through a 14-cm column with sodium sulphate buffered at pH 5.5. After the passage of 300 ml of 1.50 M eluent, they decreased the

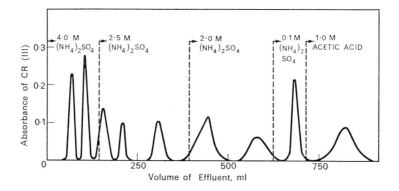

FIG. 9.6. Separation of alcohols by salting-out chromatography. Column = 32.0 cm × 2.28 cm², Dowex 1-X8, 200–400 mesh. Flow rate = 0.6 cm/min. The sequence of peaks is (1) glycerol, (2) methanol, (3) 1,2-propanediol, (4) ethanol, (5) propanol-2, (6) 1,1-dimethyl-ethanol, (7) butanol-2, (8) butanol-1, (9) pentanol-1. (Reproduced by permission from *Journal of Physical Chemistry*.)

concentration gradiently so that it was 0.64 M at $U = 1250$. The elution graph showed six peaks. The first was due to albumins, the next three to various constituents of the α-globu-lins, and the last two to other globulins.

D. Elution Chromatography with Mixed Solvents

D.I. ELUTION SOLUBILIZATION CHROMATOGRAPHY

There are many organic nonelectrolytes such as the higher homologs of pentanol that cannot be separated conveniently by elution with water either because their solubilities are too small to permit taking an adequate sample or because their distribution ratios are so large that very long times would be required to elute them from the column. The addition of salt to the eluent would, of course, aggravate the situation by decreasing the solubilities and increasing the C values. On the other hand, the addition of an organic solvent such as acetone, ethanol, or acetic acid to the eluent increases the solubilities and decreases the distribution ratios, thus making chromatographic separations possible. This method of separation is called solubilization chromatography.

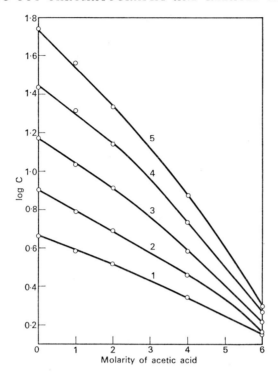

FIG. 9.7. Plots of log C vs. concentration of acetic acid for some normal primary alcohols with Dowex 50-X8. 1, amyl; 2, hexyl; 3, heptyl; 4, octyl; 5, nonyl. (Reproduced by permission from *Analytica Chimica Acta*.)

As in all separations, an eluent must be selected that does not interfere in the subsequent analysis of the eluate fractions. Aromatic compounds are readily determined by ultraviolet spectrophotometry and permit the use of a wide variety of solvents. Aldehydes and ketones can be determined by measuring the pH change caused by the liberation of hydrochloric acid on the addition of hydroxylamine hydrochloride[12, 26] and thus may be eluted with aqueous methanol or ethanol. If the conditions for the treatment of the eluate with chromic and sulphuric acid are properly chosen, aldehydes, ketones, alcohols, and ethers can be oxidized without oxidation of acetic acid;[27] therefore acetic acid can serve as the solubilizing agent in the elution of any of these compounds.

Strong-acid or strong-base resins are recommended as the stationary phase. In most cases, the counter ion of the resin is unimportant. Cation exchangers are generally used in the hydrogen form, but the sodium form should be used in the elution of esters because the hydrogen form would catalyze their hydrolysis. If the eluent contains acetic acid, the resin must be in either the hydrogen or the acetate form to avoid ion exchange. Small particle size is desirable. If the compounds of the sample have sufficiently small molecular dimensions, resins with 8% DVB are preferred, but those with 4% DVB are better for large sample molecules.

Figure 9.7 shows the graphs of log C vs. concentration of acetic acid in the eluent for the primary straight-chain alcohols from amyl through nonyl with Dowex 50-X8 as the stationary phase. Plots of log C vs. concentration of eluent for other migrant compounds, other resins, and other organic solvents generally resemble Fig. 9.7 in the following aspects. (1) Log C decreases with increasing concentration of the organic solvent. This is to be expected since the solubilities of sample compounds increase with increasing concentration of organic solvent. (2) The plots are linear for only a part of their length. At the present state of development of the theory, this is not surprising. (3) The graphs tend to converge with increasing concentration of the organic solvent in the eluent.

Equations (6.7) and (6.29) through (6.46) of the plate theory are also applicable to solubilization chromatography.

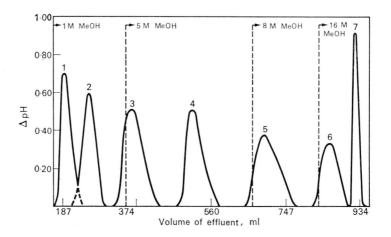

Fig. 9.8. Separation of seven ketones by solubilization chromatography with a column, 54.5 cm × 2.28 cm² of Dowex 50-X8, 200–400 mesh, hydrogen form. Flow rate, 0.4 cm/min. 1, methyl isobutyl ketone; 2, hexanone-2; 3, heptanone-2; 4, octanone-2; 5, nonanone-2; 6, decanone-2; 7, undecanone-2. (Reproduced by permission from *Analytica Chimica Acta*.)

Figure 9.8 shows the separation of seven ketones by solubilization chromatography.[28] The ordinate is the decrease of pH caused by the addition of hydroxylamine hydrochloride to the eluate fraction and is nearly proportional to the amount of ketone in the fraction. The eluent was aqueous methanol whose concentration was increased stepwise as indicated in the figure. Phenols,[27] alcohols,[27] ethers,[29] carboxylic acids,[29] and aromatic hydrocarbons[29] (Fig. 9.9) have been separated by this method.

D.II. ELUTIONS OF VERY HYDROPHYLIC NONELECTROLYTES WITH MIXED SOLVENTS

The effect of adding an organic solvent such as ethanol or acetone to a two-phase mixture of water and a rather hydrophobic compound such as butanol-1 is quite different from the effect of adding the same solvent to a mixture of water and a very hydrophylic compound such as sucrose. The solubility is increased in the former case but decreased in the latter. When a strong-base or strong-acid ion-exchange resin is equilibrated with a mixture of ethanol and water, the concentration of the ethanol in the internal solution of the resin is less than in the external solution. Table 9.5 illustrates this phenomenon.[30] It follows from the foregoing facts that the distribution ratios of very hydrophylic compounds will be greater in ethanol-water mixtures than in pure water. This is illustrated in Table 9.6. The values of C were calculated from the values[31] of U^* with the assumption that $V = 0.400\ V_b$.

Samuelson and Swenson[31] separated several mixtures of these sugars (no more than five sugars in any one sample) by elution through Dowex 1-X8, R_2SO_4, with aqueous alcohol solution between 65% and 74% by weight. They experienced difficulty because of the very slow diffusion of the sugars inside of the resin. They solved this problem by hydraulic size-grading of the resin and using the fraction between 45 and 75 μ. Elution with aqueous alcohol solutions has also been used to separate other sugar mixtures.[32]

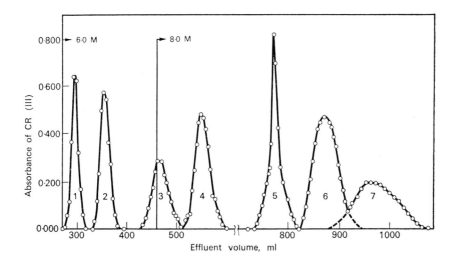

FIG. 9.9. Separation of aromatic hydrocarbons by solubilization chromatography with a column, 60 cm × 2.28 cm², of Dowex 1-X4, 200–400 mesh. Eluent, aqueous aceticacid. Flow rate, 0.4 cm/min. 1, benzene; 2, toluene; 3, m-xylene; 4, mesitylene; 5, naphthalene; 6, β-methylnaphthalene; 7, 1,4-dimethylnaphthalene. (Reproduced by permission from *Analytica Chimica Acta*.)

TABLE 9.5. DISTRIBUTION OF ETHANOL BETWEEN DOWEX 50, NaR, AND THE EXTERNAL AQUEOUS SOLUTION

Mole % EtOH		Mmol absorbed per meq of resin	
External solution	Internal solution	Water	Ethanol
0.00	0.00	10.86	0.00
1.52	1.10	10.33	0.12
33.3	12.3	6.40	0.90
55.7	17.4	4.40	0.92
73.0	25.9	2.63	0.92
84.1	38.4	1.58	0.99

(Used by permission from *Acta Chemica Scandinavica*.)

TABLE 9.6. VALUES OF LOG C FOR SEVERAL SUGARS WITH VARIOUS CONCEN-
TRATIONS OF ETHANOL IN WATER AS ELUENTS AND WITH DOWEX 1-X8, R$_2$SO$_4$

Weight % of EtOH:	65	70	72	74	82
Xylose					1.107
Glucose	0.991	1.134	1.233	1.288	1.754
Cellobiose		1.270		1.474	
Lactose			1.387	1.450	2.057
Maltose				1.450	
Sucrose			1.358	1.438	
Melisitose					1.600
Raffinose	1.212				1.816
Stachyose	1.462				
Verbascose	1.697				

(Reproduced by permission from *Analytica Chimica Acta.*)

D.III. ELUTIONS WITH DILUTE BUFFERS
IN NONAQUEOUS OR MIXED SOLVENTS

Just as the separation of weak electrolytes is improved by using a dilute aqueous buffer
as eluent rather than pure water (p. 177), so also the separation of less soluble weak electrolytes with a mixed solvent is improved by buffering the solvent. Seki[5] prepared 0.10 M
aqueous sodium citrate buffer at pH 4.00 and added 25 ml of benzyl alcohol to 1 l. of the
buffer. He used this mixture as eluent in the separation of six nitrophenols with a 64-cm
column of Amberlite IR-112. Table 9.7 lists the nitrophenols in the order of their emergence
from the column and their pK values in aqueous solution.

TABLE 9.7. ELUTION OF NITROPHENOLS WITH BUFFERED MIXED SOLVENTS

Compound	Order of emergence	pK
Picric acid	1	0.29
2,6-Dinitrophenol	2	3.71
2,4-Dinitrophenol	3	4.11
o-Nitrophenol	4	7.23
m-Nitrophenol	5	8.40
p-Nitrophenol	6	7.15

At the pH of the buffer, picric acid is almost entirely ionized, is therefore excluded from
resin by Donnan forces, and is eluted first. The two dinitrophenols are partly ionized in the
eluent, and the one with the smaller pK precedes the other down the column. The three
mononitrophenols are negligibly ionized in this buffer and their order of emergence depends
on their relative van der Waals attractions for the resin, not on their ionization constants.

D.IV. SALTING-OUT ELUTION CHROMATOGRAPHY WITH MIXED SOLVENTS

Sherma *et al.*[33] studied the effect of eluents containing both an organic solvent and a
strong electrolyte on the chromatographic behavior of water-insoluble alcohols and ketones.

With magnesium sulphate as salting-out agent in solutions of 8 M methanol, linear graphs of log C vs. the concentration of salt were obtained for elutions of hexanone-2, heptanone-2, and octanone-2 through Dowex 50-X8. Linear plots were also obtained with solutions of sodium acetate in 7 M acetic acid as eluent, Dowex 1-X8 as stationary phase, and hexanol-1, heptanol-1, and octanol-1 as migrants. However, many other combinations of solvent, salt, resin, and migrant gave curved plots of log C vs. salt concentration. These authors concluded that elution with a mixture of organic solvent and water without salt (solubilization chromatography) gives better separation of these ketones and alcohols than elution with salt solutions in mixed solvents.

On the other hand, Halmekoski and Hannikainen[34] studied the elution of phenacetine and related compounds through Dowex 50-X8 with solutions of tripotassium phosphate in 4 M methanol. They found linear graphs of log C vs. the concentration of salt. They recommend this method for the separation of phenacetine from acetanilide.

E. Comparison with Gas–Liquid Chromatography

Except for the section on ion exclusion, this chapter has dealt with the separation of organic compounds from each other by elution through ion-exchange resins with various types of eluents. Since gas–liquid chromatography is used mainly for the separation of organic compounds, a comparison of gas–liquid chromatography with the several methods of this chapter is in order. The following discussion applies to the separation of organic compounds by ion-exchange chromatography as well as to the methods of this chapter.

E.I. ADVANTAGES OF GAS–LIQUID CHROMATOGRAPHY

The greatest advantage of gas–liquid over liquid–resin chromatography is the very great economy of time. Once the proper apparatus is set up, the time required for a separation by the former method is measured in minutes, whereas the time needed for an elution through a column of resin is a matter of hours or days or even weeks. For example, the separation of sugar, malic, tartaric, and citric acids (p. 167) requires about 7 hr; the elution of the nine alcohols of Fig. 9.6 lasted 10 hr; and the separation of the seven ketones of Fig. 9.8 took 17 hr. On the other hand, the operator's time in an automated chromatographic separation with resins is very little more than in gas–liquid chromatography.

Another important advantage of gas–liquid chromatography is its ability to deal with larger and more hydrophobic compounds. Elution with water or an aqueous salt solution is obviously limited to water-soluble compounds. The use of mixtures of an organic solvent with water as eluents in solubilization chromatography and related methods extends the application of resin columns to slightly larger and more hydrophobic compounds. However, the unfortunate fact that graphs of log C vs. the concentration of organic solvent converge at large concentrations (p. 187) limits the applicability of solubilization chromatography. For example, alcohols much higher than n-amyl would require 6 M acetic acid to dissolve them; and in this eluent, their C values would differ so little (Fig. 9.7) that separation would be impracticable.

In summary, the advantages of gas–liquid chromatography are so great that elution through ion-exchange resins can compete only under special circumstances where the columns of resin offer certain advantages.

E.II. ADVANTAGES OF CHROMATOGRAPHY THROUGH
ION-EXCHANGE RESINS

The ability of this method to separate nonvolatile compounds is probably its greatest advantage. Examples of nonvolatile compounds that have been separated with resin columns are the amino acids (p. 164), organic acids of phosphorus (p. 185), sulphonic acids (p. 185), and sugars (p. 188). The exponents of gas–liquid chromatography may argue that the acids listed above can be separated by gas chromatography after conversion to volatile esters. However, the requirement of the additional step robs gas chromatography of its greatest asset— economy of time. Furthermore, for a quantitative analysis, each acid would have to be completely converted to its ester, or the percentage conversion would have to be constant and known. Moreover, compounds that can be volatilized and condensed without decomposition under ordinary conditions may decompose or react under the catalytic influence of the hot absorbant in gas chromatography.

Samples that are received as dilute solutions in water or water-soluble organic solvents must often be treated to remove the large quantity of solvent before a satisfactory gas chromatogram can be obtained. On the other hand, such samples require no or very little preparation for elution through a resin.

Although a careful test of the accuracy obtainable by the two chromatographic methods has not been made, it is probable that the use of resin columns will offer a slight advantage in this respect.

The apparatus for chromatography through ion-exchange resins is less expensive. When a large number of separations is to be performed on a repetitive basis, the cost of the apparatus for gas–liquid chromatography will soon pay for itself in man-hours saved. For the occasional separation, however, it is generally more economical to spend the extra time required by liquid–resin chromatography.

For preparative purposes, columns of resin are often preferable to those of gas chromatography. In both methods, the maximum allowable sample and hence the yield of any sample constituent are proportional to the cross-sectional area of the column. For any given column diameter, larger samples can be taken in liquid–resin than in gas–liquid chromatography. If it is desired to collect only the first constituent of a sample, frontal chromatography is preferable to elution chromatography, whatever the mobile and stationary phases may be.

F. Frontal Liquid–Resin Chromatography

The frontal technique can be applied to any of the types of chromatography discussed in this chapter. Figure 9.10 shows both the elution and the frontal graphs for the chromatography of acetophenone and β-naphthol with aqueous ethanol as eluent and solvent respectively. The frontal graph is the integral of the elution graph, i.e. the slope of the frontal graph is proportional to the ordinate of the elution curve. The proportionality factor is different for each of the solutes because the ratio of their amounts in the elution is not the same as the ratio of the concentrations in the solution used in the frontal experiment. In the frontal plot, the ordinate is the sum of the concentrations of the two solutes, i.e. the curve for naphthol rises from the plateau of the acetophenone graph.

Wheaton and Bauman[1a] recognized that the frontal curve in ion exclusion is the integral of the elution graph because they identified the frontal breakthrough with the peak of the

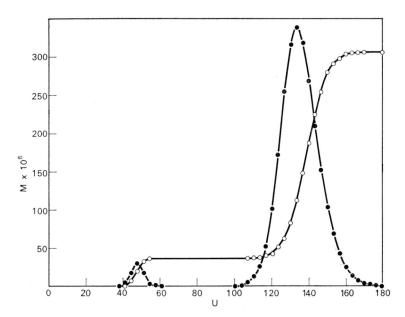

FIG. 9.10. Elution graph (filled circles) of 0.000825 mmol of acetophenone and 0.0000242 mmol of β-naphthol and frontal graph (open circles) of 0.0000380 M acetophenone and 0.000280 M β-naphthol. For both experiments: solvent = 30% by volume of 95% ethanol in water; column = 6.25 cm × 3.9 cm² of Dowex 50-X4, HR, 200–400 mesh; V = 9.0 ml; flow rate = 0.18 cm/min. (Reproduced by permission from *Talanta*.)

elution graph. The integral relationship has also been observed in gas–liquid chromatography.[35-37]

The equation for the frontal graph in solubilization chromatography has been derived.[38]

$$Y = \frac{1}{\sqrt{(2\pi)}} \int_{U=0}^{U=U} \exp \frac{-t^2}{2} \, d\acute{t},\tag{9.6}$$

$$t = \frac{U - U^*}{U^*} \Big/ \sqrt{\left\{\frac{p(1 + C)}{C}\right\}},\tag{9.7}$$

where U is the volume of effluent, U^* is the value of U when $Y = 1/2$; and Y is the concentration of sample constituent in the fraction of effluent divided by the concentration on the plateau. The other symbols have their usual meanings. In a multicomponent mixture, the concentration of each solute after the first is measured from the plateau of the previous solute. In using these equations it is not necessary to perform the integration because values of the integral can be found in probability tables[39] after conversion of the desired value of U to the corresponding value of t by eqn. (9.7). When $U = 0$,

$$t = -\sqrt{\left\{\frac{p(1 + C)}{C}\right\}} \cong -\infty.$$

This approximation is valid within the accuracy of the data in probability tables. Most probability tables give the area under the Gaussian curve from $t = 0$ to $t = t$. To find the area from $t = -\infty$ to $t = t$ from these tables, 0.5000 is added to the listed value.

The value of U^* is readily found from a frontal graph. Then C can be calculated from eqn. (6.7). The value of p is calculated by reading the value of U corresponding to any selected value Y. Probability tables are used to find the corresponding value of t. Finally, these values of t and U are substituted in eqn. (9.7) to find p. Actually, it is advisable to select two values of Y, one before and one after the mid-point of the breakthrough, and to calculate the corresponding values of p. They will agree if the curve follows eqn. (9.6) perfectly. The values $Y = 0.100$ and 0.900 are recommended because the slope of the frontal graph is moderate at these points, permitting an accurate reading of the graph. The corresponding values of t are respectively -1.282 and $+1.282$.

An experimental test of eqns. (9.6) and (9.7) was performed as follows.[38] Acetophenone, β-naphthol, and nitrobenzene were eluted, singly and as binary mixtures, through a column, $6.25\,\text{cm} \times 3.80\,\text{cm}^2$, of Dowex 50-X4, HR, 200–400 mesh, with an eluent prepared by mixing 300 ml of 95% ethanol with 700 ml of water. Flow rates were varied between 0.18 and 0.70 cm/min. From each elution graph, the values of p and C were calculated. Then fourteen frontal experiments were performed with the same compounds, singly or in pairs, through the same column at the same flow rates; the eluent of the elution experiments served as solvent for the frontal work. Again C and p were calculated.

At sufficiently slow flow rates, frontal and elution methods give the same values both for C and for P within the experimental errors. With increasing flow rates, P decreases; but the values by elution and frontal method; still check each other. The values of C also decrease with increasing flow rates, but the decrease is more rapid with the elution method. These relationships are illustrated in Table 9.8.

TABLE 9.8. CHROMATOGRAPHIC PARAMETERS; COMPARISON OF ELUTION AND FRONTAL CHROMATOGRAPHY

Solute	Flow rate (cm/min)	Values of C		Values of P	
		Elution	Frontal	Elution	Frontal
Acetophenone	0.18	5.69	5.54	29	32
,,	0.70	5.08	5.28	13	12
Nitrobenzene	0.18	8.05	7.96	42	38
β-Naphthol	0.18	17.7	18.0	30	32
,,	0.70	15.2	16.6	10	10

(Data used by permission from *Talanta*.)

Under the experimental conditions of the chromatograms summarized in Table 9.8, the parameters of any one solute are not altered by the presence of others. For example, the distribution ratios of β-naphthol in the frontal experiments at 1.8 cm/min as recorded in the table are the mean of four determinations: 18.3 when naphthol was the only solute, 18.3 when the solution contained acetophenone, and 17.9 and 17.6 when nitrobenzene was present. The concentrations of the solutions used in the frontal experiments never exceeded 0.004 M. These low concentrations were selected for convenience in the spectrophotometric analysis of the effluent fractions. The concentrations could probably be increased by a factor of about 10 without change in the parameters of one solute due to the presence of other solutes.

Theoretically, the concentration on the plateau of a frontal graph should equal the concentration of the influent. For example, the ordinates of the plateaus of Fig. 9.10 should be 38 and 318. Actually, they are 36 and 306. Similarly, in most of the frontal graphs summarized in Table 9.8, the plateau heights were a little less than the concentrations in the feed solution. A satisfactory explanation of this phenomenon cannot be given. It may possibly be due to the presence in each of the three solutes of a small amount of very strongly absorbed impurity. Since Y in eqn. (9.6) is defined as the fraction of the plateau height (not of the influent concentration), the failure of the graph to reach the concentration of the influent does not introduce an error into the calculation of C or p.

When the second plateau is reached in Fig. 9.10, the entire resin is at equilibrium with the feed solution. If the pure solvent (aqueous ethanol) is now passed through the column, the acetophenone will move down the column faster than the naphthol by virtue of its smaller C value. When all the acetophenone has been removed from the column, the graph reaches a plateau at an ordinate of $306 - 36 = 270$. Then the effluent contains β-naphthol free from acetophenone. Finally, the naphthol is entirely removed from the column; and the ordinate of the graph sinks to zero.

Since the experiments summarized in Table 9.8 were done with aqueous ethanol as solvent, they are properly classified as solubilization chromatography. However, eqns. (9.6) and (9.7) would very likely apply equally well to frontal experiments with water or an aqueous salt solution as solvent, i.e. to frontal salting-out chromatography.

Frontal separations of nonelectrolytes with ion-exchange resins as the stationary phase have found little use at present. However, frontal solubilization chromatography has been used in the separation of diastereoisomeric esters.[40]

References

1. D. W. SIMPSON and R. M. WHEATON, *Chem. Eng. Progr.*, **50**, 45 (1954).
1a. R. M. WHEATON and W. C. BAUMAN, *Ind. Eng. Chem.*, **45**, 228 (1953).
2. T. YOSHIMO, M. MATSUSHITA and M. SUGIHARA, *Kagaku to Kogyo.* **34**, 164 (1960); *Chem. Abs.*, **55**, 11029b (1961).
3. R. M. WHEATON and W. C. BAUMAN, *Ann. N.Y. Acad. Sci.*, **57**, 159 (1953).
4. H. MAEHR and C. P. SCHAFFNER, *Anal. Chem.*, **36**, 105 (1964).
5. T. SEKI, *J. Chromatog.*, **4**, 6 (1960).
6. V. E. BOWER and R. G. BATES, Equilibrium constants of proton-transfer reactions in water, in L. MEITES, *Handbook of Analytical Chemistry*, McGraw-Hill, New York, 1963, pp. 1–20.
7. R. SARGENT and W. RIEMAN, *J. Phys. Chem.*, **61**, 354 (1957).
8. J. SETSCHENOW, *Z. physik. Chem.*, **4**, 117 (1889).
9. W. RIEMAN and R. SARGENT, Ion exchange, in W. G. BERL's *Physical Methods in Chemical Analysis*, Academic Press, New York, 1961, vol. 4, p. 138.
10. A. BREYER and W. RIEMAN, *Talanta*, **4**, 67 (1960).
11. R. SARGENT and W. RIEMAN, *Anal. Chim. Acta*, **18**, 197 (1958).
12. A. BREYER and W. RIEMAN, *Anal. Chim. Acta*, **18**, 204 (1958).
13. R. SARGENT and W. RIEMAN, *Anal. Chim. Acta*, **17**, 408 (1957).
14. R. SARGENT and W. RIEMAN, *Anal. Chim. Acta*, **14**, 381 (1956).
15. A. VARON, F. JAKOB, K. C. PARK, J. CIRIC and W. RIEMAN, *Talanta*, **9**, 573 (1962).
16. G. D. MANALO, A. BREYER, J. SHERMA and W. RIEMAN, *J. Phys. Chem.*, **63**, 1511 (1959).
17. W. RIEMAN, unpublished work.
18. J. SHERMA, *Chemist-Analyst*, **52**, 114 (1963).
19. R. B. LEW, H. GARD and F. JAKOB, *Talanta*, **10**, 911 (1963).
20. W. FUNASAKA, T. KOJIMA and K. FUJIMARA, *Bunseki Kakagu*, **13**, 42 (1964); *Chem. Abs.*, **60**, 8646c (1964).
21. W. FUNASAKA, T. KOJIMA and K. FUJIMARA, *Bunseki Kakagu*, **10**, 374 (1961); *Chem. Abs.*, **55**, 23188b (1961).
22. W. FUNASAKA, T. KOJIMA, K. FUJIMARA and T. MINAMI, *Bunseki Kakagu*, **12**, 446 (1963); *Chem. Abs.*, **59**, 6968g (1963).

23. W. Funasaka, T. Kojima, K. Fujimara and S. Koshida, *Bunseki Kakagu*, **12,** 1170 (1963); *Chem. Abs.*, **60,** 4757e (1964).
24. H. J. Keily, A. L. Garcia and R. N. Peterson, *Chem. and Eng. News*, **42** (March 8, 1963).
25. R. N. Sargent and D. L. Graham, *Anal. Chim. Acta*, **30,** 101 (1964).
26. H. Roe and J. Mitchell, *Anal. Chem.*, **23,** 1758 (1951).
27. J. Sherma and W. Rieman, *Anal. Chim. Acta*, **18,** 214 (1958).
28. J. Sherma and W. Rieman, *Anal. Chim. Acta*, **19,** 134 (1958).
29. J. Sherma and W. Rieman, *Anal. Chim. Acta*, **20,** 357 (1959).
30. H. Rückert and O. Samuelson, *Acta Chem. Scand.*, **11,** 303 (1957).
31. O. Samuelson and B. Swenson, *Anal. Chim. Acta*, **28,** 426 (1963).
32. J. Dahlberg and O. Samuelson, *Acta Chem. Scand.*, **17,** 2136 (1963).
33. J. Sherma, D. Locke and D. Bassett, *J. Chromatog.*, **7,** 273 (1962).
34. J. Halmekoski and H. Hannikainen, *Suomen Kem.*, **35B,** 221 (1962); *Chem. Abs.*, **58,** 7350a (1963).
35. J. Boecke, in R. P. W. Scott's *Gas Chromatography*, Butterworths, London, 1960, p. 88.
36. C. N. Reilley, J. P. Hildebrand and J. W. Ashley, *Anal. Chem.*, **34,** 1198 (1962).
37. G. J. Krige and V. Pretorius, *J. Gas Chromatog.*, **2,** 115 (1964).
38. H. D. Spitz, H. L. Rothbart and W. Rieman, *Talanta*, **12,** 395 (1965).
39. R. C. Weast, *Handbook of Chemistry and Physics*, 45th edn., Chemical Rubber Publishing Co., Cleveland, 1964, p. A108.
40. R. E. Leitch, H. L. Rothbart and W. Rieman, *Talanta*, **15,** 213 (1968); H. D. Spitz, H. L. Rothbart and W. Rieman, *J. Chromatog.*, **29,** 94 (1967).

CHAPTER 10

LESS COMMON ION EXCHANGERS

THIS chapter is devoted mainly to a discussion of the less common ion-exchange materials such as inorganic ion exchangers and liquid ion exchangers. The properties and uses of the common ion exchangers in unusual forms (as membranes or impregnants of paper) are also considered.

A. Porous Resins

It is well known that the usual strong-acid and strong-base ion-exchange resins with small DVB content undergo considerable swelling when immersed in water or a dilute aqueous solution. In this swollen condition, these resins may be said to be porous. This section, however, is devoted to resins of a different type that retain their porous structure even in the dry condition.

A.I. DOWEX 21K

In the preparation of this resin, a hydrocarbon copolymer is first made by the reaction between styrene and a small percentage of DVB. This slightly crosslinked resin is then treated in the swollen state with a Friedel–Crafts reagent to introduce methylene bridges which serve as additional crosslinks. Then the resin is chloromethylated and treated with trimethylamine to introduce the usual type of strongly basic anion-exchange group, $—CH_2-NMe_3^+$. The methylene bridges increase the rigidity of the hydrocarbon network so that the resin resembles Dowex 1-X8 in regard to its swelling and shrinking, also in regard to its selectivity. Nevertheless, it is more similar to Dowex 1-X4 in regard to permeability and faster approach to equilibrium. It is unfortunate that the resin is not commercially available in sizes smaller than 50–100 mesh.

A.II. MACRORETICULAR RESINS

A.II.a. *Preparation of Macroreticular Resins*

Regarding the preparation of these resins, Kunin[1] *et al.* wrote: "The polymerization technique involves the suspension polymerization of styrene-divinylbenzene copolymers in the presence of a substance that is a good solvent for the monomer but a poor swelling agent for the polymer." The formation of the copolymer starts at numerous points in the droplet. The polymer networks grow in all directions from these centers until they join and occupy most of the volume of the bead. Most of the solvent is driven from regions occupied by the polymer. When the process is completed, the bead has a skeleton of copolymer consisting of some regions of high crosslinking connected by fibers of copolymers of lesser crosslinking. The solvent (usually toluene) occupies the interconnecting regions among the centers and strands of copolymer. If sufficient DVB was used in the original mixture, the polymer structure is sufficiently rigid so that it does not collapse (i.e. the bead does not shrink) when the solvent is evaporated. Then air occupies the macropores previously occupied by the solvent. Figure 10.1 shows the difference between the structure of an ordinary resin and that of a macroreticular resin as revealed by the electron microscope.[2]

The macroreticular hydrocarbon polymer may then be sulphonated or chloromethylated and treated with an amine to introduce the usual strong-acid or strong- or weak-base ionogenic groups. The resulting macroreticular ion-exchange resin retains the porous structure of the hydrocarbon polymer from which it was made.

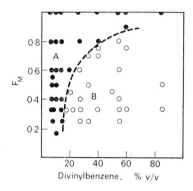

FIG. 10.2. Dependence of macroporosity on the composition of the polymerization mixture. Open circles represent macroporous resins; filled circles represent expanded networks. (Reproduced by permission from *Journal of the Chemical Society*.)

Millar *et al.*[3] prepared forty-five hydrocarbon copolymers from mixtures of varying percentages of DVB, styrene, and toluene, also eight resins with only DVB and styrene. They studied the properties of these polymers in relation to the composition of the original mixture. They express this composition in terms of the volume percentage of DVB in the mixture of monomers (i.e. DVB and styrene) and F_M, the fraction by volume of monomers in the total mixture.

They found that a large percentage of toluene in the reaction mixture (i.e. small F_M) does not produce a macroreticular resin unless the percentage of DVB is also large. This is illustrated in Table 10.1. In regard to the absorption of cyclohexane, and other properties, resin B differed very little from resin A in spite of the 40% of toluene used in the preparation of the former. It swelled slightly on absorption of the cyclohexane and shrank again on drying. The authors designate such a resin as having an expanded network. A comparison of C and D, on the other hand, reveals a striking difference in the absorption of cyclohexane and in other properties. Resin D is macroreticular.

Resins with the ability to absorb more than 0.1 ml of cyclohexane per g showed the macroreticular structure while the resins absorbing less than this amount had the expanded network. The authors used this criterion to classify their resins. Figure 10.2 shows how the macroreticular structure depends on both F_M and the percentage of DVB.

FIG. 10.1. Electron micrographs of an ordinary (left) and a macroreticular (right) resin. (Reproduced by permission from *Journal of Polymer Science*.)

TABLE 10.1. EFFECT OF TOLUENE ON THE ABSORPTION OF
CYCLOHEXANE BY STYRENE-DVB COPOLYMERS

Resin	F_M	% DVB	Absorption of cyclohexane (ml/g)
A	1.00	7	0.05
B	0.60	7	0.04
C	1.00	27	0.03
D	0.60	27	2.0

(Data used by permission from *Journal of the Chemical Society*.)

The introduction of ionogenic groups into the macroreticular or expanded-network copolymers produces ion-exchange resins having the same structure as the hydrocarbon copolymer. Table 10.2 shows the essential features in the preparation of these resins and some of their properties.[4] Resin A is obviously an ordinary resin; E had the expanded network; F is on the border line between the expanded-network and macroporous structure; G and H are clearly macroreticular.

A.II.b. *Kinetics of Exchange of Macroreticular Resins*

Millar *et al.*[4] studied the kinetics of ion exchange of these resins using the indicator method (p. 57). Since all of these resins had about the same average diameter (between 511 and 549 μ), the differences in the rate of exchange are due mainly to differences in structure. With all of the resins, the exchange between the hydrogen-form resin and 0.020 M tetraalkylammonium ion was controlled by bead diffusion. Table 10.3 gives the values of the

TABLE 10.2. PREPARATION AND PROPERTIES OF SOME SULPHONIC CATION-EXCHANGE RESINS

	A	E	F	G	H
% DVB in mixture of monomers	7	15	27	34	55
F_M	1.00	0.62	0.51	0.33	0.33
Absorption of toluene by hydrocarbon copolymer (ml/meq)	0.79	0.87	1.14	1.95	2.21
Sulphonated resin:					
Specific capacity (meq/g)	4.98	4.90	4.36	4.40	3.83
Absorption of H_2O (ml/meq)	0.20	0.20	0.22	0.40	0.54

(Data used by permission from *Journal of the Chemical Society*.)

diffusion coefficient $\bar{D}$ inside the resin as calculated from the kinetic data. Pr and Bu stand for the normal propyl and butyl groups. Resin A shows a marked decrease in $\bar{D}$ with increasing size of the cation; this indicates the difficulty encountered by large ions in diffusing through an ordinary resin with about 7% DVB. In spite of the larger percentage of DVB in resin E, the expanded network permits easier diffusion through the resin. Ion size is still less important in resin F and much less important in the truly macroreticular resins G and H.

TABLE 10.3. DIFFUSION COEFFICIENTS INSIDE THE RESIN

Resin	$\bar{D} \times 10^6$			
	NMe_4^+	NEt_4^+	NPr_4^+	NBu_4^+
A	1.08	0.26	0.10	0.015
E		0.29		0.032
F		0.23		0.088
G	0.49	0.32	0.31	0.24
H	0.62	0.42	0.48	0.32

(Reproduced by permission from *Journal of the Chemical Society*.)

Table 10.4 shows the comparison in exchange rates between the hydrogen forms of resins A and H and various 0.02 N cations. For any one resin, the exchange becomes slower as the size of the exchanging ion is increased. This is to be expected. For the smaller exchanging cations, Na^+, NMe_4^+, and NEt_4^+, the ordinary resin A with 7% DVB reacts faster

TABLE 10.4. EXCHANGE RATES

Time	Q_t = fractional approach to equilibrium									
	Na^+		NMe_4^+		NEt_4^+		NPr_4^+		NBu_4^+	
	Resin A	Resin H	Resin A	Resin H	Resin A	Resin H	Resin A	Resin H	Resin A	Resin H
2	1.00	0.92	0.81	0.68	0.55	0.55	0.38	0.47	0.17	0.38
4	1.00	1.00	0.98	0.76	0.69	0.63	0.48	0.56	0.23	0.46
6	1.00	1.00	1.00	0.78	0.78	0.64	0.56	0.58	0.26	0.47
8	1.00	1.00	1.00	0.79	0.83	0.65	0.63	0.58	0.28	0.48

(Reproduced by permission of the Chemical Society.)

because of the lower degree of crosslinking. However, with the larger cations, the existence of the macropores causes a faster exchange with resin H.

A.II.c. *Selectivity Coefficients of Macroreticular Resins*

Millar *et al.*[5] also studied the selectivity coefficients of macroreticular resins. They found that a resin with 15% DVB (based on the monomers) and $F_M = 0.40$ had $K_H^{Na} = 1.4$ at $[NaR] = 0.20$ and $K_H^{Na} = 1.1$ at $[NaR] = 0.80$. This rather small decrease in K_H^{Na} with increasing $[NaR]$ is similar to ordinary resins (Chapter 3). On the other hand, a highly macroreticular resin (27% DVB based on monomers, $F_M = 0.40$) had K_H^{Na} values of 2.9 and 1.2 at $[NaR] = 0.20$ and 0.80 respectively. The marked effect of the mole fraction of sodium in the resin on the selectivity coefficient is due to the highly heterogeneous structure of the macroreticular resin. The value of K_H^{Na} is greatest in the most crosslinked portions of the resin. At small values of $[NaR]$, the sodium ion is almost entirely in the regions of great crosslinking; therefore the selectivity coefficient is large. The nuclei of great crosslinking constitute only a small part of the resin. Therefore at large values of $[NaR]$, the majority of the sodium is in the regions of lesser crosslinking, and the selectivity is small.

The selectivity coefficients of other alkali metals with respect to hydrogen and with respect to each other show a similar large dependence on the mole fraction of any one cation in the resin.

A.II.d. *Donnan Invasion of Macroreticular Resins*

In the same paper,[5] the British authors also discuss the invasion of macroporous resins by electrolytes. It seems reasonable that the Donnan forces would be very ineffective in the large pores and that therefore the invasion of these resins would be large. In confirmation of this expectation, the average internal concentration of hydrochloric acid was found to be 0.67 M when a highly macroreticular resin prepared with a mixture of monomers containing 34% of DVB and with $F_M = 0.33$ was equilibrated with 1.00 M hydrochloric acid. In an ordinary resin with 7% DVB, the internal concentration of hydrochloric acid was only 0.11 M under the same conditions.

A.II.e. *Other Properties of Macroreticular Resins*

The properties of the macroreticular resins[2] manufactured by Rohm and Haas are given in Table 10.5. The distribution of pore sizes (columns 2, 3, and 4) was found by measuring the volume of mercury that enters the resin at known pressures and applying the equation $P = -4 \, \sigma \cos\theta/d$ where P is the pressure necessary to drive the mercury into pores of diameter d (Angstrom units), σ is the surface tension, and θ is the contact angle. Apparent and skeletal densities were determined by displacement of mercury and helium respectively.

Since dry macroreticular resins swell very much less than dry ordinary ion-exchange resins, it is not surprising that they suffer less damage from sudden immersion of the dry resin in water. A series of alternate drying and sudden wetting[6] that broke 87% of the beads of Amberlite IR-120, 16–20 mesh, broke only 10% of the beads of Amberlyst 15 of the same size. The macroreticular resins are also more resistant against chemical attack. Treatment of the sodium-form resins, 20–30 mesh, with 3% hydrogen peroxide at 45° for 24 hr, dissolved only 2% of the Amberlyst 15 but 24% of the Amberlite IR-120.

A.III. APPLICATIONS OF POROUS RESINS

Porous resins have been used much more extensively for industrial operations than in the laboratory. Among the industrial uses are the purification of sugar and the deionization of natural waters, especially those containing appreciable organic matter. In both of these applications the porous resins have the advantage of being able to absorb large organic molecules or ions better than ordinary resins. They are also used as catalysts in organic reactions.

In the laboratory, the porous resins are used principally with nonaqueous or mixed solvents, where the ordinary ion-exchange resins are very slightly swollen and hence very slow in reaching equilibrium. After methods had been developed for the separation of sugars by elution through an ordinary anion-exchange resin (p. 189), Dahlberg and Samuelson[7] studied the use of porous resins in the chromatographic separation of other sugar mixtures. They wrote: "These [porous] resins offer advantages over conventional resins in the chromatographic separations of sugars, and probably also in many other ion exchange separations."

TABLE 10.5. SOME PROPERTIES OF MACRORETICULAR RESINS

Resin	Ionogenic group	Average diam. d for given % of total pores			Density		Pores (ml/g)	Spec. surf. (m²/g)	Spec. capac. (meq/g)	Sorption of H_2O (g/g)
		10%	50%	90%	Apparent	Skeletal				
Amberlyst 15	—SO₃H	300	180	120	0.982	1.527	0.36	47.2	4.8	0.96
Amberlyst XN-1005	—SO₃H	350	<120	<120	0.795	1.359	0.52	117.9	3.5	0.79
Amberlyst A-27[a]	—CH₂NMe₃Cl	1140	690	660	0.555	1.114	0.91	42.2	2.6	1.50
Amberlyst A-29	—CH₂NME₂(C₂H₄OH)Cl	475	130	<120	0.836	1.237	0.39	53.0	2.7	0.79
Amberlite IR-120[b]	—SO₃H				1.483	1.488	0.00	<0.1	4.6	0.85
Amberlite IRA-401[b]	—CH₂—NMe₃Cl				1.136	1.131	0.00	<0.1	4.0	1.28

[a] Previously designated Amberlyst XN-1001.
[b] Not macroreticular.

(Used by permission from *Journal of Polymer Science*.)

Fritz and Waki[8] separated calcium and magnesium by elution through a 10-cm column of Amberlyst XN-1002, an anion-exchange resin, with 0.5 M nitric acid in 90% propanol-2. Fritz and Greene[9] separated rare earths from Al, Co, Ga, In, Fe(III), Mg, Mn, Ni, V(IV) and Zn by elution with 1.5 M nitric acid in 85% propanol-2 through a 16-cm column of the same resin. Ordinary resins are not sufficiently swollen in these eluents to be used for these separations.

An interesting separation of the esters of *cis-* and *trans-*fatty acids has been accomplished with a 160-cm column of Amberlyst XN-1005 in the silver form.[10] A sample of about 0.6 mmol each of methyl oleate and methyl elaidate was eluted with methanol. The silver ion forms complexes with the double bond (see p. 172). Since *cis-*compounds form more stable complexes than their *trans-*isomers, the elaidic ester was eluted first.

Cassidy and Streuli[11] studied the behavior of acetanilide and its derivatives on Amberlyst 15 and Dowex 50W-X2. With methanol, acetonitrile or their mixtures as eluents, they were unable to obtain satisfactory separations with Dowex 50-X2. With Amberlyst 15 they obtained a nearly quantitative separation of acetanilide from its *N*-n-propyl derivative, using 10 volume % methanol in acetonitrile as eluent.

A.III.a. *Summary*

The foregoing examples illustrate the superiority of porous resins over conventional resins in chromatographic separations with nonaqueous or largely nonaqueous eluents. The porous resins may also have an advantage in the separation of solutes of large molecular (or ionic) dimensions even with aqueous eluents. On the other hand, the macroreticular resins usually have values of P about one or less; hence they give wide graphs and require large values of $\Delta \log C$ for satisfactory separations.

B. Resins with Interpenetrating Polymer Networks

Millar[12] prepared beads of copolymerized styrene and DVB by the conventional method. After drying the beads, he soaked them in a mixture of styrene, DVB, and catalyst of the same composition as that used for the preparation of the beads. They absorbed the hydrocarbon mixture and catalyst. Then he suspended them in water and heated to 70° to induce copolymerization of the absorbed hydrocarbons. Thus a secondary copolymer network was formed interpenetrating the primary network. He repeated the procedure to produce a tertiary interpenetrating network. With the formation of each additional interpenetrating network, the density of the dry beads increased and their ability to absorb toluene decreased. These changes in properties are illustrated in Table 10.6.

Millar *et al.*[13] also prepared a resin with two interpenetrating networks each containing 4.5% DVB. They sulphonated this resin and compared the properties of the resulting cation-exchange resin with those of three cation-exchange resins prepared from a single hydrocarbon network with various percentages of DVB. These properties are summarized in Table 10.7. In regard to swelling properties (column 3), the intermeshed exchanger resembles most closely the single-network resin with 7% DVB. Its kinetic behavior, as judged from the interdiffusion coefficient, is closest to that of a single-network resin with 4% DVB. The selectivity coefficients of the intermeshed exchanger resemble most closely those of an ordinary resin with much higher DVB content.

TABLE 10.6. PROPERTIES OF HYDROCARBON COPOLYMERS WITH INTERPENETRATING NETWORKS

DVB (%)	One network		Two networks		Three networks	
	Density (g/ml)	Absorption of toluene (g/g)	Density (g/ml)	Absorption of toluene (g/g)	Density (g/ml)	Absorption of toluene (g/g)
1	1.0491	3.38	1.0507	2.44	1.0541	1.57
2	1.0494	1.93	1.0512	1.33	1.0544	0.79
4	1.0494	1.17	1.0512	0.73	1.0561	0.44
7	1.0500	0.74	1.0531	0.44		
10	1.0520					

(Data used by permission from *Journal of the Chemical Society*.)

The foregoing comparison indicates that the exchangers with interpenetrating networks should be very useful in chromatography. Although the secondary network has very little adverse effect on the kinetic properties, it decreases the undesirable swelling and shrinking and causes a marked increase in selectivity coefficients. It is unfortunate that resins with interpenetrating networks are not commercially available.

TABLE 10.7. PROPERTIES OF A CATION-EXCHANGE RESIN WITH INTERPENETRATING NETWORKS

No. of networks	DVB (%)	Absorption of water (g/g)	$\bar{D} \times 10^6$ for Na–H exchange	Log K_H^{Na} at indicated [NaR]			
				0.27	0.40	0.60	0.80
1	4.0	1.74	8.2				
1	7.0	1.04	5.9	0.15	0.17	0.18	0.15
1	15.0	0.60	2.5	0.39	0.37	0.28	0.15
2	4.5	1.08	7.3	0.42	0.32	0.25	0.22

(Data used by permission from *Journal of the Chemical Society*.)

C. Ion-retardation Resins

Retardion 11-A-8, marketed by the Dow Chemical Co., is an example of an ion-retardation resin. It is prepared[14] by immersing the chloride form of Dowex 1-X8 in an aqueous solution of acrylic acid until the latter penetrates into the resin. Then persulphate is added to catalyze the formation of polyacrylic acid. The linear chains of polyacrylic acid, although not linked covalently to the polystyrene resin, are so enmeshed in the resin that they cannot be washed out. Finally, the polyacrylic acid on the surface of the beads is removed by washing with sodium hydroxide. The quantities of Dowex 1 and acrylic acid are chosen so that the final resin contains nearly equivalent quantities of the two ionic groups, —$CH_2NMe_3^+$ and —COO^-. The ion-retardation resin is also called a "snake-cage" resin because the "snakes" of polyacrylic acid can not escape from the "cage" of polystyrene resin.

These resins have several unusual and interesting properties. When the dry resin is immersed in water, it swells very little because the two types of ionic groups neutralize each other and do not become extensively hydrated. However, when sodium chloride or other salt is added to the system, the close association of the carboxylic and tetra-alkylammonium ions is broken. The former associates with sodium or other cation, the latter with chloride or other anion. Then the resin swells. The absorption of salt can be reversed by passing water through the resin bed. Then the resin shrinks.

If a mixture of salt and a nonelectrolyte is eluted with water through such a resin, a separation is achieved; the nonelectrolyte emerges first because the mobile ions are "retarded" by the electrostatic attraction between them and the ionic groups of the resin. It should be recalled that ion exclusion (p. 175) may be used for the same purpose but that the order of emergence is different in the two procedures. Ion retardation has been found[15] to be more efficient in the industrial separation of glycerol from sodium chloride as obtained in the manufacture of soap. Ion retardation is particularly useful in the separation of salts from nonelectrolytes of large molecular size; in such a case, ion exclusion is ineffective as a method of separation because the large molecules are also excluded by the resin.

The fixed ions of Retardion 11-A-8 retain their characteristic selectivities toward the appropriate counter ions. For example, if a mixture of sodium chloride, bromide, and iodide is eluted with water through Retardion, the salts emerge in the order chloride, bromide, and, finally, iodide. This has an advantage over the separation of these anions by ion-exchange chromatography because in the latter procedure the halides are mixed with the eluent salt. Another example of the separation of a salt mixture by ion retardation is the elution of potassium, sodium, and lithium chloride. The salts emerge in the order of the increasing selectivity of a carboxylic resin toward these cations, i.e. potassium first and lithium last.

If equivalent amounts of lithium chloride and potassium acetate are dissolved in water and if the solution is then evaporated, the first of the four possible salts to crystallize will be potassium chloride and lithium acetate, these being less soluble than lithium chloride and potassium acetate.   Thus a metathesis of lithium chloride and potassium acetate can be accomplished very simply by mixing and partial evaporation. It follows that lithium acetate and potassium chloride can not be converted to potassium acetate and lithium chloride by the same procedure. This metathesis can be achieved by dissolving equivalent amounts of these salts in water and eluting with water through Retardion 11-A-8. The faster-moving cation and anion in the column are the potassium and acetate ions. Hence potassium acetate emerges from the column before lithium chloride.

It is possible to prepare many other snake-cage resins. For example, if Dowex 50 is permitted to absorb vinylbenzyltrimethylammonium chloride, $CH_2 = CH-C_6H_4CH_2-NMe_3Cl$, which is then polymerized, the product will be a snake of polyvinylbenzyltrimethylammonium chloride in a cage of Dowex 50.

Ion retardation resins have received much more attention from industrial chemists than from analysts. The following procedure[16] for the separation of amino acids and polypeptides from sodium chloride indicates the potential usefulness of these resins in the analytical laboratory. Some gelatine was partly hydrolyzed by boiling 2 hr in a solution of sodium chloride and hydrochloric acid. After neutralization of the acid with sodium hydroxide a 5-ml sample of the solution containing 50 mg of hydrolysis products and 5 mmol of sodium chloride was eluted through a column, 41 cm × 1.02 cm², of AG 11-A-8. A quantitative separation of the hydrolysis products ($C = 1.06$) from the sodium chloride ($C = 2.2$) was obtained.

D. Ion-exchange Membranes

D.I. PREPARATION OF ION-EXCHANGE MEMBRANES

Phenolsulphonic acid and formaldehyde can be condensed in a wide, shallow vessel to give a homogeneous cation-exchange membrane containing both sulphonate and phenolic hydroxyl groups. The presence of two ionogenic groups in the membrane makes it less desirable for some purposes than a membrane of sulphonated crosslinked polystyrene. On the other hand, the sulphonation of a membrane of crosslinked polystyrene generally yields a product that is too easily torn or cracked to be satisfactory. Analogously, homogeneous anion-exchange membranes can be made by the condensation of amines with formaldehyde, but the treatment of a membrane of crosslinked polystyrene with chloromethyl ether and then with a tertiary amine fails to yield a satisfactory anion-exchange membrane.

Most ion-exchange membranes are *mixtures*, either homogeneous or heterogeneous, of an ion-exchange substance and an inert, more plastic resin. For example, a mixture of linear polystyrene-sulphonic acid and a copolymer of vinyl chloride and acrylonitrile is dissolved in a common solvent. Then the solvent is evaporated so as to leave a thin homogeneous film of the two polymers. Apparently the chains of the two polymers are so intimately intermeshed that subsequent treatment with water fails to dissolve the sulphonated linear polymer. Homogeneous anion-exchange membranes may be made analogously.

Heterogeneous ion-exchange membranes may be prepared by mixing finely ground ion-exchange resin of any type with an inert material such as polyethylene and then shaping the mixture into a film of the desired thickness (0.1–0.6 mm) by heat and pressure. The fraction of ion-exchange resin in the mixture must be high enough so that it is possible for an ion to move from one surface of the membrane to the other either by diffusion or migration in an electric field; indeed, there must be very many paths by which this motion can occur. This requires that particles of ion-exchange resin be exposed on both surfaces and that the internal particles touch several neighbouring particles. On the other hand, too large a fraction of ion-exchange resin in the mixture leads to a brittle membrane.

Ion-exchange membranes can also be made of inorganic exchangers such as clay, but only organic membranes will be discussed in this book.

D.II. PROPERTIES OF ION-EXCHANGE MEMBRANES

D.II.a. *Membrane Potentials*

Since ion-exchange membranes consist in whole or in large measure of ion-exchange material, they resemble ion-exchange resins in many of their properties. When a dry ion-exchange membrane is immersed in water it swells by imbibition. In a solution of an electrolyte, it swells less than in water. If the solution contains a cation (anion) other than the counter ion in the cation-exchange (anion-exchange) membrane, ion exchange occurs between the solution and the membrane. If the cation or anion of the solution is the same as the exchangeable ion of the resin, Donnan invasion occurs. In general, the equilibrium condition in the absorption of water, in ion exchange, or in Donnan invasion, is reached more slowly in an ion-exchange membrane than in beads of resin because diffusion must occur over greater distances.

Let us consider a cation-exchange membrane in the potassium form separating two solutions of potassium chloride of different concentrations. Inside the membrane there will be water, potassium ions, and chloride ions. The concentration of chloride ions in the internal water is much less than that of the potassium ions. Water will diffuse slowly through the membrane under the influence of osmotic and/or hydrostatic pressure. Usually the latter is small, and osmotic pressure drives the water from the dilute to the concentrated solution. The potassium and chloride ions diffuse from the concentrated to the dilute solution. Because of the Donnan principle, there will be more potassium ions than chloride ions inside the membrane. Also the mobility of the potassium ions inside the membrane is greater than that of the chloride ions. Hence, there is a tendency for more potassium ions than chloride ions to diffuse through the resin. This would violate the principle of electroneutrality. Actually the potassium ions do diffuse in slightly greater quantity and thus create a potential difference between the solutions, the dilute solution being more positive. The potential difference, in turn, increases the diffusion of chloride ions and decreases the diffusion of potassium ions, thus avoiding an appreciable violation of electroneutrality.

An accurate equation for the potential difference between the two solutions is very complicated, taking into account such considerations as the transport of water through the membrane, the hydrostatic pressure, the difference in osmotic pressure, and the concentration gradients of potassium chloride inside the resin. An approximate form of the equation is satisfactory for our purpose.

$$E_M = E_A + (2t - 1)\,\frac{2.303\,RT}{zF}\,\log\frac{a_2}{a_1}. \tag{10.1}$$

This approximation can be derived from the equations of Meyer and Sievers,[17] Teorell,[18] or Helfferich.[19] Equation (10.1) is the general equation for the potential difference between two solutions separated by a cation-exchange membrane if each solution contains only one species of cation, the same as the counterion of the membrane. E_M is the membrane potential in millivolts, i.e. the potential difference between the two solutions; E_A is the asymmetry potential characteristic of each individual membrane and analogous to the asymmetry potential of the glass electrode; t is the transport number of the cation inside of the membrane; z is the valence of the cation; T is the absolute temperature; and a_1 and a_2 are the activities of the cation in the two solutions. The transport number t depends primarily on the ratio of mobile (i.e. not fixed) cations to anions in the resin. Since the concentration of anions inside the resin is governed by Donnan's principle, t approaches unity as the concentration of the external solution approaches zero. The value of t is greater in membranes of high crosslinkage than in resins of low crosslinkage. Hence for sufficiently dilute solutions and membranes of moderate or high crosslinkage, eqn. (10.1) can be simplified to

$$E_M = E_A + \frac{2.303\,RT}{zF}\,\log\frac{a_2}{a_1}. \tag{10.2}$$

If two cations are present in the solution on one side of the membrane and only one of these cations on the other side, the equation is more complicated. In the case of a solution with hydrogen and sodium ions on one side and only hydrogen ions on the other, the approximate equation becomes[20]

$$E_M = E_A + 0.1983T\,\log\frac{(\mathrm{H}^+)_2\,\bar{D}_\mathrm{H}\,\bar{\gamma}_\mathrm{Na} + (\mathrm{Na}^+)_2\,\bar{D}_\mathrm{Na}\,\bar{\gamma}_\mathrm{H}}{(\mathrm{H}^+)_1\,\bar{D}_\mathrm{H}\,\bar{\gamma}_\mathrm{Na}}, \tag{10.3}$$

where $\bar{D}$ represents the diffusion coefficients inside the membrane, $\bar{\gamma}$ represents the activity coefficients inside the membrane, and the numerical subscripts refer to the two solutions.

Equations (10.1) and (10.2) are also applicable to anion-exchange membranes. Then t, z, and a refer to the anion, and the more dilute solution is negative with respect to the other.

D.II.b. *Permselectivity*

Let us consider an experiment in which a cation-exchange membrane and an anion-exchange membrane separate three electrolytic solutions. Electrodes are inserted in the two end solutions and an e.m.f. is applied (Fig. 10.3) sufficient to cause electrochemical oxidation and reduction at the electrodes. The electric field drives anions to the left and cations to the right. The cations from the central compartment migrate to the cation-exchange membrane, enter the membrane, and eventually enter the cathode compartment. Analogously anions migrate from the central compartment through the anion-exchange membrane into the anode compartment. The anions of the cathode compartment tend to migrate to the left.

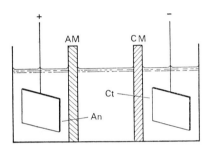

Fig. 10.3. Simple cell for the deionization of a solution. AM = anion-exchange membrane; CM = cation-exchange membrane.

However, since the Donnan forces maintain a small concentration of anions in the cation-exchange membrane, comparatively few anions migrate through this membrane. Analogously, few cations migrate from the anode compartment to the central compartment. Thus the effect of the current is to decrease the concentration of electrolyte in the central compartment. The principle can be used to remove salts from solutions of proteins.

D.III. APPLICATIONS OF ION-EXCHANGE MEMBRANES
D.III.a. *Determination of Ion Activities*

Early work on the determination of the activity or concentration of ions by the measurement of membrane potentials was done by Marshall[21] with clay membranes. Since membranes of ion-exchange resins are more satisfactory, only the latter will be discussed in this book.

The accuracy and limitations of eqn. (10.2) are illustrated by the work of Basu.[22] He prepared membranes by incorporating three commercial anion-exchange resins in an inert plastic. He rejected all those membranes for which the asymmetry potential was not zero. Then he set up cells of the type

$$\text{Hg} + \text{Hg}_2\text{Cl}_2 \mid \text{sat'd KCl} \parallel \text{sol'n A} \mid \text{Membrane} \mid \text{sol'n B} \parallel \text{sat'd KCl} \mid \text{Hg}_2\text{Cl}_2 + \text{Hg}$$

The e.m.f. of this cell is equal to the membrane potential plus the sum of the liquid-junction potentials, which is conventionally taken as zero. He prepared a series of solutions of sodium or potassium nitrate, chloride, bromide, bromate, and iodate of such concentrations that the mean activities of the two solutions stood in a ratio of 3.00.

Table 10.8 summarizes the results. The entries in the columns designated E_M were the average values obtained with each of the three types of membranes. Basu fails to state the temperature of the systems. However, he states that the theoretical value of E_M was 28.1 mV. From eqn. (10.2) it can then be calculated that the temperature was 24°.

Nitrate and chloride ions seem to follow the equation better than the other anions, but the differences between the observed and theoretical potentials exceed the experimental errors with all anions if one of the solutions is 0.026 M or more.

TABLE 10.8. MEMBRANE POTENTIALS

Larger activity	NO_3^-		Cl^-		Br^-		BrO_3^-		IO_3^-	
	E_M	SD	E_M	SD	E_M	SD	E_M	SD	E_M	SD
0.0029	28.4	0.5	28.0	0.4	27.4	0.4	27.8	0.2	27.4	0.5
0.0087	28.0	0.1	28.0	0.4	27.7	0.3	27.8	0.7	26.8	0.3
0.026	27.6	0.1	27.4	0.4	27.3	0.2	26.5	0.5	26.0	0.2
0.079	26.5	0.4	27.5	0.5	26.5	0.5	25.4	0.5	24.8	0.9

(Data used by permission of the Indian Chemical Society.)   SD = standard deviation.

On the other hand, Schindewolf and Bonhoeffer[23] found that a membrane of poly-phenolsulphonic acid gave the theoretical potential for hydrogen ion up to 1 M, for sodium ion up to 0.1 M, and for barium ion up to 0.03 M.

Within the limitations noted above, eqn. (10.2) can be used to determine the activity or concentration of an anion (or cation) in an unknown solution that contains only one anion (or cation). At first thought, this last restriction would seem to rob eqn. (10.2) of all of its usefulness in analytical chemistry. Obviously, there is no advantage in determining the concentration of a solution of pure sodium nitrate by the use of either a cation-exchange membrane or an anion-exchange membrane; the concentration could be determined more accurately and with less trouble by any one of several classical analytical procedures.

Nevertheless, ion-exchange membranes have been used to determine ion concentrations or activities in cases where no other method was available. For example, solutions of the linear condensed phosphates, $Na_{n+2}P_nO_{3n+1}$, are not completely ionized. Schindewolf and Bonhoeffer[23] prepared a series of these solutions with the mean degree of polymerization n, varying from 2 to 10,000. Each solution was only 0.01 M with total sodium so that eqn. (10.2) was applicable. They measured the membrane potentials of these solutions against 0.0100 M sodium chloride. The results indicated that the ratio of ionized sodium to total sodium fell from 0.9 when $n = 2$ to 0.22 when $n = 160$.

Carr[24] has used cation-exchange membranes with solutions of pure salts of sodium, potassium, magnesium, calcium, strontium, or barium on one side of the membrane and the same salt plus protein on the other side to measure the extent to which various proteins form complexes with these cations.

D.III.b. *Membranes as Indicator Electrodes in Titrations*

Sinha[25] separated two solutions of hydrochloric acid by a cation-exchange membrane and then titrated one of the solutions with sodium hydroxide, measuring the membrane potential at intervals during the titration. Equation (10.2) is not applicable because two kinds of cations are present in the titrated solution up to the equivalence point. Nevertheless, the equivalence point is marked by a jump in the titration graph. This type of titration is not recommended because it is more troublesome and less accurate than the use of a glass electrode or an indicator. Whereas the pH changed during the titration from about 1 to 12 corresponding to a change in e.m.f. of a glass electrode of about 650 mV, the change in the membrane potential was only about 55 mV. Sinha[25] also titrated sodium chloride with silver nitrate using an anion-exchange membrane. This titration suffers from the same disadvantages as the acid–base titration with the cation-exchange membrane.

D.III.c. *Membranes as Monitors in Chromatography*

Spencer and Lindstrom[26] used a cell with a cation-exchange membrane to monitor the effluent in the separation of sodium from potassium by ion-exchange chromatography. The eluent, either 0.1 M or 0.4 M hydrochloric acid, flowed through one compartment of the small cell before entering the column. The eluate flowed through the other compartment. Two silver–silver chloride electrodes, one in each compartment of the cell, were connected to a recording potentiometer. Thus graphs of e.m.f. vs. time were plotted automatically.

Since the ionic strength of the eluate is the same as that of the eluent, concentrations may be substituted for activities in eqn. (10.3). If the further assumption is made that $\bar{\gamma}_{Na} = \bar{\gamma}_{H}$, the equation becomes

$$E_M - E_A = 0.1983T \log \frac{\bar{D}_H[H^+]_1}{\bar{D}_H [H^+]_2 + \bar{D}_{Na} [Na^+]_2}, \tag{10.4}$$

where the subscripts 1 and 2 refer to the eluent and eluate, respectively. Equation (10.4) indicates that $E_M - E_A$ will be zero when the eluate contains neither sodium nor potassium. Furthermore, if $[H^+]_2 > 100[Na^+]_2$, as is true in a properly conducted elution, the value of $E_M - E_A$ varies linearly with $[Na^+]_2$ within the experimental error. This same statement is true of potassium. Therefore if a constant flow rate is maintained, the automatically plotted graph can be interpreted as a graph of the concentration of sodium (or potassium) ion vs. the volume of eluate, i.e. as an elution curve. Furthermore, the area under the sodium graph should be proportional to the quantity of sodium in the sample. Similarly, the area under the potassium graph should be proportional to the amount of potassium; the proportionality factor of potassium will differ from that of sodium because $\bar{D}_{Na} \neq \bar{D}_K$. The maximum value of $E_M - E_A$ was only 4.4 mV.

The authors[26] show the graphs of three separations. All of the six curves have the typical Gaussian shape. On the other hand, the proportionality factors, area/quantity, show relative standard deviations of about 10% for both sodium and potassium.

Rothbart and Lindsay[27] have used an anion-exchange membrane to monitor the eluate in the elution of condensed phosphates with potassium chloride. They pass the eluate through one compartment of the cell and a stream of potassium chloride identical in composition to the eluent through the other compartment. This stream flows into the sink. They believe that this arrangement is better than that of Spencer and Lindstrom[26] because in the latter case there exists the possibility that sodium or potassium ions might diffuse through the membrane, enter the eluent, and be recycled through the column. Rothbart[27] also found a poor constancy in the ratio of area/quantity, caused, in his opinion, by drifts in the base line of the graph.

It is obvious that the automatically plotted graph can not be used for an accurate quantitative determination of any constituent unless the ratio of area to quantity can be made constant. Nevertheless, monitoring with membranes may be used in the development of methods of chromatographic separation by enabling the chemist to know at a glance the $U*$ values of the various sample constituents and whether the several curves overlap. It may also be useful in quantitative analysis by indicating which fractions of eluate may be discarded and which should be subjected to quantitative determination.

D.III.d. *Separation of Amino Acids into Groups*

Hara[28] set up the cell

| + | 1 | 2 | 3 | 4 | 5 | — |

Pt | Na₂SO₄ | AM | NaCl | CM | amino acids | AM | NaCl | CM | Na₂SO₄ | Pt

$$\text{Pt} \mid \text{Na}_2\text{SO}_4 \mid \text{AM} \mid \text{NaCl} \mid \text{CM} \mid \text{amino acids} \mid \text{AM} \mid \text{NaCl} \mid \text{CM} \mid \text{Na}_2\text{SO}_4 \mid \text{Pt}$$

The amino acids, dissolved in buffer solution of pH 5.6, were placed in compartment 3. After 2 hr of electrolysis, each of the basic amino acids (arginine, histidine, and lysine) had migrated quantitatively into compartment 2. Similarly, each of the acidic amino acids (glutamic and aspartic) had migrated quantitatively to compartment 4. The neutral amino acids (glycine, alanine, leusine, isoleusine, valine, methionine, proline, serine, threonine, and phenylglycine) remained for the most part in compartment 3; but percentages of the compounds varying from 5% to 16% had migrated into compartment 2 while 4–10% had migrated into compartment 4.

D.III.e. *Coulometric Titrations*

Feldberg and Bricker[29] used a cation-exchange membrane to separate the cathode and anode compartments in the coulometric titration of bases.

The anode compartment consisted of a polyethylene bottle containing 100 ml of 0.25 M potassium sulphate (the supporting electrolyte). Into this they immersed the glass and calomel indicating electrodes, a platinum generating anode, and a tube which served as the cathode compartment. This tube contained 15 ml of 0.25 M potassium sulphate and the generating cathode. A cation-exchange membrane at the bottom of this tube separated the two solutions. After adjusting the pH of the cathode compartment to that of the equivalence point (7.6 in the titration of sodium hydroxide), they added the sample of base and passed current through the generating circuit until the pH of the equivalence point was reached again.

When current is passed through the generating circuit, hydrogen ions are liberated at the anode

$$2H_2O \rightarrow 4H^+ + O_2 + 4e$$

Electroneutrality is maintained in the anode compartment by the migration of potassium ions from the anode compartment through the membrane into the cathode compartment. The quantity of hydroxide ion passing through the membrane is negligible if the concentration of supporting electrolyte is sufficient. The quantity of hydrogen ion migrating through the membrane is also insignificant provided that the anode solution is well stirred and also that its pH does not drop below 6.

The relative mean deviation in the titration of fourteen samples of sodium hydroxide was only 0.16% with no bias toward high or low results. Titrations of sodium carbonate in aqueous solutions could be carried only up to the first equivalence point (NaHCO₃). As the second equivalence point (H₂CO₃) is approached, significant quantities of hydrogen ion migrate through the membrane. Nevertheless, an average relative error of only -0.28% was encountered when the supporting electrolyte was 0.5 M sodium perchlorate in methanol.

They also tried to determine acids by an analogous procedure using a membrane containing fixed quaternary ammonium ions. These experiments failed because of the presence of some weak-base groups in the strong-base resin.

Ho and Marsh[30] elaborated the method of Feldberg and Bricker[29] by providing a recorder to draw the titration curve and a control to stop the generating current at the equivalence point. With samples of 1.0 ml of sodium hydroxide, their accuracy and precision are even better than those of Feldberg and Bricker.

It should be noted that the foregoing authors[29, 30] generated the titrant H^+ at the anode and that they used the permselective membrane merely to avoid the escape of either hydroxide or hydrogen ions from the anolyte. Hanselman and Rogers,[31] on the other hand, used ion-exchange membranes quite differently in the coulometric titration of silver nitrate with halides. They put the sample of silver nitrate into the anode compartment, which contained either sodium perchlorate or sodium naphthalene sulphonate as the supporting electrolyte and an inert electrode. The catholyte was a sodium halide. An anion-exchange membrane separated the two solutions. Ideally, when the current flows through the generating circuit, one equivalent of halide anion should enter the anolyte via the membrane for every faraday of electricity. This represents a current efficiency of 100%. The observed current efficiencies varied from 96.0% to 97.6% when 0.5 M sodium chloride was the catholyte. The low results are caused by the imperfect permselectivity of the membrane. That is, an appreciable part of the current was carried through the membrane by cations. With 0.5 M sodium iodide at catholyte, the current efficiencies were between 50.5% and 62.7%. Presumably, iodide ions formed ion pairs with the fixed quaternary ammonium ions of the membrane, thus decreasing tremendously the permselectivity. With 0.005 M or 0.01 M sodium iodide as catholyte, the current efficiencies varied between 94.9% and 101.3%. The values above 100% are caused by the reaction

$$RI + ClO_4^- \text{ (anolyte)} \rightarrow RClO_4 + I^- \text{ (anolyte)}$$

It is obvious that this technique will have to be greatly improved before it can be used for accurate titrations.

D.III.f. *Continuous Electrochromatography*

Caplan[32] modified the usual apparatus for continuous electrochromatography and applied it to the quantitative analysis of mixtures of the alkali metals. Instead of the usual thick sheet of filter paper he used a pack of 13 sheets of thin filter paper (0.18 mm) and 12 cation-exchange membranes (0.10 mm), alternately located. The membranes were in the dimethylammonium form. Platinum wires, extending vertically along opposite edges of the pack, served as electrodes, the anode being on the left. A slow stream of 0.1 M ammonia was fed to the upper left-hand corner of the pack and flowed from a large lead-off tab in the lower left-hand corner into a small beaker. A similar stream of 0.1 M hydrochloric acid entered the pack at the upper right-hand corner and left via a large lead-off tab in the lower right-hand corner. The eluent, 0.014 M uramildiacetic acid adjusted to a pH of 9.5 with dimethylamine, was fed along the upper edge of the pack and dropped from 22 smaller lead-off tabs into 11 beakers. These tabs were equally spaced between the two larger ones. Two adjacent tabs fed into one beaker. The receiving beakers were numbered 1 to 13, starting at the left. The sample, dissolved in eluent, was fed to the pack at a point on the upper edge directly above the beaker number 3.

In one experiment a solution containing 0.50 meq each of lithium, sodium, and potassium entered the pack during 8 hr. The flow of eluent and electric current was continued for another 16 hr so that all of the alkali-metal cations were eluted into the receiving beakers. Then the amount of metal in each beaker was determined by flame photometry. The results are given in Table 10.9.

The lithium was almost entirely converted by uramildiacetate into an uncharged complex and did not migrate either to the left or to the right while being eluted down the pack. Sodium, on the other hand, migrated toward the cathode so that most of it was collected in beakers 6 and 7. This indicates that the formation of the uncharged sodium complex was incomplete and reservable. The potassium ion was complexed to a smaller extent since it moved further toward the cathode.

TABLE 10.9. SEPARATION OF THE ALKALI METALS
BY CONTINUOUS ELECTROCHROMATOGRAPHY

Beaker number	Alkali metals found (meq)		
	Li	Na	K
1	0.00		
2	.00		
3	.54		
4	.00	0.00	
5		.01	
6		.24	
7		.24	0.00
8		.00	0.01
9			.03
10			.24
11			.21
12			.07
13			.00
Total	0.54	0.49	0.56

(Data used by permission of the Electrochemical Society.)

The inclusion of the cation-exchange membranes in the pack gives a better separation of the metals for two reasons. In the first place, a considerable fraction of the alkali metals (lithium excepted) that are in the pack at any given moment during the experiment are located in the resin membranes. Here they are not subjected to the downward flow of the eluent. Hence, the metals are eluted more slowly and are subjected to the separating influence of the electric field for longer periods. In the second place, the membranes improve the efficiency of the current as a separating agent. In the paper, the current is carried by both cations and anions; but in the membranes the current is carried almost entirely by the cations. A mathematical treatment of the theory of the separation is given in the paper.[32]

D.III.g. *X-ray Emission Spectrography*

The use of ion-exchange resins for concentrating trace constituents from solution was discussed on pp. 82–85. Ion-exchange membranes may also be used for this purpose and have the added advantage that the collected ions are in an ideal condition for determination by X-ray emission spectrography.[33]

Zemany et al.[34] applied this method to the determination of potassium extracted from mica. In preliminary work they suspended hydrogen-form cation-exchange membranes $22 \times 35 \times 0.088$ mm, in 75 ml of water containing known amounts of potassium chloride. After drying the membranes in a vacuum desiccator at room temperature, they mounted them in the X-ray apparatus and counted the number of X-rays emitted by the potassium. Plots of counts per second against the quantity of potassium chloride taken were linear provided that this quantity did not exceed 2 μmols. Although the total capacity of the membranes was 58 μeq, the exchange was incomplete with more than 2 μeq of potassium. Then, they applied the same method to 75 ml of water in which 1-g samples of mica, 40–60 mesh, had been soaked.

E. Inorganic Ion Exchangers

As discussed in Chapter 1, the first ion exchangers studied and used were inorganic compounds, the natural zeolites or synthetic compounds of similar composition. Soon after the first synthesis of ion-exchange resins in 1935, the organic exchangers largely displaced the zeolites both in industrial and analytical applications and as objects of scientific investigation. Interest in nuclear power shortly after the midpoint of the twentieth century focused attention on two weaknesses of the ion-exchange resins and caused a revival of interest in inorganic exchangers. The weaknesses of the resinous exchangers have already been mentioned (pp. 16–20). (1) The instability at high temperatures of the hydrogen form of strong-acid cation-exchange resins and (more importantly) of the hydroxide form of strong-base anion-exchange resins limits the use of these resins in the removal of radioactive ions from the water used as coolant or as a heat-transfering agent in nuclear reactors. (2) Decomposition of the resins by radiation limits the aforementioned application and also the use of resins in the treatment of fission products, whether for the disposal of the radioactive wastes or for the recovery of radioisotopes.

Dissatisfaction with both the resinous and old inorganic exchangers led to a search for new and better inorganic ion-exchange compounds. Several have been found that are unaffected by the intensities of radiation to which they are exposed in nuclear work. In addition, most of them have a very large selectivity coefficient for cesium, a very useful property in nuclear laboratories. Also, since they swell and shrink with changes in the external solutions much less than the resinous exchangers, they furnish a method for the theoretical study of ion-exchange equilibrium where the swelling term is very small. On the other hand, they undergo exchange reactions more slowly than the resins and have a much smaller capacity.

An excellent survey of the modern inorganic exchangers is available.[35]

E.I. HYDROUS OXIDES

The hydrous oxides of chromium(III), zirconium(IV), tin(IV), and thorium(IV) are prepared[36] by adding ammonia to the appropriate salt solution, filtering or centrifuging the precipitate, drying at room temperature, grinding, and sieving.

A very interesting comparison between ion-exchange resins and hydrous oxides is given by Kraus.[36] The latter are polymers crosslinked by oxygen bonds between the metal atoms. In acid medium, the hydroxide groups are partly ionized. Thus the polymer network becomes positively charged and functions as a weak-base anion exchanger. The exchange capacity increases with decreasing pH. In basic solution, the polymer acquires a negative charge by the sorption of hydroxide ions and functions as a weak-acid cation exchanger, the capacity increasing with increasing pH. For many hydrous oxides, the pH limits of the anion-exchange and cation-exchange regions overlap. Hydrous zirconium oxide has appreciable exchange capacity for both cations and anions between pH 5 and 8. For hydrous stannic oxide, the limits are 4 and 7. Each type of hydrous oxide has its own characteristic isoelectric point where the exchange capacities for cations and anions are equal. This value is 6.7 for zirconium oxide and 4.8 for stannic oxide.

Hydrous oxides and ion-exchange resins are both water-swollen gels. In both gels, the exchanging ions diffuse through the imbibed water to and from the exchange sites. The rate of exchange is therefore strongly dependent on the degree of hydration of the gel. The resins may be dried almost completely and rehydrated with practically no loss of exchange capacity. Hydrous oxides in general require higher temperatures than the resins to reach any given degree of dehydration. Furthermore, complete or nearly complete dehydration of the hydrous oxides is an irreversible process and causes almost complete loss of exchange properties.

Maeck et al.[37] determined the D values of nearly all the metals between hydrous zirconium oxide and nitrate solutions of various pH values. From the results, they predict that several chromatographic separations should be possible with this exchanger. Among them is the separation of lead from germanium, tin, arsenic, antimony, bismuth, selenium, and polonium.

The ion-exchange properties of silica gel have also been studied. Ahrland et al.[38] found that three commercial samples gave identical results within the experimental error in equilibrium experiments. In a titration with sodium hydroxide, the gel behaved like a resin with very weakly acidic groups such as those of phenol. The pH rose rapidly at first, reaching a value of 11 when 0.8 mmol of base had been added per g of gel. However, 2.2 meq/g were required to raise the pH to 12. A plot of log D of uranyl ion vs. pH was linear but had a slope of 1.6 instead of 2.0. The authors attribute this discrepancy to the invasion of the gel by the hydrochloric acid, and hence a change in the internal activity coefficients. They compared the liberation of hydrogen ion from the gel (or loss of hydroxide ion from the solution) with the sorption of sodium ion when the gel was equilibrated with sodium hydroxide. Within the limits of 0.07–5 meq of sodium hydroxide to 1 g of gel, the molar ratio of hydrogen ion liberated to sodium ion sorbed was 0.98 with a standard deviation of 0.02. These facts are in accord with the hypothesis that the gel functions as a weak-acid cation exchanger.

Kinetic experiments showed that the exchanges between the hydrogen-form gel and uranyl, gadolinium, calcium, or sodium ions were virtually complete in 5 min or less. On the other hand, ions that undergo hydrolysis and polymerization in dilute acid solutions reach equilibrium much more slowly, presumably because the polymerized ions diffuse slowly through the gel. The exchange is particularly slow with aged solutions or solutions of low acid concentration. For example, a freshly prepared solution of plutonium(IV) in 0.100 M nitric acid reached 79% of the equilibrium exchange in 30 min whereas a similar solution 8 days old reached only 67% under similar conditions. In 0.050 M nitric acid, the respective figures are 74% and 48%.

The same authors[39] described the separation of zirconium, plutonium, and uranium by means of silica gel.

They passed 50 ml of 4 M nitric acid containing these elements into a column, 100 cm $\times$ 0.79 cm², at 3.5 cm/hr. This very slow flow rate was necessary because of the slow rate of exchange of the zirconium and plutonium. They washed the column with 4 M nitric acid. At this point most of the zirconium was in the column and all of the plutonium and uranium in the eluate. They removed the zirconium from the column with 0.1 M oxalic acid. They evaporated to dryness the solution containing the plutonium and uranium and redissolved it in 100 ml 0.1 M nitric acid. They passed this solution through the column again and washed with 0.1 M nitric acid. Now all the plutonium was on the column, also the small amount of zirconium that had gone into the first eluate. They removed the plutonium from the column with 4.5 M nitric acid, leaving the zirconium still in the gel. They adjusted the eluate containing the uranium to a pH of 3 with sodium hydroxide and passed this solution through the column. Thus the uranium was retained by the gel. Finally, they removed the uranium by treating the column with "strong acid".

They also describe a similar procedure for separating plutonium and uranium from each other and from long-lived fission products.

Although aluminum oxide prepared according to Brockman can hardly be considered a hydrous oxide, Nydahl and Gustafsson[40] found that it has an anion-exchange capacity of 0.17 meq/ml of column bed. They attribute this exchange to the structural group $(-AlO)_2Al^+An^-$. Nydahl[41] obtained excellent results in the determination of sulphur in steel after separating the sulphate from the iron and phosphate by means of a column of alumina.

E.II. SALTS OF MULTIVALENT METALS

E.II.a. Zirconium Phosphate

According to Kraus et al.,[36] if the hydrous oxide of phosphorus(V) were insoluble, it would have a very low isoelectric point and hence have a larger cation-exchange capacity and a smaller anion-exchange capacity than the hydrous oxides discussed in the previous section. In order to utilize the advantages of the nonexistent hydrous phosphorus oxide for cation exchange, it is possible to precipitate a gelatinous compound of the formula $xZrO_2 \cdot yP_2O_5 \cdot zH_2O$, i.e. a zirconium phosphate. As expected, this compound has good cation-exchange properties and very small anion-exchange capacity, even in alkaline solutions. In fact, 0.5 M sodium hydroxide changes the composition of the precipitate by extracting P_2O_5.

The composition and ion-exchange properties of zirconium phosphate depend on the method of preparing and drying the precipitate. Since there are almost as many procedures for preparing this compound as there are investigators in this field, it is not surprising that the literature contains some conflicting statements. Nevertheless, there is substantial agreement on the following points:

(1) The molar ratio of P/Zr in the exchanger is generally not the same as in the reagents from which it was prepared. That is, complete precipitation of both zirconium and phosphate does not ordinarily occur in the preparation of this exchanger.

(2) There is an upper limit of the molar ratio of P/Zr in a satisfactory exchanger. If this limit is exceeded, the gel loses phosphate ion on contact with water. The upper limit has been given as 1.67 by Baetslé and Pelsmaekers[42] and as 2.00 by Clearfield and Stynes.[43]

(3) A large molar ratio of P/Zr favors a large capacity.

(4) A large molar ratio of H_2O/Zr in the exchanger, i.e. a large degree of hydration, favors rapid exchange. In this respect, drying zirconium phosphate is analogous to increasing the crosslinkage in ion-exchange resins.[44] Zirconium phosphate resembles the hydrous oxides inasmuch as excessive dehydration is irreversible.

(5) To continue this analogy, high drying temperature sometimes increases the selectivity coefficients.

(6) The drying temperature has no effect on the capacity of the exchanger (in terms of milliequivalents per gram of anhydrous exchanger) unless the dehydration is so great as to prevent diffusion of ions to some of the exchange sites.

(7) Selectivity coefficients are generally greater on zirconium phosphate than on the resinous exchangers.

(8) The temperature coefficients of selectivity coefficients are much greater with zirconium phosphate than with resinous exchangers.

The foregoing statements will now be illustrated by data from the literature.

Baetslé and Pelsmaekers[42] prepared samples of zirconium phosphate by the dropwise addition of 500 ml of 2.2 M zirconyl chloride in 1 M hydrochloric acid to 1550 ml of vigorously stirred phosphoric acid of various concentrations in 6 M hydrochloric acid. After the precipitates had settled they determined zirconium and phosphorus in the supernatant liquids so that they could calculate the molar ratios of P/Zr in the unwashed precipitates. They washed the precipitates several times with water, dried them at 50°, and washed them again until the pH values of the wash water were between 3 and 4. Then they determined the molar ratio of P/Zr in each precipitate. They also determined the specific cation-exchange capacities of each precipitate at several pH values. The results are summarized in Table 10.10.

TABLE 10.10. COMPOSITION AND SPECIFIC CAPACITY OF SEVERAL SAMPLES OF ZIRCONIUM PHOSPHATE[42]

Sample number	Molar ratio of P/Zr			Specific cation-exchange capacity			
	In reagents	In unwashed ppts.	In washed ppts.	pH = 3.15	pH = 5.75	pH = 7.55	pH = 9.15
1	3.00	1.84	1.68	2.9	3.5	4.3	4.2
2	2.00	1.75	1.66	2.8	3.7	4.3	4.3
3	1.75	1.68	1.67	2.9	3.6	4.4	
4	1.50	1.48	1.50	2.7	3.3	4.1	4.2
5	1.25	1.24	1.25	1.8	2.6	3.3	3.8
6	1.00	1.04	1.00	1.2	2.0		3.5
7	0.90	0.94	0.94	0.47	1.4	2.5	3.3

The ratio of P/Zr in the washed precipitate No. 3 is 1.67 with the ratio in the reagents equal to 1.75. Larger ratios in the reagents (No. 1 and No. 2) give larger ratios in the unwashed precipitates but not in the washed precipitates. For any given pH, the capacity decreases with decreasing ratios of P/Zr in the washed precipitates. For any given sample of exchanger, the capacity increases with increasing pH; this behavior is characteristic of weak-acid cation exchangers.

Somewhat different results were obtained by Clearfield and Stynes.[43] They precipitated samples of zirconium phosphate and washed and dried them several days over calcium chloride or phosphorus pentoxide. (The original article must be consulted for the rather complicated details.) Then they refluxed the precipitates 1–91 hr in aqueous solutions of phosphoric acid, 2.6–10 M. This treatment increased the crystallinity of the precipitates and caused slight changes in their composition. The analysis of six samples of these exchangers indicated that the molar ratio of P/Zr was 2.01 with a standard deviation of 0.02. Heating at 110° caused a reversible loss of water corresponding to 0.97 ($\sigma = 0.03$) mole of water per mole of zirconium. Heating at 1000° caused an irreversible loss of water corresponding to 2.04 moles of water per mole of zirconium with a standard deviation of 0.08. After treatment with sodium chloride and sodium hydroxide, the sodium content of the exchanger corresponded to a molar ratio of Na/Zr equal to 2.01 with a standard deviation of 0.02. These facts point to the formula $Zr(HPO_4)_2 . H_2O$. The water of hydration is lost reversibly at 110°. At 1000°, zirconium pyrophosphate, ZrP_2O_7, is formed with the loss of ion-exchange properties. The theoretical capacity calculated from the formula $Zr(HPO_4)_2 . H_2O$ is 9.9 meq/g.

The effect of the degree of hydration of the exchanger on the rate of exchange is illustrated in Table 10.11. The investigators[45] mixed 1.0 g of hydrogen-form zirconium phosphate

TABLE 10.11. RATE OF SORPTION OF URANYL ION BY
ZIRCONIUM PHOSPHATE[45]

% H$_2$O in exchanger	Uranium absorbed (% of total uranium)			
	15 min	1 hr	4 hr	24 hr
50	98	99	99	99
10	82	94	96	99

(Used by permission from *Acta Chemica Scandinavica*.)

with 10 ml of a solution containing 0.1 mmol each of uranyl and hydrogen ions and determined the extent of the sorption of uranium at various time intervals. The mole ratio of P/Zr in the exchanger was 1.95, and the water content was controlled by a previous heating of the exchanger.

In spite of their slower exchange rates, samples of zirconium phosphate with small water content sometimes have an advantage in separation processes over the more highly hydrated samples. The reason is that the selectivity coefficients may be larger in the less hydrous exchangers. For example, Alberti and Conte[44] obtained a better chromatographic separation of strontium from cesium with an exchanger dried at 260° than with one dried at room temperature. Also, a gel dried at 260° gave a better separation of lithium from sodium than a gel dried at 50°.

Drying a sample of zirconium phosphate at moderate temperatures causes an increase in the specific exchange capacity if this is defined in terms of milliequivalents per gram of moist gel because the partial dehydration yields a gel with more "monomeric units" of $Zr(HPO_4)_2$ per gram. The capacity as calculated on the basis of the completely dry gel should not be affected by moderate drying. The use of higher temperatures in the dehydration causes a change in the gel structure that prevents diffusion to some of the otherwise exchangeable ions. In addition, orthophosphate groups condense to form pyrophosphate with loss of replaceable hydrogen ions and hence decrease in capacity. These facts are illustrated in Table 10.12. The apparent rise in the capacity on a dry basis between 50° and 260° may be due to experimental error.

TABLE 10.12. SPECIFIC EXCHANGE CAPACITY[44] OF A
ZIRCONIUM PHOSPHATE FOR POTASSIUM AT pH 2

Drying temp.	% H$_2$O in exchanger	Specific capacity	
		Wet basis	Dry basis
50°	13.9	0.75	0.87
260°	5.3	0.92	0.97
500°	2.4	0.40	0.41

(Used by permission of Elsevier Publishing Co.)

The separation of two (or more) ions by ion-exchange chromatography is facilitated by a large ratio of the two selectivity coefficients, vs. the exchanging ion of the eluent. In the case of the alkali metals, as discussed in Chapter 6, changes in eluent are not effective in changing the selectivity coefficients. It is still possible, however, to select an exchanger that gives a large ratio of the selectivity coefficients and hence of the C values. Since selectivity coefficients may change markedly with the mole fraction of any ion in the resin and since a small sample is taken in elution chromatography, the important values of the selectivity coefficients are those that apply when the exchanger is almost entirely the eluent-ion form. Table 10.13 shows the selectivity coefficients at 25° of rubidium and cesium ions vs. hydrogen ion for several exchangers when the mole fraction of hydrogen ion in the resin is almost 1.00. The values for zirconium phosphate were calculated from the data of Baetslé[46] and those for Dowex 50 are taken from Bonner.[47] These data illustrate the superiority of zirconium

TABLE 10.13. SOME SELECTIVITY COEFFICIENTS AT 25°

Exchanger	Dowex 50-X4	Dowex 50-X8	Dowex 50-X16	Zirconium phosphate
E_H	2.7	4.9	11.7	73
E_H^{Rb}	2.3	4.0	8.3	37
Ratio	1.17	1.22	1.41	1.97

(Used by permission from *Journal of Physical Chemistry*.)

phosphate for the separation of rubidium from cesium. It also has advantages in several other chromatographic separations.

Although changes in temperature have only a slight effect on the selectivity coefficients of ion-exchange resins, the exchange equilibria on zirconium phosphate are markedly dependent on temperature. For example, the selectivity coefficient of cesium vs. hydrogen under the conditions given in the last paragraph changes from 150 at 5° to 17 at 72°.

Kraus *et al.*[36] obtained a quantitative chromatographic separation of the alkali metals (Na, K, Cs) from the alkaline earths (Mg, Ca, Sr, Ba) with a column, 2.0 cm × 0.2 cm², of zirconium phosphate.

Osterrid[48] used this exchanger in the determination of traces of cesium in rocks. He dissolved a 1-g sample in hydrofluoric acid and expelled the excess by evaporation with perchloric acid. Then he passed the solution of the sample through an exchange column, $3.0 \text{ cm} \times 0.79 \text{ cm}^2$ and washed it with 150 ml of 1.1 M hydrochloric acid and 20 ml of water. Of all the cations and anions of the sample, only cesium remained on the column at this point. Since he found it impracticable, to elute the cesium from the column, he dissolved the exchanger in hydrofluoric acid, diluted the solution and passed it through a column of anion exchanger. This retained all the phosphate and zirconium (as ZrF_6^{2-}), leaving only hydrofluoric acid and cesium fluoride in the eluate. He finally determined the cesium by flame photometry.

Analysis of known solutions resembling the solutions of the rock always yielded a recovery of cesium greater than 90%, usually greater than 95%. He reports replicate analyses of six rocks with results ranging from 0.8 to 26 ppm of cesium. For the rocks with the small cesium contents, the standard deviations were between 0.07 and 0.5 ppm. For the rocks with larger cesium contents the relative standard deviations were between 6% and 8%.

Gal and Ruvarac[49] used zirconium phosphate to separate plutonium from uranium in fission products.

E.II.b. *Other Salts of Multivalent Metals*

Several other gelatinous salts in which titanium, tin, or antimony(V) are substituted for zirconium and antimonate, arsenate, molybdate, or tungstate are substituted for the phosphate have been studied as cation exchangers. Although some of these may be preferable to zirconium phosphate for specific purposes, they generally compare unfavorably with this exchanger in regard to exchange capacity and stability toward acids and bases. For example, Ahrland et al.[45] prepared some titanium phosphate by the same method as they used for the preparation of zirconium phosphate except that they used a soluble titanium salt instead of the zirconium salt. The molar ratio of P/Ti was less than that of P/Zr. Therefore, the specific exchange capacity of the titanium phosphate was less than that of the zirconium phosphate.

Kraus et al.[36] performed chromatographic separations of cobalt from iron and of the five alkali metals on columns of zirconium tungstate and of calcium, strontium, barium, and radium on zirconium molybdate. Phillips and Kraus[50] used zirconium antimonate for the separation of potassium, rubidium, and sodium, which emerged from the column in this order. Campbell[51] separated magnesium, calcium, strontium, and barium on a column of zirconium molybdate. All of these separations were quantitative.

Inoue et al.[52] have developed a method for separating the various metals from fission products.

If zirconium is absent they pass 10 ml of sample solution into a column of stannic phosphate, $4.0 \text{ cm} \times 0.38 \text{ cm}^2$ Then they pass 10 ml of water through the column and elute the strontium with 20 ml of 0.5 M ammonium chloride. This leaves considerable hydrogen ion still on the column, which would cause contamination of the cesium fraction by uranium and rare earths. Therefore, they pass 10 ml of 0.1 M ammonium acetate through the column, followed by 60 ml of 3.0 M ammonium chloride. These reagents remove all the cesium. Then they pass 1 M phosphoric acid through the column. This brings out the uranium in a very badly tailing graph. After 70 ml of phosphoric acid have been used, about 8% of the uranium still remains on the column but its concentration in the eluate is so small that the investigators change the eluent to 5 M nitric acid. Sixty milliliters of this solution removes all of the rare earths, which are contaminated with some uranium. The procedure is more involved if zirconium is present in the sample.

E.III. SALTS OF HETEROPOLY ACIDS

Ammonium molybdophosphate (or phosphomolybdate) is the best known salt of the heteropoly acids. Although the formula is generally written as $(NH_4)_3PO_4 \cdot 12MoO_3 \cdot 3H_2O$ or as $(NH_4)_3PMo_{12}O_{40} \cdot 3H_2O$, the actual composition is subject to some variation. There may be somewhat more or less than 12 units of MoO_3 per phosphorus atom and more or less than three molecules of water of hydration. Furthermore, part of the ammonium ions may be replaced by hydrogen or other ions. For example, Krtil[53] found that the composition of one of his precipitates corresponded to the formula $(NH_4)_{2.38}H_{0.62}PMo_{12}O_{40} \cdot 9.3H_2O$. The cation-exchange properties of ammonium molybdophosphate depend on the ability of other cations to replace ammonium ion at rates of the same order of magnitude as the exchange reactions of the common resins.

There are insoluble salts of heteropoly acids in which the central phosphorus atom is replaced by arsenic or silicon or in which the surrounding molybdenum atoms are replaced, wholly or partly, by tungsten. Also cations such as thallium(I) or the alkylammonium ions, can be used to precipitate the heteropoly anion. Most of these salts also show interesting ion-exchange properties.

Broadbank[54] et al. studied the molybdophosphates of potassium, rubidium, cesium, ammonium, and alkylammonium cations in regard to their ion-exchange properties and solubility. They concluded that none of these compounds has any advantage over the ammonium salt.

The very large selectivity of several heteropoly salts for the heavy alkali metals is shown in Table 10.14. Aside from having the greatest affinity for cesium, ammonium molybdophosphate has two other advantages:[55] (1) it has the smallest solubility; (2) its properties are least affected by minor variations in the method of preparation; in contrast, the great difference between two preparations of the molybdoarsenate are given in the table.

TABLE 10.14. DISTRIBUTION COEFFICIENTS OF TRACE AMOUNTS OF CATIONS IN 0.1 M AMMONIUM NITRATE

Exchanger approximate formula[a]	Distribution coefficients and standard deviations			Ref.
	Cs	Rb	K	
$(NH_4)_3AsMo_{12}O_{40}$[b]	4160 ± 40	187 ± 1		55
$(NH_4)_3AsMo_{12}O_{40}$[b]	1178 ± 12	62.6 ± 1.0	5.6 ± 1.2	55
$(NH_4)_3PMo_{12}O_{40}$	5500 ± 60	192 ± 1	4	55
$(NH_4)_3PMo_8W_4O_{40}$	4100			56
$(NH_4)_3PMo_6W_6O_{40}$	3200			56
$(NH_4)_3PMo_4W_8O_{40}$	2700			56
$(NH_4)_3PW_{12}O_{40}$	3500 ± 20	136	5.0 ± 0.6	55
Dowex 50	62	52	46	55

[a] Water of hydration omitted.
[b] Both samples supposedly prepared by the same method but by two different persons.

Plots of log D of traces of cesium and rubidium on the ammonium molybdotungstophosphates of Table 10.14, vs. log concentration of ammonium nitrate are linear with a slope very nearly equal to the theoretical value of -1.00. This indicates that the sorption of the alkali metal by these compounds is truly an ion-exchange phenomenon. The effect of nitric acid on the values of D at constant concentrations of ammonium nitrate is very small provided that the hydrogen ion is less than 2 M. This indicates that the hydrogen ion plays only a minor role in the exchange unless its concentration is above 2 M.

The precipitates of the heteropoly salts consist of very fine crystals. This entails the disadvantage that eluents flow through columns of these exchangers at intolerably slow rates. At first, this difficulty was overcome by mixing the exchangers with roughly an equal amount of asbestos. Smit[57] succeeded in preparing reasonably large particles of the precipitate. He carefully crushed crystals of molybdophosphoric acid to break up the aggregates and dropped these crystals into an aqueous solution of ammonium nitrate at least 1 M. Before the particles of the acid dissolve, they are converted to the insoluble ammonium salt. Under low magnification, the crystals of the acid seem to have been changed into crystals of the ammonium salt of the same size and shape. However, since the ammonium salt is not isomorphous with the free acid, it is not surprising that high magnification reveals that each particle of the ammonium molybdophosphate consists of an aggregate of much smaller crystals. These aggregates have sufficient mechanical stability to permit their use in columns that give satisfactory flow rates. Smit subjected a sample of his coarse-grained ammonium molybdophosphate to sieve analysis and found 52% in the 40–80mesh range.

Not all of the ammonium ions in a heteropoly salt can undergo exchange. For example, the theoretical specific exchange capacity calculated from the formula $(NH_4)_3PMo_{12}O_{40} \cdot 2H_2O$ is 1.57 meq/g; but the observed[58] capacities, measured by the breakthroughs of rubidium and cesium, varied between 0.84 and 1.22. It is suggested[58] that the exchange takes place between unhydrated ions and that after sufficient amounts of these large alkali cations have entered the crystals, the channels become blocked so that further exchange is prevented. Similarly, the theoretical specific exchange capacity of thallous tungstophosphate is 0.86 meq/g, but the observed[59] values run from 0.11 to 0.50.

E.III.a. *Applications of Heteropoly Exchangers*

Several chromatographic separations have been performed on rather short columns (3–12 cm) of heteropoly exchangers. These include the use of ammonium molybdophosphate for the separation of sodium from potasium,[58] rubidium from cesium,[58, 60] strontium from yttrium,[61] and cadmium from indium,[61] also the use of $(NH_4)_{2.55}H_{0.45}PMo_4W_8O_{40} \cdot 6.5H_2O$ for the separation of rubidium from cesium.[56] In almost all of these separations, the elution graphs tailed considerably. Furthermore sodium and yttrium had two peaks each, the second appearing in each case after an increase in the concentration of the eluent. These abnormalities are probably due to slow rates of exchange.

Broadbank *et al.*[62] used ammonium molybdophosphate to determine cesium-137 in rain water. They acidified a litre of sample with nitric acid and passed it through a layer of the exchanger supported by filter paper on a demountable Buchner funnel, 1 in. in diameter. After drying the exchanger they measured the cesium in it by radioactive counting.

Caron and Sugihara[59] determined radioactive cesium in samples containing other fission products by the use of thallous tungstophosphate.

They passed the sample dissolved in no more than 2 l. of nitric acid at least 0.4 M through a column of the exchanger mixed with paper pulp. Then they passed 90 ml of 0.005 M thallous nitrate in 4 M nitric acid through the column to elute all cations except cesium. Next, they eluted the cesium with 90 ml of 0.15 M thallous nitrate, also in 0.4 M nitric acid. They treated this fraction of eluate with chlorine to oxidize the thallium to the valence number of 3. Then they made the solution 3 M with hydrochloric acid and passed it through a column of Dowex 1 to remove the thallium. They determined the cesium in the final eluate by counting.

E.IV. MOLECULAR SIEVES

Some zeolites have the unusual property of retaining their original crystalline structure when they are dehydrated. Unfortunately, the natural zeolites with this very valuable property are found in only small amounts; but synthetic crystalline zeolites with this property can be prepared in the laboratory. When such a zeolite is dehydrated, void spaces previously occupied by the water are left inside the crystals. These *internal* voids, which should not be confused with the void volume between the crystals, constitute about 25% of the volume of the crystals.

The crystal surface exerts strong adsorptive forces especially on polar and polarizable molecules. By virtue of the myriads of internal voids, these dehydrated (or activated) zeolites have very large specific surfaces, about 600 m²/g, almost all of which is internal surface. Hence these zeolites have very large sorptive capacity. Indeed, the internal voids can become filled with the liquefied sorbate, even when the crystal is in equilibrium with the vapor of the sorbate.

These dehydrated zeolites possess still another very unusual and very valuable property, which we shall illustrate by considering briefly the crystal structure of the Linde Molecular Sieve Type 4A. Each internal void space is roughly spherical with a diameter of 11.4 Å. Each individual void is connected with other voids and eventually with the surface of the crystal by "windows" that consist of eight-membered oxygen rings, 4.2 Å in diameter. Therefore, no substance can be sorbed by this zeolite (except to an insignificant extent on the surface of the crystals) unless its molecules are small enough to diffuse through these rings. For this reason these dehydrated zeolites are called *molecular sieves*. Other types of molecular sieves have "windows" of different diameters leading to the inner voids and hence set different limits to the molecular size of substances that can be sorbed.

The exchangeable cation in Linde Molecular Sieve Type 4A is sodium. Some of these sodium ions are held by electrostatic forces close to the oxygen atoms that form the window between the voids. Hence the diffusion of sorbate molecules into the voids is further restricted by the presence of these counter ions. If the sodium ions are replaced by the larger potassium ions the effective size of the windows is diminished. (Radii of unhydrated ions are considered here.) If the sodium ions are replaced by half as many calcium ions of about the same diameter, the window size is usually increased.

Table 10.15 presents some Linde Molecular Sieves with a list of some compounds excluded by each.[63] Of course, any compound excluded by a given sieve will be excluded also by sieves of smaller diameter. Furthermore, the substances that are listed as not sorbed do undergo slight sorption on the external surfaces of the crystals and on the clay (20%) that is used to bind the small crystals into pellets. Table 10.16 illustrates numerically the dramatic difference in sorption caused by a small change in the structure of the sorbate or of the replaceable cation of the sieve.[64]

A mixture of gases or vapors of different molecular sizes can readily be separated by passage through a series of molecular sieves so chosen that only the smallest molecules are sorbed by the first zeolite, only the smallest of the remaining compounds by the second zeolite, etc. Later the separately sorbed compounds can be desorbed by application of a

TABLE 10.15. LINDE MOLECULAR SIEVES

Sieve	Approximate formula of hydrated zeolite	Critical diam. (Å)	Substances not sorbed[a]
13X	$Na_{26}Ca_{30}(AlO_2)_{86}(SiO_2)_{106} \cdot 276H_2O$	10	$N(C_4F_9)_3$
10X	$Na_{86}(AlO_2)_{86}(SiO_2)_{106} \cdot 276H_2O$	9	$NH(CH_2CH_2CH_3)_2$, $NH(CH_2CH_2CH_2CH_3)_2$
5A	$Na_4Ca_4(AlO_2)_{12}(SiO_2)_{12} \cdot 27H_2O$	5	*iso*-Paraffins, cyclic cmpds. with 4 or more atoms in ring
4A	$Na_{12}(AlO_2)_{12}(SiO_2)_{12} \cdot 27H_2O$	4	*n*-BuOH, butene, C_3H_8 and higher homologs, cyclopropane
3A	$Na_4K_8(AlO_2)_{12}(SiO_2)_{12} \cdot 27H_2O$	3	C_2H_4, C_2H_6, O_2

[a] Substances sorbed by all these sieves include H_2O, MeOH, and CO_2.
(Data used by permission of Academic Press.)

vacuum or a stream of inert gas such as helium at high temperature. Most moleculer sieves are stable up to 500°.

Because of their strong attraction for water, these compounds are excellent drying agents for both gases and liquids. Water is sorbed in preference to organic compounds because of its very polar nature even when the organic molecules are small enough to enter the inner voids.

Molecular sieves are also used as the stationary phase in gas–solid chromatography. In this application they generally function simply as adsorbents, not rejecting any of the migrant gases by the sieve effect.

TABLE 10.16. SORPTION OF SOME COMPOUNDS ON DEHYDRATED ZEOLITES

Zeolite	Form	Sorption (g per 100 g of anhydrous zeolite)		
		n-Hexane	3-Methylpentane	Water
A	K	0.2	0.2	22.2
A	Na	0.4	0.3	28.9
A	Ca	12.6	0.2	30.5
ZK-4	K	0.4	0.2	19.5
ZK-4	Na	12.5	0.2	24.8

(Used by permission from *Journal of the American Chemical Society*.)

E.V. OTHER INORGANIC EXCHANGERS

It was mentioned on p. 2 that the sodium in analcite, $NaAlSi_2O_6 \cdot H_2O$, was completely replaced by potassium, when the mineral was kept in contact with a hot solution of potassium chloride for 3 months. Barrer and Sammon[65] prepared the silver form of analcite by heating synthetic analcite with ten times its weight of silver nitrate at 225° for 4 hr. Then they separated the silicate from the water-soluble salts and repeated the treatment. X-ray examination revealed that the product had the same structure as the sodium analcite.

They used this material for the quantitative removal of sodium from cesium by virtue of the exchange reaction

$$AgAlSi_2O_6 . H_2O + Na^+ + Cl^- \rightarrow NaAlSi_2O_6 . H_2O + AgCl$$

Cesium does not exchange under the conditions of their experiments. They heated 334.2 mg (1.095 mmol) of silver analcite, 35.7 mg (0.611 mmol) of sodium chloride, 51.7 mg (0.307 mmol) of cesium chloride, and 2 ml of water in a sealed tube at 95° for 18 hr. They filtered the mixture, washed the residue, and evaporated the filtrate to dryness. The residue from the evaporation weighed 52.8 mg. Subtracting from this the weight of cesium chloride taken and a blank of 1.4 mg ($\sigma = 0.3$) representing mostly some dissolution of the glass tube, they found -0.3 mg representing the amount of sodium chloride left after the foregoing exchange reaction. Since this is zero within the experimental error, they concluded that all of the sodium and none of the cesium had exchanged with the silver. Four similar experiments supported this conclusion.

In ten experiments similar to those above except that no sodium chloride was taken, 1–6% of the silver ion was removed from the silicate and precipitated as the chloride. However, the exchange occurred between silver and hydrogen ions.

$$AgAlSi_2O_6 . H_2O + H_2O + Cl^- \rightarrow HAlSiO_2 . H_2O + AgCl + OH^-$$

The unhydrated cesium ion is too large to enter the analcite structure. For this reason, Barrer[65] refers to the silver analcite as an *ion-sieve* reagent.

The ferrocyanides[66–68] of several metals have been studied as possible cation exchangers The most promising of these seem to be compounds that approximate to the formulas[69, 7]

$$[H_4Fe(CN)_6]_5[MoO_3(H_2O)_x]_{12}$$

and

$$[H_4Fe(CN)_6]_4[MoO_3(H_2O)_x]_{16}$$

Both of these compounds, especially the former, have unusually great selectivity for cesium, the distribution coefficient in 1 M nitric acid being 3.5×10^4 and 5.1×10^3 respectively. Huys and Baetslé passed a solution of 2 M nitric acid containing sulphuric, phosphoric and molybdic acids in smaller concentrations and the following cations:

$$UO^{2+}, Fe^{3+}, Ni^{2+}, Cr^{3+}, Al^{3+}, Na^+, NH_4^+, Mg^{2+}, Ce^{3+}, Sr^{2+}, \text{ and } Cs^+$$

(total concentration of metal ions $= 0.85$ M) through a bed of this exchanger with a molar ratio of Mo/Fe of 2. The cesium was 0.0046 M. No cesium was detected in the effluent until the volume of effluent was equal to 82 bed volumes.[69]

Phillips and Kraus[71] precipitated cadmium sulphide by pouring ammonium sulphide into cadmium nitrate. They prepared columns of the washed precipitate and passed metallic salts through the columns. Copper salts displace cadmium from the column, mole for mole. Silver and mercuric salts displace 1 mole of cadmium for 2 moles. In the case of mercuric salts, this unexpected result is probably due to the formation of a salt such as HgS . HgCl$_2$. The theoretical specific capacity of the exchanger is 14 meq/g. Even when 0.052 M copper nitrate in 0.004 M nitric acid was passed through a 1.4-cm column at 5 cm/min, 70% of the cadmium was displaced from the column at the midpoint of the breakthrough graph of the copper.

Lieser and Hild[72] partly converted a silica gel into barium silicate by treatment with aqueous barium chloride. After washing the gel, they converted the barium silicate into barium sulphate with aqueous sulphuric acid. The product contained about 35% of barium sulphate, whose specific surface was about ten times that of ordinarily precipitated barium sulphate. They passed solutions of 0.018 M calcium, 0.005 M magnesium, and more concentrated solutions of sodium, all contaminated with tracer-free radioactive strontium, through a column 15 cm × 1 cm² at 1 cm/min. The column retained all of the strontium, very little of the calcium, and none of the magnesium or sodium. It also sorbed some iron(III) and yttrium, probably as basic salts.

F. Liquid Ion Exchangers

F.I. LIQUID ANION EXCHANGERS

In 1948, Smith and Page[73] used long-chain amines dissolved in organic liquids such as chloroform and nitrobenzene for the selective extraction of acids from aqueous solutions.

$$H^+(a) + Cl^-(a) + MeN(C_8H_{17})_2(o) \rightarrow Me(C_8H_{17})_2NHCl(o) \qquad (10.a)$$

The symbols (a) and (o) denote that the species is in the aqueous or organic phase. In analogy with the absorption of acids by weak-base anion-exchange resins, there is a marked increase in the extraction of various acids with increasing ionization constants. Another similarity of these amines to resinous anion exchangers is seen in the reactions such as

$$Br^-(a) + (C_8H_{17})_3NHCl(o) \rightleftharpoons Cl^-(a) + (C_8H_{17})_3NHBr(o) \qquad (10.b)$$

Also, these amines can extract metals as anionic complexes such as $Th(NO_3)_6^{2-}$, $UO_2(SO_4)_3^{4-}$, $FeCl_4^-$, and $CuCl_4^{2-}$. For these reasons, amines of this type are known as *liquid anion exchangers*.

The amines used for this purpose usually have 18–27 carbon atoms and may be primary, secondary, or tertiary.[74] In the case of the primary and secondary amines there should be two branches on at least one of the carbon chains located on the carbon next to the nitrogen. Aliphatic amines are used more than aromatic amines. Quaternary ammonium compounds can serve as liquid anion exchangers but are generally less desirable because of their greater solubility in water and their tendency to form emulsions when the two phases are mixed.

Two frequently used products are Amberlite LA-1 and Amberlite LA-2. These are not pure compounds but mixtures of several closely related compounds. The former is dodecenyl-(trialkylmethyl)amine, $RR'R''C—NH—CH_2—CH=CH—CH_2—CMe_2—CH_2—CMe_3$; the latter is lauryl(trialkylmethyl)amine, $RR'R''C—NH—CH_2—(CH_2)_{10}—CH_3$. In both products, the three alkyl groups R, R' and R'' have a total of 11–14 carbon atoms.

F.II. LIQUID CATION EXCHANGERS

Acids that are insoluble in water and soluble in organic liquids serve as liquid cation exchangers. The most useful of these are the dialkyl esters of phosphoric acid and the mono-alkyl esters of alkanephosphonic acids. The total number of carbon atoms is generally 10–17.

Although the actual reactions are more complex, the following simplified equations illustrate the formal similarity between the liquid and resinous cation exchangers.

$$Th^{4+}(a) + 4(C_8H_{17})PO(OC_8H_{17})OH(o) \rightleftharpoons 4H^+(a) + [(C_8H_{17})PO(OC_8H_{17}O)]_4Th(o) \quad (10.c)$$

$$Na^+(a) + (C_8H_{17}O)_2POOLi(o) \rightleftharpoons Li^+(a) + (C_8H_{17}O)_2POONa(o) \qquad (10.d)$$

In most processes of liquid ion exchange, whether anions or cations are involved, one species of ions moves across a liquid–liquid interface while an equivalent quantity of similarly charged ion moves in the opposite direction across the same boundary. Although the absorption of acetic acid by a liquid anion exchanger in the free-base form may be considered to involve the transfer of only nonionized molecules across the interface, it is generally regarded as an example of liquid ion exchange.

All examples of liquid ion exchange involve the transfer of some species, usually ionic, across a liquid–liquid interface and hence are properly included in the more general category of liquid–liquid extraction. On the other hand, all cases of liquid–liquid extraction do not involve liquid ion exchange. For example, the classical method of separating iron(III) from most other metals by shaking the solution of the metals in aqueous 6 M hydrochloric acid with ether involves the extraction of an uncharged complex, $HFeCl_4$ (probably solvated), without the transfer of any solute species in the opposite direction across the interface.

In many cases it is difficult to decide whether a given extraction is or is not an example of liquid ion exchange. For example, let us consider the extraction of copper(II) by a solution of dithizone, HDz, in carbon tetrachloride. The phenomenon can be considered to occur in the following four steps:

$$HDz(o) \rightarrow \qquad\qquad HDz(a)$$

$$HDz(a) \rightarrow \qquad\qquad H^+(a) + Dz^-(a)$$

$$Cu^{2+}(a) + 2Dz^-(a) \rightarrow CuDz_2(a)$$

$$CuDz_2(a) \rightarrow CuDz_2(o)$$

Since no ions are transferred according to this mechanism, the process is usually classified as liquid–liquid extraction. However, if the foregoing four reactions are combined the resulting overall reaction

$$2HDz(o) + Cu^{2+}(a) \rightarrow CuDz_2(o) + 2H^+(a)$$

must be classified as liquid ion exchange.

F.III. DESIRABLE PROPERTIES

The properties of a good liquid ion exchanger follow:[74]

Small solubility in water. Amines or acids that are too soluble in water undergo appreciable loss in large-scale or counter-current processes. Solubility in water may be decreased by increasing the number of carbon atoms in the amine but this involves a larger molecular weight and hence a smaller specific exchange capacity. Most sulphonic acids and carboxylic acids are too soluble to serve well as liquid exchangers.

Large specific exchange capacity. Since the usual anion exchangers have only one functional group per molecule, the specific exchange capacity is inversely proportional to the molecular weight.

Large solubility in a water-immiscible organic liquid. Undiluted liquid ion exchangers are very rarely used; they are almost always dissolved in an organic solvent known as the *diluent*. The concentrations of these solutions are usually between 2% and 12%. Difficulties that are often encountered in the absence of a diluent are (a) the formation of three liquid phases or of a precipitate, (b) emulsification, and (c) greater loss of the liquid exchanger by dissolution in the aqueous phase. Frequently used diluents are kerosene, benzene, toluene, xylene, and chloroform.

High selectivity. The advantage of high selectivity is obvious.

Low cost. This is less important in analytical than in industrial applications.

Stability. No difficulty is experienced because of decomposition of the liquid exchangers except when they are used with strong oxidizing agents.

Small surface activity. Great surface activity leads to the formation of emulsions and hence difficulty in separating the two phases. Quaternary ammonium salts and sulphonic and carboxylic acids are undesirable in this respect. The addition of a few per cent of a long-chain alcohol or ketone to the diluent decreases the stability of the emulsions.

F.IV. COMPARISON WITH ION-EXCHANGE RESINS

There are several advantages of liquid ion exchangers in comparison with the resins.

Greater selectivity. The selectivity sequence of liquid ion exchangers can generally be predicted from the order in analogous ion-exchange resins. For example, the selectivity sequence of both Amberlite LA-1 and LA-2 in kerosene is $NO_3^- > Cl^- > C_2H_3O_2^-$. Nevertheless the selectivity coefficients are usually much greater with the liquid exchangers. Furthermore, the selectivity coefficient of a given amine or acid toward a given pair of ions can be altered by changing the diluent or the concentration of the organic solution.

Faster exchange. Diffusion is faster in the organic solution of a liquid exchanger than in a swollen ion-exchange resin. This leads to faster exchange reactions. When a liquid ion exchanger (absorbed by an inert porous solid) is used as the stationary phase in chromatography for the separation of traces, the number of plates per centimeter of column is usually greater than with ion-exchange resins. See, for example, p. 234.

No interstitial water. When an ion previously absorbed by a column of ion-exchange resin is removed by passing an appropriate solution through the column, the volume of effluent containing the desired ion is equal to the volume of solution necessary to remove this ion plus the interstitial volume of the column. Except in chromatographic separations, no such dilution by interstitial water occurs in the case of the liquid ion exchangers.

No difficulty with suspended matter. When a solution containing a suspended solid is passed through an ion-exchange column, the suspended matter is likely to clog the column. If the suspended material is of colloidal dimensions, it will not clog the column but may "poison" the resin (decrease the exchange rate by settling on the surface of the resin beads). Neither of these difficulties is encountered with liquid ion exchangers.

Facility of true countercurrent processes. Although true countercurrent methods (where the ion exchanger and the aqueous solution both move through the column in opposite directions) can be performed with both resinous and liquid ion exchangers, this is accomplished much more easily with the latter.

Greater permselectivity. Sollner and Shean[75] passed direct current through cells consisting of two aqueous solutions separated by a 20% solution of Amberlite LA-2 in benzene, xylene, or nitrobenzene. From the changes in concentration of the aqueous solutions, they calculated the ratios of the amounts of anion to the amounts of cation that had passed through the organic solution. Except for very dilute aqueous solutions, these ratios were ten to several thousand times greater than the analogous ratios with membranes of ion-exchange resins. Bonner and Lunney[76] have measured the membrane potential of cells with liquid ion-exchanger membranes.

On the other side of the coin, liquid ion exchangers have the disadvantage of being appreciably soluble in water. Brown et al.[77] report the following losses in terms of milligrams of amine per litre of sulphuric acid (pH = 1.0): Amberlite LA-1, 12; 1-(3-ethylhexyl)-4-ethyloctyl amine, 20; tris(2-ethylhexyl)amine, 1,500. Changes in the properties of the columns were reported[78] after 20 elutions of rare earths had been performed through columns of di-(2-ethylhexyl) phosphate absorbed by diatomaceous earth previously treated with a silicone. It is likely that this change in properties was caused by the loss of some liquid cation exchanger.

The formation of emulsions and a method of decreasing the stability of the emulsion was mentioned on p. 228.

F.V. THEORY

F.V.a. *Theory of Liquid Anion Exchangers*

Neither the extraction of an acid from an aqueous solution by a solution of a long-chain amine in an organic solvent nor the exchange of ions with liquid ion exchangers is as simple as the chemical eqns (10.a) to (10.d) indicate. It is beyond the scope of this book to discuss or even to summarize the large amount of research that has been done to elucidate the mechanism of the extraction and the nature of the compound or complex of the extracted ion in the organic phase, but a few papers dealing with these subjects will be discussed.

Prior to 1962, several investigators had observed that solutions of long-chain amines in organic solvents extracted more nitric acid from aqueous solutions than could be explained by the reaction

$$Amn(o) + H^+(a) + NO_3^-(a) \rightleftharpoons AmnHNO_3(o) \qquad (10.e)$$

Amn represents a primary, secondary, or tertiary amine. Considerable disagreement existed regarding the mechanism by which the excess acid was extracted. The careful work of Kertes and Platzner[79] contributed much to the clarification of this confused situation. They shook 0.89 M solutions of Amberlite LA-1 in carbon tetrachloride with equal volumes of aqueous nitric acid of various concentrations for various lengths of time. Then they determined the total concentration of nitrate in the organic phase. This concentration depended not only on the initial concentration of amine in the organic phase but also on the duration of the agitation. The relative differences between the amounts extracted by 1 and 7 days of gentle agitation amounted in some cases to 6%. Fifteen minutes of violent shaking always gave intermediate results. They measured the viscosity of the organic phase immediately after the 15 min of vigorous agitation and separation of the phases, also at intervals thereafter. This viscosity increased 20% during 5 days, indicating either a slow oxidation of some organic constituent by the extracted nitric acid or a slow reaction involving the various extracted species.

In the subsequent experiments, they used the 15 min of vigorous mixing, separated the phases as quickly as possible, and analyzed both phases. The concentrations of total nitrate in the carbon tetrachloride were in good agreement with the assumption that the following three equilibria occur:

$$H^+(a) + NO_3^-(a) + Amn(o) \rightleftharpoons AmnHNO_3(o) \qquad (10.f)$$
$$K_1 = 3.8 \times 10^5$$

$$HNO_3(a) + AmnHNO_3(o) \rightleftharpoons Amn \cdot 2HNO_3(o) \qquad (10.g)$$
$$K_2 = 1.8$$

$$HNO_3(a) \rightleftharpoons HNO_3(o) \qquad (10.h)$$
$$K = 0.10$$

Amn represents the amine. $HNO_3(a)$ represents nonionized nitric acid (or ion pairs) in the aqueous phase. To calculate the concentration of this species they used some measurements of Raman spectra.[80] $HNO_3(o)$ represents nonionized nitric acid (or ion pairs) in the organic phase extracted by virtue of the $AmnHNO_3$ or $Amn \cdot 2HNO_3$ in the organic solution. Carbon tetrachloride alone does not extract nitric acid from aqueous solution.

Additional evidence that nitrate exists in the organic phase in three states is found in measurements of the viscosity of the organic phase. A plot of the viscosity vs. the ratio of total nitrate to total amine in the organic phase consists of three straight lines with intersections at ratios of 1.0 and 2.0. This indicates that almost all the nitrate enters the organic liquid by reaction (10.f) until this reaction is practically complete. Then reaction (10.g) occurs almost exclusively until it is essentially complete. Finally, reaction (10.h) takes place. Furthermore, the graph of the conductivity of the organic phase vs. the molar ratio of total nitrate to total amine in this phase coincides with the X-axis up to ratio of 1.0 and then rises. This indicates that $AmnHNO_3$ is a much weaker electrolite than $Amn \cdot 2HNO_3$.

The system is even more complicated when benzene is used as the diluent instead of carbon tetrachloride because benzene alone does extract appreciable amounts of nitric acid from concentrated aqueous solutions. For example, in equilibrium with 6.10 M and 14.35 M aqueous nitric acid, the concentration in the benzene[81] is 6.7×10^{-3} M and 0.26 M, respectively. However, Kertes and Platzner[79] found that the system with benzene as diluent behaves like the system with carbon tetrachloride with minor changes in the values of K_1 and K_2, provided that a correction is applied for the concentration of nitric acid that would be extracted in the absence of the amine.

In spite of complications introduced by reactions (10.g) and (10.h), for most purposes of analytical chemistry it is satisfactory to write eqn. (10.f) to represent the extraction of nitric acid by an amine dissolved in an organic liquid.

If we assume that thorium is extracted from an aqueous solution of thorium nitrate and nitric acid by an organic solution of an amine nitrate according to the reaction

$$AmnHNO_3(o) + Th(NO_3)_5{}^-(a) \rightleftharpoons AmnHTh(NO_3)_5(o) + NO^-_3(a) \qquad (10.i)$$

we can write

$$K = \frac{[AmnHTh(NO_3)_5]_o \, [NO_3^-]_a}{[AmnHNO_3]_o \, [Th(NO_3)_5^-]_a}.$$

The distribution ratio of thorium between the two phases is

$$C = \frac{[AmnHTh(NO_3)_5]_o}{[Th(NO_3)_5^-]_a}.$$

It is assumed that equal volumes of the two phases are taken. Therefore

$$K = C[NO_3^-]_a/[AmnHNO_3]_o.$$

If the concentration of nitrate ion in the aqueous phase is kept constant at k in a series of extractions while the concentration of the amine nitrate in the organic phase is varied, this equation becomes

$$\log C = \log K - \log k + \log [\mathrm{AmnHNO_3}]_0. \qquad (10.5)$$

On the other hand, if the extraction occurs according to reaction (10.j),

$$2\mathrm{AmnHNO_3(o)} + \mathrm{Th(NO_3)_6^{2-}(a)} \rightleftharpoons (\mathrm{AmnH})_2\mathrm{Th(NO_3)_6(o)} + 2\mathrm{NO_3^-(a)}, \qquad (10.j)$$

an analogous derivation gives

$$\log C = \log K - \log k + 2 \log [\mathrm{AmnHNO_3}]_0. \qquad (10.6)$$

Therefore in a series of extractions at constant concentration of nitrate ion in the aqueous phase, a plot of the experimentally determined values of $\log C$ versus the logarithm of the concentration of amine nitrate should be linear with a slope of 1 or 2 depending on whether the extraction proceeds according to reaction (10.i) or (10.j). In general, the slope of a log-log plot of C vs. the concentration of liquid anion exchanger at constant concentration of complexing anion indicates the number of molecules of exchanger associated with one atom of metal in the extracted compound.

Keder *et al.*[82] have published data on the distribution ratio of the actinide metals between solutions of tri-*n*-octylamine in xylene and aqueous nitric acid. As explained above, information about the probable formula of the extracted species can be obtained from plots of $\log C$ vs. log concentration of amine at constant concentration of nitric acid. The scattering of the experimental points is such that an uncertainty of about 0.2 unit exists in the slope. Within this limit, thorium(IV), neptunium(IV), and plutonium(IV) have slopes of 2.0, indicating formulas such as $(\mathrm{HAmn})_2\mathrm{Th(NO_3)_6}$. The slope of americium(III) is 1.0, indicating $\mathrm{HAmnAm(NO_3)_4}$. Anomalous slopes between 1 and 2 were found for uranium(VI), neptunium(VI), plutonium(VI), plutonium(III), and protoactinium(V). Aside from indicating the probable formula of the extracted species, data such as these are useful in planning separations by liquid anion exchange.

The slope of the log-log plot should be regarded as an indication rather than as a proof of the formula of the metallic compound in the organic phase. For example, Nelson *et al.*[83] found a slope of 0.78 for the plot of $\log C$ of indium(III) vs. log concentration of tri-*n*-octylamine in nitrobenzene in the extraction of this element from 0.97 M hydrochloric acid by solutions of the amine in nitrobenzene. These investigators also performed a series of extractions in which the amine concentration was held constant at 0.10 M while the concentration of indium(III) in an equal volume of 8 M hydrochloric acid was varied. A plot of the concentration of metal in the organic phase after equilibration vs. the concentration of indium in the aqueous phase starts at the origin, rises steeply at first and approaches asymptotically the value of 0.033. Since $0.10/0.033 \simeq 3.0$, they conclude that the metal compound in the organic phase is $(\mathrm{HAmn})_3\mathrm{InCl_6}$. A similar series of experiments with 0.01 M amine gave a similar graph with an asymptote at 0.0033, pointing again to the formula $(\mathrm{HAmn})_3\mathrm{InCl_6}$. To explain the apparent discrepancy, the authors[83] suggest that the indium is first extracted as $\mathrm{HAmnInCl_4}$ and that this compound combines by dipole forces with two molecules (or ion pairs) of HAmnCl in the organic phase to form $(\mathrm{HAmn})_3\mathrm{InCl_6}$.

Lindenbaum and Boyd[84] examined the absorption spectrum of the organic phase after nickel had been extracted from aqueous hydrochloric acid by a solution of tri-*iso*-octylamine in toluene. This spectrum is very similar to the spectra of (1) a solution of nickel chloride in molten pyridinium chloride, (2) crystals of $\mathrm{Cs_2ZnCl_4}$ containing nickel(II) in solid solution, and (3) a solution of $\mathrm{H_2NiCl_4}$ in nitromethane. Since the nickel exists in these three solutions as the complex anion $NiCl_4^{-2}$, these investigators conclude that the nickel extracted by tri-*iso*-octylamine exists as $(HAmn^+)_2NiCl_4^{-2}$. Similar work with iron(III), cobalt(III), copper(II), and manganese(II) indicated that each of these metals is extracted as the tetra-chloro anion.

F.V.b. *Theory of Liquid Cation Exchangers*

Before 1958 it was generally believed that di(2-ethylhexyl) phosphate extracts uranium-(VI) from aqueous solution according to the reaction

$$UO_2^{2+}(a) + 2HX(o) \rightleftharpoons UO_2X_2(o) + 2H^+(a) \qquad (10.k)$$

where HX represents the dialkylphosphoric acid. However, Baes *et al.*[86] measured the vapor pressure of solutions of this liquid cation exchanger in *n*-hexane. Between 0.037 and 0.89 molal (as HX), the vapor pressure, corresponded to a degree of polymerization of 2.00 ($\sigma = 0.08$). The graph of log C vs. log concentration of H_2X_2 at constant concentration of nitric acid had a slope of $+2.0$ and a graph of log C vs. log concentration of hydrogen ion in the aqueous phase at constant concentration of the phosphate had a slope of -2.0. All these observations are in harmony with the reaction.

$$UO_2^{2+}(a) + 2H_2X_2(o) \rightleftharpoons UO_2(HX_2)_2(o) + 2H^+(a) \qquad (10.l)$$

Additional evidence in support of this reaction is found in the measurement of the change in hydrogen-ion concentration that accompanies a given change in uranyl-ion concentration. This ratio was found to be 1.97 ($\sigma = 0.03$).

Reaction (10.l) applies only to extractions with moderate concentrations of the liquid cation exchanger. There is evidence that the dimerization is incomplete in 0.005 molal solution and that further polymerization occurs at 3.11 molal; both of these concentrations refer to the total monomer. Also, polymerization of the $UO_2(HX_2)_2$ occurs when nearly all of the H_2X_2 is converted to this compound.

F.V.c. *Synergistic Effect*

Uranium(VI) and plutonium(VI) are extracted from aqueous solutions not only by organic acids and amines dissolved in an organic liquid, but also by neutral phosphorus compounds such as the trialkyl phosphates and the dialkyl esters of alkanephosphonic acids (as complexes of these compounds with the salt of the metal). If the concentration of the organic acid of phosphorus in the organic liquid is M_1 and if that of the neutral phosphorus compound is M_2, the amount of metal extracted is greater than the sum that would be extracted by M_1 molar phosphorus acid and M_2 molar neutral phosphorus compound, used separately. These facts indicate not only that the uranium reacts with each type of phosphorus compound individually but that there is also formed a complex containing the uranium and *both* compounds of phosphorus. This subject is treated mathematically by Baes.[85]

F.VI. APPLICATIONS OF LIQUID ION EXCHANGERS

The most important large-scale application is probably in the recovery of uranium from poor ores.[74] The ore is treated with about 0.3 M sulphuric acid to yield a dilute solution of uranyl sulphate contaminated with large amounts of metals such as iron, aluminum, and calcium. The uranium is then extracted by the use of either a cationic or anionic liquid exchanger. It is put back into aqueous solution by treatment with sodium carbonate. This double extraction increases the concentration of the uranium and also gives a partial separation from the accompanying metals.

The determination of traces of uranium in urine[87] follows the same general outline as the foregoing industrial enrichment. The organic matter in a sample of 250 ml is destroyed by treatment with concentrated nitric acid and 30% hydrogen peroxide. The solution is evaporated to dryness, and the residue is dissolved in 50 ml of 3.5 M hydrochloric acid. Treatment with 50 ml of 10% tri-iso-octylamine in xylene extracts about 99% of the uranium. The aqueous phase is treated again with 25 ml of the amine solution. The combined organic extracts are treated with 15 ml of 3.5 M hydrochloric acid to remove traces of actinides that have been extracted. Then the uranium is stripped from the xylene by treatment with two 25-ml portions of water. These aqueous solutions are evaporated and the uranium is determined by measurement of its radioactivity. The mean recovery of uranium in twelve spiked samples of urine was 98% with a standard deviation of 10%. The quantities of uranium in these samples were so small that an average of only ten atomic decompositions per minute was observed.

The determination of traces of cadmium in zinc metal, zinc oxide, etc., illustrates the great selectivity of liquid exchangers. The older method involved extraction of the cadmium by a solution of dithizone in carbon tetrachloride; some zinc always accompanied the cadmium and was removed from the carbon tetrachloride by treatment with an aqueous solution of sodium hydroxide. Although this method seemed to give a satisfactory separation of cadmium from zinc, the results were subject to an erratic negative error probably caused by a partial decomposition of the dithizone on contact with the alkaline solution.

Knapp et al.[88] performed the determination as follows: They dissolved a sample containing 2–10 μg of cadmium in enough 9 M sulphuric acid to give a final pH of 3.0 or less. After adding 5 ml of 1 M potassium iodide, they diluted the solution to 50 ml and extracted the cadmium quantitatively as the complex iodide with 10 ml of 1% solution of Amberlite LA-2 in xylene. They treated the organic phase with 50 ml of an aqueous solution 0.1 M with potassium iodide and 0.036 M with sulphuric acid to remove the traces of zinc that had been extracted. They put the cadmium back into aqueous solution by treatment with two 10-ml portions of 1 M sodium carbonate. After combining these aqueous solutions, they extracted the cadmium with a solution of dithizone in carbon tetrachloride and determined the cadmium spectrophotometrically in that solution. The relative standard deviation was 3.5%.

Nakagawa[89] recommends the following simple method for the separation and determination of tin and lead.

Twenty milliliters of 7 or 8 M hydrochloric acid containingtin (IV) and lead(II) is treated with an equal volume of 10% Amberlite LA-2 in xylene. This extracts about 98% of the tin and a small but significant fraction of the lead. The extraction is repeated, and the combined organic phases are treated with hydrochloric acid again to remove the lead. Lead is determined polarographically in the combined aqueous solutions. Tin is returned to aqueous solution by extraction with three 30-ml portions of 0.5 M nitric acid and is also determined polarographically.

Irving and Damodaran[90] have devised a novel and rapid method for the determination of perchlorate ion based on the use of an unusual liquid anion exchanger. They treated 5.00 ml of the sample solution at a pH between 6 and 10 with 5.00 ml of 0.01 M tetrahexylammonium erdmannate in a mixture of xylene and methyl iso-butyl ketone, 4/1 by volume.

$$(C_6H_{13})_4N[Co^{III}(NH_3)_2(NO_2)_4](o) + ClO_4^-(a) \rightleftharpoons$$
$$(C_6H_{13})_4NClO_4(o) + Co^{III}(NH_3)_2(NO_2)_4^-(a) \qquad (10.m)$$

Then they measured the absorbance of the erdmannate anion in the aqueous phase at 355 mμ using as a blank the aqueous phase obtained by treating 5.00 ml of water by the same procedure.

Since the equilibrium constant of reaction (10.m) is only 17, the absorbance of the aqueous phase after equilibration is proportional to the original concentration of perchlorate only if the latter does not exceed 10^{-4} M. A calibration graph is used for greater concentrations. Fluoride, phosphate, and sulphate anions do not interfere. Chloride reacts analogously to perchlorate but to a much smaller extent; 0.02 M chloride introduces a relative error of only 3.8% in the determination of 2.5×10^{-4} M perchlorate. Many separations by liquid ion exchange are listed in Table II of Freiser's biennial review of extraction.[90a]

Liquid ion exchangers are also employed as the stationary phase in chromatography. Di(2-ethylhexyl) phosphate has been found to be very useful in the separation of the rare earths. It is supported in the column by Kel-F (polytrichlorofluoroethylene)[91] or diatomaceous earth that has been rendered hydrophobic by exposure to the vapor of dichlorodimethylsilane.[92, 93] The phosphate is applied to the support by dissolving it in a volatile solvent such as acetone or chloroform, mixing the support with this solution, and warming to evaporate the solvent. There should be about 100 mg[91,93] or 200 mg[92] of the liquid exchanger per 1.0 g of the support. Special care is needed in preparing columns of this material to avoid air bubbles and channels. Sochacka and Siekerski[93] prepared columns with various ratios of di(2-ethylhexyl) phosphate to diatomaceous earth, used them for the separation of the rare earths in elutions with nitric acid, and calculated the value of P (plates per centimeter) for each column. P had a constant value of 30 for columns with 50–100 mg of exchanger per g of support. With larger amounts of the phosphate, P decreased, being 5.1 with 500 mg/g.

Winchester[92] used a similar column, 6 cm $\times$ 0.32 cm², at 65° to separate all the rare earths as radioactive isotopes from a sample containing a total of 0.015 mg of rare earth. The eluent was nitric acid whose concentration was increased in 13 steps from 0.05 to 3.80 M. Each rare earth gave a separate peak except that erbium and yttrium came out together. Most of the separations were quantitative.

Sochacka and Siekerski[93, 94] also prefer nitric acid as an eluent for this separation. Hydrochloric acid gives slightly larger ratios of C between two adjacent rare earths, but has the disadvantage of smaller P values than nitric acid for some metals. These authors found that the mean P for all the rare earths with nitric acid as eluent was 30 with a standard deviation of only 2. With hydrochloric acid, elements 57 through 64 also had P values averaging 30 with a standard deviation of 2. The next three elements had very small P values. Elements 68 through 71 had a mean P value of 2.3 with a standard deviation of 0.5. The explanation of this curious phenomenon is not known. The authors[94] believe that the exchange reaction reaches equilibrium more slowly with the elements of small P and suggest that the sluggish equilibrium is caused by coordination with ligands other than water by the second-shell electrons.

The use of permselective membranes for the determination of the activity or concentration of a given cation or anion was discussed in section D.III.a. The experiments of Sollner and Shean[75] and of Bonner and Lunney[76] (p. 228) indicate that the solutions of liquid ion exchangers can also be interposed between a test solution and a standard solution to determine the activity or concentration of an ion in the former. Until quite recently, however, the use of permselective membranes (whether resinous or liquid) was limited to those solutions that contained no cation (or anion) other than the cation (or anion) to be determined.

It is well known that the "glass electrode" serves well for the measurement of pH because it contains a membrane that is essentially permeable only to hydrogen ions. A slight permeability to sodium (and other alkali-metal cations) causes errors only if the ratio of sodium ion to hydrogen ion in the test solution is very large, about 10^8. A great deal of research has been done to find other membranes that are permeable to only one specific cation or anion and hence could be used for the determination of the activity of that ion. A membrane cut from a single crystal of lanthanum fluoride is permeable only to fluoride ions and serves admirably in the determination of fluoride-ion concentration.[95] Other "ionically conducting" solids permeable within certain limitations to only one cation or anion, are incorporated into electrodes and marketed by Orion Research, Inc., Cambridge, Massachusetts 02139. The ions whose activities can be determined by the appropriate electrode of this type are fluoride, chloride, bromide, iodide, and sulphide.

If an ion-exchange material—resinous, liquid, inorganic, or of any other type—could be found whose selectivity coefficient for one counterion relative to all other counterions is infinity, a membrane of that exchanger would serve as an ideal "electrode" for that ion because no other ion could migrate through the membrane (Donnan invasion disregarded). In that case, eqn. (10.2) would be applicable. Although infinite selectivity coefficients are out of the question, exchangers can be found with selectivity coefficients so large that eqn. (10.2) is applicable over wide ranges of concentration.

Since liquid ion exchangers generally have greater specificity than the resins, it is not surprising that the former have been used in the development of "specific-ion electrodes" based on ion-exchange membranes. For example, a solution of didecylphosphate in dioctyl benzenephosphonate, $C_6H_5PO(OC_8H_{17})_2$ has a very great specificity for calcium ion. Ross[96] has described a calcium-ion electrode based on this fact. A porous membrane near the bottom of a tube 15 cm in length and 1 cm in diameter supports a 2-cm layer of 0.1 M calcium didecylphosphate in dioctyl benzenephosphonate. Contacting this solution from above is an aqueous 2% agar gel, 0.1 M with calcium chloride. A silver–silver chloride electrode in the gel provides a lead to the potentiometer. When the tube is immersed in an unknown solution containing calcium ions, the electrolytic contact between the liquid ion exchanger and the unknown is made at the porous membrane. Of course, a calomel electrode is inserted in the unknown solution to complete the cell.

In solutions containing no cation except calcium, the electrode responds in accordance with the Nernst equation between 10^{-4} and 1 M calcium ion

$$E_{Ca} = E° + 0.0992T \log (Ca^{2+}). \qquad (10.7)$$

Below 10^{-4} M calcium ion, an error occurs because appreciable calcium didecyl phosphonate is extracted from the organic by the aqueous phase.

In solutions containing other cations, the electrode potential is given by the equation

$$E_{Ca} = E° + 0.0992T \log [(Ca^{2+}) + \Sigma k_i (Ct^{z+})^{2/z}]. \qquad (10.8)$$

The second term inside the brackets is the contribution to the electrode potential of the foreign cations. If the activities of the other cations are known, the second term can be evaluated; and thus the unknown calcium-ion activity can be calculated from the observed electrode potential. The values of k_i are characteristic of each foreign cation. For hydrogen, sodium, potassium, magnesium, and barium ions, their respective values are 10^{-5}, 10^{-4}, 10^{-4}, 0.014, and 0.010. These k_i values may vary 20%, depending on the concentration of the solution. They can be used, however, to calculate the upper limit of concentration of any foreign ion above which the simpler eqn. (10.7) is subject to an appreciable error.

A chloride-ion electrode based on a liquid ion-exchange membrane is also available. The organic phase is a solution of a tetraalkylammonium chloride of large molecular weight in decanol.[97] Specific-ion electrodes based on liquid ion exchangers are also available for cupric, nitrate, and perchlorate ions, also one that responds about equally to most bivalent cations and is used in determinations of the hardness of water. All of these electrodes are the products of Orion Research, Inc., Cambridge, Massachusetts 02139.

G. Ion-exchange Paper

Ion-exchange paper is paper containing ion-exchange groups. One method of preparing such paper is to treat the cellulose with chemical reagents that introduce ion-exchange groups into the cellulose molecule. Another method is simply to impregnate the paper with an ion-exchange material (resin, inorganic exchanger, or liquid exchanger). Additional information on the preparation is given later in the discussion of each of the four main types of ion-exchange paper.

G.I. THEORY OF CHROMATOGRAPHY WITH ION-EXCHANGE PAPER

The results of paper chromatography with both ordinary and ion-exchange paper are generally expressed as R_f values. R_f is defined as the distance traveled by the spot of migrant divided by the distance traveled by the eluent (also called developer or solvent). Both distances are measured from the origin, i.e. the position on the paper where the spot of sample is applied. In many cases, a more useful parameter is R_m, which is defined by eqn. (10.9)

$$R_m = \log \left(\frac{1}{R_f} - 1 \right). \tag{10.9}$$

Thus R_f decreases and R_m increases with increasing sorption of the migrant by the stationary phase (ion exchanger or paper). Curve C of Fig. 10.4 shows R_m as a function of R_f. Curves A and B of this figure indicate that small experimental errors in R_f cause larger errors in R_m, especially at very small or very large values of R_f.

Lederer and Kertes[98] derived a relationship between R_m of a cation Ct^{z+} and the concentration of the sulphuric acid used as eluent in chromatography with ion-exchange-resin paper. A more general equation is derived more rigorously as follows:

The ion-exchange equilibrium is

$$z \mathrm{ElR} + Ct^{+z} \rightleftharpoons CtR_z + z \mathrm{El}^+$$

where El is the cation of the eluent, assumed to be univalent, and R is the anionic part of the exchanger, which may be a resin, a liquid or inorganic exchanger, or modified cellulose. Therefore,

$$\log E = \log \frac{[CtR_z]}{[Ct^{z+}]} + z \log[\mathrm{El}] - z \log[\mathrm{ElR}]. \tag{10.10}$$

Since the cation can move only when it is in the mobile phase, R_f is equal to the fraction of the sample cation in the mobile phase

$$R_f = \frac{[Ct^{z+}]\, v}{[Ct^{z+}]\, v + [CtR_z]\, w\, Q},$$

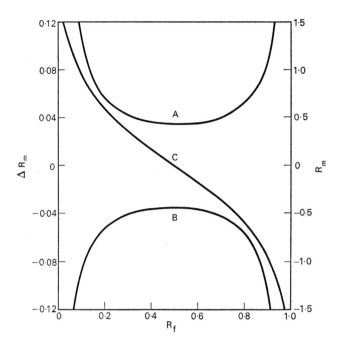

Fig. 10.4. Relationship between R_m and R_f. Curve A: ΔR_m corresponding to $\Delta R_f = -0.020$. Curve B: ΔR_m corresponding to $\Delta R_f = +0.020$. Curve C: Plot of R_m vs. R_f.

where v is the volume of mobile phase in a unit quantity of the wet ion-exchange paper, w is the weight of ion-exchange material in this same quantity of paper, and Q is the specific exchange capacity of the exchanger. From the last two equations

$$\frac{1}{R_f} - 1 = \frac{[CtR_z]\, w\, Q}{[Ct^{z+}]\, v}.$$

Therefore by eqn. (10.9)

$$R_m = \log \frac{[CtR_z]}{[Ct^{z+}]} + \log \frac{wQ}{v}.$$

The last equation may be combined with eqn. (10.10) to eliminate $\log [CtR_z]/[Ct^{z+}]$.

$$R_m = \log E - z \log [El] + z \log [ElR] + \log \frac{wQ}{v}. \tag{10.11}$$

Since E, w, Q, and v are constant for any given paper, cation, and eluent, the first and fourth terms on the right may be combined into a single constant k. Furthermore, the third term on the right can be taken as zero since $[ElR]$ is approximately equal to unity. Thus

$$R_m = k - z \log [El]. \tag{10.12}$$

In a series of experiments in which all conditions are maintained constant except that

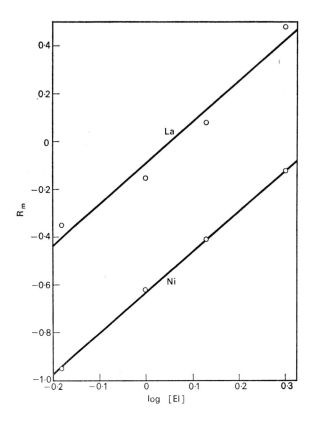

FIG. 10.5. Plots of R_m vs. log concentration of eluent. (Reproduced by permission from *Analytica Chimica Acta.*)

the concentration of eluent is varied, eqn. (10.12) predicts that a plot of R_m vs. $-\log [El]$ will be a straight line with a slope equal to z.

Many experimentally investigated systems do not conform to this prediction. For example Fig. 10.5 is a plot of the data of Lederer and Kertes[98] with paper impregnated with colloidal Dowex 50 and sulphuric acid solutions as eluents. The authors' data with eluents below 0.5 M ($-\log [H^+] > 0.3$) have been excluded from the plot because the R_f values in these solutions are too small to yield reliable values of R_m, also because of the complications caused by the second ionization of sulphuric acid in these dilute solutions. The points for nickel lie on a good straight line, but the slope is 1.70 instead of the predicted 2.00. The points for lanthanum show distinct curvature, and the best linear graph for these points has the same slope as the graph for nickel. The major cause for the failure of eqn. (10.12) in these cases is probably the formation of complexes between the cations and sulphate or bisulphate ion. Sorption of the metal ions by the cellulose and the variation of the activity coefficients in the aqueous solutions (0.5–1.5 M) probably contribute to the discrepancies.

Using paper impregnated with zirconium phosphate and solutions of potassium chloride as eluents, Alberti *et al.*[99] found slopes of R_m vs. $-\log [K^+]$ of 0.80 for sodium, 1.72 for nickel, and 2.49 for lanthanum.

G.II. ION-EXCHANGE CELLULOSE

Cellulose is a polymer of anhydroglucose with the structure

$$-O-CH-(CHOH)_2-CH-O-CH-(CHOH)_2-CH-$$
$$| \qquad\qquad | \qquad\qquad\qquad |$$
$$CH\!-\!\!-\!\!-\!\!-\!\!-\!O \qquad CH\!-\!\!-\!\!-\!\!-\!\!-\!O$$
$$| \qquad\qquad\qquad\qquad |$$
$$CH_2OH \qquad\qquad\qquad CH_2OH$$

Natural cellulose fibres have about 3000 monomers per molecule. Suitable reagents can be made to react with the various functional groups of cellulose to introduce ionogenic groups into the polymer. For example, oxidation of the primary-alcohol group produces *carboxy-cellulose*, which is a weak-acid cation exchanger.[100] However, this reaction is difficult to control; and the product has little importance as an ion-exchange material.

When cellulose is treated with sodium hydroxide and chloroacetate, the carboxylmethyl group is introduced into some of the monomeric units.[101]

$$(C_6H_9O_4)OH + NaOH + ClCH_2COONa \rightarrow (C_6H_9O_4)OCH_2COONa + NaCl + H_2O$$

The resulting carboxymethylcellulose is a weak-acid cation exchanger. Other carboxylic exchangers can be made by esterifying a diprotic acid with cellulose. Succinic, phthalic, maleic, and glutaric anhydrides have been used for this purpose.[102]

$$(C_6H_9O_4)OH + (CH_2CO)_2O \rightarrow (C_6H_9O_4)OOCC_2H_4COOH$$

These products have specific capacities in the range of 0.8–3.2. Since the ester linkage is more reactive than the ether linkage, these cellulose esters are less stable than carboxymethyl cellulose.

Before chemists became interested in the ion-exchange derivatives of cellulose, phosphorylated cotton or *cellulose orthophosphate*, $(C_6H_9O_4)OPO(OH)_2$, was important in the textile industry because it is much less flammable than ordinary cellulose.[103] It is also a cation exchanger of rather unusual properties. Each ionogenic group has two ionizable hydrogen atoms, one rather strong the other weak. It is prepared by the action of phosphoric acid on cellulose in the presence of pyridine or preferably urea.[104]

When cellulose is treated with sodium hydroxide and sodium chloroethanesulphonate. *sulphoethylcellulose* is produced.[105]

$$(C_6H_9O_4)OH + ClCH_2CH_2SO_3Na + NaOH \rightarrow NaCl + H_2O +$$
$$(C_6H_9O_4)OCH_2CH_2SO_3Na$$

This is a strong-acid cation exchanger.

The simplest example of an anion-exchange derivative of cellulose is *aminoethylcellulose*, prepared according to the reaction[106]

$$(C_6H_9O_4)OH + NaOH + NaO_3SCH_2CH_2NH_2 \rightarrow Na_2SO_3 + H_2O +$$
$$(C_6H_9O_4)OCH_2CH_2NH_2$$

As is to be expected, this is a weak-base exchanger. An analogous reaction with diethylamino-2-chloroethane yields *diethylaminoethylcellulose*[106] $(C_6H_9O_4)OCH_2CH_2NEt_2$. If this compound is refluxed with a solution of ethyl iodide in ethanol, a strong-base anion exchanger is produced $(C_6H_9O_4)OCH_2CH_2NEt_3I$.

TABLE 10.17. SOME COMMERCIALLY AVAILABLE CELLULOSIC
ION EXCHANGERS

Product	Approximate specific capacity Q	
	Paper	Powder or flocs
Cation exchangers		
Sulphoethylcellulose		0.2
Cellulose phosphate	2.1[a]	0.8[a]
Carboxymethylcellulose	0.5	0.7
Anion exchangers		
Guanidoethylcellulose		0.4
Triethylaminoethylcellulose		0.5
Diethylaminoethylcellulose	0.4	0.4–0.9
Ecteola cellulose	0.2	0.3
Aminoethylcellulose	0.6	0.9
Polyethylenediaminecellulose		0.2
p-Aminobenzylcellulose		0.2

[a] These are the capacities when both hydrogen ions of the phosphate group are replaced.

Table 10.17 lists some of the ion-exchange derivatives of cellulose that are available from laboratory supply houses. Both the cation and anion exchangers are listed in the order of decreasing acid or basic strength. Guanidoethylcellulose is $(C_6H_9O_4)OCH_2CH_2NHC\text{-}(NH_2) = NH_2Cl$. Ecteola cellulose is prepared by the action of epichlorohydrin and triethanolamine on cellulose. Although its structure is somewhat ill defined, it is known to contain weakly basic amino groups. Polyethylenediaminecellulose contains the groups —$(CH_2CH_2NH—)_x$—$CH_2CH_2NH_2$ where x is variable. It should be emphasized that only a fraction of the $(C_6H_9O_4)OH$ units of cellulose are linked to ionogenic groups in the foregoing products.

The properties and applications of the powder and floc types of ion-exchange cellulose are discussed in section H.I of this chapter.

G.II.a. *Chromatography with Paper Made from Ion-exchange Cellulose*

Wieland and Berg[107] reasoned that electrolytes would probably be sorbed more and with greater selectivity by cellulose containing ionogenic groups than by ordinary cellulose. In collaboration with Zelstoff–Fabrik Waldhof, they prepared papers of carboxycellulose with various percentages of COOH groups. Then they used these papers for a study of the chromatographic behavior of some amino acids and some metallic cations. Table 10.18 indicates some of the results. As is to be expected, decreasing the concentration of the eluent and increasing the specific exchange capacity of the paper increases the sorption of arginine and decreases its R_f. With the paper containing 1% of carboxyl groups and with 0.05 M ammonium formate adjusted to pH 7, 8, 9, and 10, the R_f values of arginine were respectively, 0.26, 0.27, 0.30, and 0.35. The increases in pH caused decreasing fractions of the amino acid to exist in the cationic form and hence less sorption and larger R_f values.

With this eluent at pH 9 and with the same paper, the R_f values of lysine and histidine are 0.51 and 0.72 respectively. Thus the three amino acids are easily separated. The same investigators also separated lead, bismuth, and mercury on this paper using 0.05 M ammonium formate adjusted to pH 3. These were the first separations accomplished by paper chromatography on ion-exchange cellulose.

TABLE 10.18. R_f VALUES OF ARGININE ON CARBOXYCELLULOSE PAPER
ELUENT = HCOONH$_4$ ADJUSTED TO pH 9.0

Molarity of HCOONH$_4$	Percentage of COOH groups in paper			
	1	2	3	4
0.05	0.30	0.14	0.04	0.01
0.1	0.47	0.33	0.28	0.18
0.5	0.66	0.58	0.53	0.52
1	0.76	0.71	0.67	0.65

Kember and Wells[104] studied the behavior of metals on cellulose phosphate paper with sodium chloride as eluent. The spots were diffuse, and they overlapped if the sample contained more than three metals.

In working with paper consisting of unaltered cellulose, mixtures of several constituents that can not be separated by the usual techniques of paper chromatography can often be separated by the two-dimensional method. This consists of eluting the sample spot in one direction with a suitable eluent, drying the paper and then eluting in a direction perpendicular to the first with a different eluent. Thus, two constituents that give overlapping spots with the first eluent may be separated by the second.

Knight[108] wrote: "In the separation of a series of related substances . . . it is common in conventional two-dimensional paper chromatography to obtain separations in which most of the spots are concentrated in one diagonal half of the paper, since in many cases the relative R_f values of the series of substances are similar in a whole range of different organic solvents." He then pointed out that this disadvantage is avoided if ion-exchange paper is used and if the eluents are chosen so that the ionogenic group is ionized in one and non-ionized in the other. For example, he separated mixtures of mercury, cadmium, cobalt, manganese, zinc, copper, bismuth, and iron, eluting first with the organic phase obtained by shaking butanol with 2 M aqueous hydrochloric acid and then with 0.5 M aqueous magnesium chloride. Sufficient hydrogen chloride dissolved in the butanol to repress the ionization of the —O—PO(OH)$_2$ groups of the paper. Thus the metals migrated through the paper almost as if it had been unmodified cellulose. With the second eluent, ion exchange occurred and the selectivity of the phosphate group toward the several metals played an important role in governing their migration. Knight also separated other mixtures of six or seven metals by similar methods.

Moore[109] found carboxymethylcellulose paper very helpful in the qualitative detection of the anions in one drop of a solution of mixed electrolyte.

He put the drop of unknown solution on a 9-mm circle of unmodified chromatographic paper. He placed this paper in the center of a 5.5-cm circle of carboxymethylcellulose paper in the sodium form. Then he let water drop upon the small paper disk. As the water flowed outward through the larger paper, it carried the electrolytes with it; and the cations of the sample were retained by ion exchange. Thus only the sodium salts of the anions of the sample reached the circumference of the larger paper. This operation occurred in a ring oven, a device which heats the circumference of the paper and thus evaporates the water as it reaches the circumference. He then cut a narrow band of the paper including the circumference. He cut this into several fractions, dissolved the sodium salt from each fraction in a small volume of water, and applied qualitative tests for the various anions in the solutions thus obtained.

The advantage of this method is that the tests for the anions are not subject to the possible interference of the several cations in the sample.

The separation of the lighter rare earths by paper chromatography with unmodified cellulose is very unsatisfactory.

Cerrai and Triulzi[110] found that the R_f values of the eight lightest of these metals (La through Gd) on ordinary chromatographic paper with eluents of 0–3 M nitric acid in aqueous methanol containing 0–99.9 volume % of the alcohol were close together and always between 0.54 and 0.86. However, these ternary eluents give better separations on diethyl-aminocellulose paper. Increasing the concentrations of either methanol or nitric acid causes decreases in R_f values and better separation. The reason is that the methanol and nitric acid favor the formation of anionic complexes. The authors give the composition of eight eluents, each consisting of water, methanol, and nitric acid, each of which is able to isolate three elements of the mixture. The behavior of these metals on this anion-exchange paper is similar to their behavior in a column of anion-exchange resin. It is a curious fact, however, that the impregnation of ordinary chromatographic paper with trioctylamine, a liquid anion exchanger, does not decrease the R_f values of these metals with these ternary eluents.

Paolini and Serlupi-Crescenzi[111] studied the separation of the nucleotides on diethyl-aminoethylcellulose paper. Elution with acetate–citrate buffers served to separate mixtures of the orthophosphates of adenosine, guanosine, cytidine, and uridine. Furthermore, mixtures of the ortho-, pyro-, and triphosphates of any one of these compounds could be separated either by chromatography or electrophoresis on this paper. They obtained a nearly perfect separation of all twelve compounds by applying first electrophoresis and subsequently chromatography with the eluent migrating at a right angle to the previous direction of the electric current.

Hartel and Pleumeekers[112] used a paper with quaternary nitrogen groups in the quantitative determination of cysteic acid in protein hydrolysates. The eluent in ascending chromatography was 0.34 M chloroacetic acid. After the elution, they treated the dried paper with ninhydrin to develop the red spots; and they used a densitometer to determine the quantity of cysteic acid in its spot. According to the authors, no amino acid found in protein gives an overlapping spot with cysteic acid. They tested the precision of the method by doing sixteen determinations of this acid in the hydrolysate of 1 g of wool. The mean value was 1.25% with a relative mean deviation of 3%.

G.III. ION-EXCHANGE-RESIN PAPER

The successful applications of ion-exchange cellulose in paper chromatography led M. Lederer to try paper in which an ion-exchange resin was incorporated. In 1955, after unsuccessful attempts to persuade English and French paper manufacturers to prepare such a product, he made his own by drawing strips of Whatman No. 1 paper through aqueous suspensions of carefully purified colloidal aggregates of Dowex 50 in the ammonium form.[113] He dried these strips overnight at room temperature on sheets of Whatman No. 1. With these strips he succeeded in separating yttrium, cerium, and lanthanum, eluting with 3% citrate adjusted to pH 3.

A few years later, Rohm and Haas started to market chromatographic paper containing about 45% by weight of their ion-exchange resins. This company now manufactures four kinds of ion-exchange-resin paper: SA-2 containing a strong-acid cation-exchange resin in the sodium form; SB-2 containing a strong-base anion-exchange resin in the chloride form; WA-2 containing a hydrogen-form carboxylic resin; and WB-2 containing a weak-base resin in the free-base form. In 1959 Peterson[114] used the paper impregnated with IRA-400 to separate small amounts of uranium from 250 times as much bismuth.

Ion-exchange-resin papers are used frequently for the qualitative separation and identification of metals. In a series of papers,[115–118] Sherma has recommended about fifteen eluents, each one of which in conjunction with the appropriate ion-exchange-resin paper, will isolate one or two metals from mixtures of all or nearly all of the common metals. For example, in an eluent of 0.0125 M nitrilotriacetic acid in 3.0 M ammonia, silver exists as $Ag(NH_3)_2^+$ and migrates on SB-2 paper with an R_f of 0.77. Thallium(I) exists as a complex anion with little affinity for the resin and has an R_f of 0.21. All of the other thirteen metals studied under these conditions formed anionic complexes that were tightly bound by the resin. They had R_f values between 0.00 and 0.10. Therefore, this paper and eluent serve to detect both silver and thallium in a mixture of all of the common metals.[115]

Another example is the isolation of arsenic(III) and zinc from a mixture of twenty-seven cations[116] with SA-2 paper and 0.50 M sodium hydroxide as eluent. Arsenic(III) and zinc are anionic in this solvent and migrate with R_f values of 0.85 and 0.57 respectively. Many of the other metals are precipitated as hydroxides and remain at the origin. Others such as barium are cations and migrate only slightly because they undergo ion-exchange absorption.

In another paper,[119] Sherma and Cline discuss qualitatively the theory of the migration of metals on SB-2 paper. They studied the migration of twenty-six metals with five different aqueous eluents, each containing a complexing anion. They also did electrophoretic experiments with each of the metals in each of the solvents to ascertain whether the metal had a positive or negative charge or no charge. When no electromigration occurred, they did test-tube experiments to ascertain if the metal was precipitated by the eluent. Their conclusions may be summarized as follows:

(1) In an eluent where the metal bears a positive charge, it generally migrates with a large R_f. However, they observed eleven exceptions to this statement. For example, silver and aluminum have R_f values of 0.00 in 0.05 M tartaric acid. These cations are probably adsorbed by the cellulose.

(2) Negatively charged metals ordinarily have small R_f values. The largest R_f that they observed in such a case was 0.55 for aluminum with 0.020 M EDTA at pH 11. In this case, the selectivity coefficient of the anionic metal complex is small.

(3) Uncharged soluble metal complexes usually have large R_f values, but fourteen exceptions were observed. For example, in the 0.50 M tartaric acid, both mercury(I) and mercury(II) had R_f values of 0.00. This may be due to adsorption by the paper. It may also be due to an equilibrium in solution involving cationic, neutral, and anionic species of mercury; then an ion-exchange reaction on the part of the anionic mercury could displace the equilibrium so that the mercury was quantitatively retained as an anion by the resin.

(4) Precipitates would be expected to remain at the origin, but cobalt migrated (presumably as a colloidal precipitate) with an R_f of 1.00 in an eluent of 0.020 M EDTA at pH 11. A few other precipitates migrated with much smaller R_f values.

In the same paper,[119] the authors discussed the anomalous behavior of tin(IV) with 0.50 M tartaric acid as eluent. The tin is anionic in this solvent and is rapidly eluted from a column of strong-acid cation-exchange resin. Nevertheless, with the same solvent and SA-2 paper the R_f value of tin is 0.00. The retention of the tin can not be caused by adsorption by the cellulose because R_f is 1.00 on a paper of pure cellulose with the same solvent. The authors suggested that the tin may be adsorbed by the binder used to hold the resin to the paper, but Sherma rejected this explanation in a later paper.[120] In conclusion, the authors warn that there are striking exceptions to the oft-repeated statement that the behavior of metals on ion-exchange-resin paper is analogous to their behavior in a column with the same resin and eluent. On the other hand, Grimaldi[121] points out that misleading results may be obtained if the spots are applied to the dry-resin-impregnated paper. The correct technique, he notes, is to apply the metal solutions to the wet paper behind the solvent front.

Sherma[122] also discusses the effect of the impregnation of ion-exchange-resin papers with water-insoluble organic reagents. For example, he treated SA-2 and a pure-cellulose paper with an ethanolic solution of phenylbenzohydroxamic acid and let them dry in the air. He studied the behavior of several metals on these papers, also on unimpregnated SA-2 paper, using 0.50 M hydrochloric acid as eluent. The results are summarized in Table 10.19. Note that cadmium is isolated only with the paper that contains both the resin and the hydroxamic acid. Other examples are given for papers impregnated with 8-hydroxyquinoline and dimethylglyoxime.

TABLE 10.19. EFFECT OF IMPREGNATION OF ION-EXCHANGE-RESIN
PAPERS WITH PHENYLBENZOHYDROXAMIC ACID
Eluent = 0.50 M hydrochloric acid

Metal	R_f Values $\times$ 100		
	SA-2 paper alone	Cellulose paper + impregnant	SA-2 paper + impregnant
Cd	25–46	91–100	25–42
V(V)	8–25	0–98	< 10
Sn(IV)	35–45	0	0
Sb(III)	11–39	0–56	< 10
Hg(II)	35–45	100	0–18

Heininger and Lanzafama[123] studied the behavior of the five lightest rare earths and yttrium in centrifugal paper chromatography on SA-2 paper with glycolate buffers as eluents. They used a circular sheet of this paper, 15 cm in radius. They drew a pencilled circle 4 cm in radius concentric with the circumference of the sheet. They spotted the samples along this line. Then they sprayed eluent on the center of the paper at a rate of 5 ml/min while rotating the disc at 1000 rev/min. Since the eluent traveled beyond the edge of the paper, R_f values have no significance in this type of chromatography. Therefore they express their results as the distances moved by the spot in a given time. As is to be expected, the distance increased with increasing eluent concentration, eluent pH, and atomic number of the rare earth. For example, centrifugation for 9 min with 0.40 M glycollate at pH 3.75 separated lanthanum, cerium, and neodymium, their spots moving 3.7, 6.1, and 8.5 cm respectively.

Ossicini[124] determined the R_f values of twenty-one anions on SB-2, WB-2, and cellulose paper with three different concentrations of potassium nitrate as eluent.

Although ion-exchange-resin papers have been used mostly for the separation of inorganic substances, they may also serve for the separation of organic compounds. Hüttenrauch and Klotz[125] separated several vitamins of the B group with WA-2 paper in the acetate form using water as eluent. Sherma and his students have extended the principles of solubilization chromatography (p. 186) to the separation of phenols,[126] alcohols[127], and ketones[128] on ion-exchange-resin paper.

One of the few quantitative applications of ion-exchange-resin paper is the determination of ϵ-aminocaproic acid.[129] When a mixture of amino acids is spotted on SB-2 paper and eluted with an acetate buffer, this acid gives a spot isolated from all of the natural amino acids. After development of the spot by the ninhydrin-copper method, McNicol[129] et al. determined the quantity of ϵ-aminocaproic acid with a recording densitometer. The mean relative error was less than 20%.

Lewandowski and Jarczewski[130] used cation-exchange resin paper not to separate amines but to determine the amount of a qualitatively known pure amine. They put the unknown solution into one depression of a spot plate and four standard amounts of the same amine into other depressions of the same spot plate. (The amine was dissolved in water or in dilute aqueous hydrochloric acid or ethanol.) They diluted the liquid in each depression to 1.00 ml and inserted in each a strip of Whatman No. 1 paper, 2 mm wide, the upper end of which touched a piece, 80 × 10 mm, of paper containing 10% of phenolsulphonic resin. The liquid ascended through the strip of ordinary paper and more slowly through the strip of ion-exchange paper. A nearly quantitative ion-exchange reaction occurred at the bottom of the ion-exchange paper so that the amine formed a sharp band at the bottom. They put some lignin at the top of each strip of ion-exchange paper to absorb the liquid as it migrated to the top. When no more liquid remained in the depression of the spot plate, they added two more drops of solvent to wash into the ion-exchange paper the amine remaining in the ordinary paper. Then they developed the color in the ion-exchange-resin paper and measured the area of the colored zone at the bottom of the strip.

Plots of this area vs. the amount of amine in the sample were always linear, but the extrapolation of the curves did not pass through the origin (Fig. 10.6). This means that the linearity must fail near the origin. They measured the area of the colored zone of the paper strip containing the unknown amount of amine and read the amount of this amine from the graph. The relative errors were 1–3%.

This paper[130] also has references to other publications describing the determination of other organic compounds by analogous methods.

G.IV. PAPER IMPREGNATED WITH INORGANIC EXCHANGERS

Alberti and Grassini[131] prepared a filter paper impregnated with zirconium phosphate. Thereafter papers were prepared with other inorganic ion exchangers. These include zirconium and titanium oxides,[132] zirconium molybdate,[133] and zirconium tungstate.[134] In general, the procedure for preparing such a paper consists in dipping the paper into a solution of the cation of the desired precipitate, letting the paper dry without washing it, dipping the paper into a solution that will produce the desired precipitate, and washing and drying it.

Alberti and Grassini[135] used paper impregnated with ammonium phosphomolybdate to separate the alkali cations. Although they could not find an eluent capable of separating all five cations in one step, they succeeded in separating them by the following method. The first eluent in ascending chromatography was 0.1 M with nitric acid and 0.2 M with ammonium nitrate. Cesium ($R_f = 0.00$) and rubidium ($R_f = 0.06$) overlapped, potassium was isolated ($R_f = 0.27$), while sodium ($R_f = 0.73$) and lithium ($R_f = 0.78$) overlapped. They cut the strip into three parts and used the central (smallest) part for the potassium test. They immersed the lower edge of the lowest part in 0.2 M nitric acid in 3.5 M ammonium nitrate. This moved cesium and rubidium with R_f values of 0.10 and 0.60 respectively, thus separating them. Similarly, they used 95% ethanol on the highest part to separate the sodium and lithium.

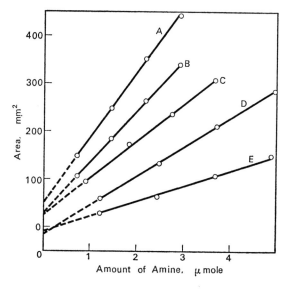

FIG. 10.6. Area of zone as a function of quantity of amine. A = o-aminobenzoic acid, B = p-aminobenzoic acid, C = phenylhydroxylamine, D = triethanolamine, E = methylamine.

Adloff[134] gives a list of eight binary mixtures of metals that can be separated from each other with paper impregnated with inorganic exchangers. Sastri and Rao[136] studied the separation of the two valence states of one element. The elements were iron, uranium, cesium, arsenic, chromium, vanadium, molybdenum, and mercury. Cabral[133] separated radioactive calcium, strontium, and barium. After the elution, he dried the paper and measured the radioactivity at various distances from the origin. A plot of these activities vs. the distance gives three nearly Gaussian curves with very small overlap.

Organic compounds have also been separated by chromatography with paper impregnated with inorganic exchangers. Catelli[137] gives the R_f values of twelve amino acids on zirconium phosphate paper with acetate buffers of three different pH's. He separated mixtures of two or three amino acids. Coussio et al.[138] conducted similar studies with the alkaloids.

G.V. PAPER IMPREGNATED WITH LIQUID ION EXCHANGERS

Testa[139] was the first to perform paper chromatography on paper impregnated with liquid ion exchangers. In his first experiments he applied the spots of aqueous solutions of the unknowns and standards to the paper, then dried the paper, soaked it in a solution of 0.2 M tri-*n*-octylamine in kerosene, dried it again, and finally dipped the end into the aqueous eluent. In later experiments he immersed the paper first in the nonaqueous solution of the liquid ion exchanger, dried it, and then used it as in the usual methods of paper chromatography. Many other liquid ion exchangers have been used in paper chromatography. These include tri-*n*-octylphosphine oxide,[140] polyethylenimine,[141] di-(2-ethylhexyl) phosphate,[142] and dinonylnaphthalene sulphonic acid;[143] they are generally applied to the paper before the spots of sample or standard.

Cerrai and Testa[140] prepared paper strips that contained both an anion and a cation exchanger. They dipped an end of the strip into a solution of tri-*n*-octylphosphine oxide and allowed the solution to progress up to a certain point. After letting the strip drain and dry, they dipped the opposite end into a solution of di-(2-ethylhexyl) phosphate until this solution had risen to the front of the phosphine oxide. After a second draining and drying, the strip was ready for use. The authors used these strips to separate some rare-earth mixtures that could not be separated with paper containing only one liquid ion exchanger.

Although liquid-ion-exchanger papers have been used for the separation of a wide variety of metals, and also of some organic compounds, their principle application is probably the separation of the rare earths. Testa studied the behavior of these metals on paper impregnated with tri-*n*-octylamine[144] using eluents of lithium nitrate containing 0.002 M nitric acid to prevent the hydrolysis of the cations. The plot of R_m for any given eluent vs. the atomic number of the rare earth was approximately linear with a slope of -0.151, i.e. the average value of ΔR_m between two adjacent rare earths was 0.151. Thus we may write

$$R_{mA} - R_{mB} = 0.151,$$

where A represents a rare earth of atomic number one unit less than that of rare earth B. Substitution of eqn. (10.11) in the foregoing equation yields

$$\log \frac{E_A}{E_B} = 0.151,$$

$$E_A/E_B = 1.42.$$

This rather large ratio of the selectivity coefficients of adjacent rare earths indicates that chromatography with paper impregnated with this amine is an efficient method of separating very small quantities of rare earths.

Cerrai and Ghersini[145] studied the separation of the alkaline earths on paper impregnated with di-(2-ethylhexyl) phosphate, using acetic acid as the eluent. Plots of R_m of any one cation vs. the logarithm of the hydrogen-ion concentration were generally linear with slopes of -2, corresponding to the reaction

$$Ct^{2+} + 3H_2Y_2 \rightleftharpoons CtY_2.4HY + 2H^+$$

However, with concentrations of acetic acid above about 0.6 M and with large contents of the liquid exchanger on the paper, the slopes were -1, corresponding to the equilibrium

$$CtC_2H_3O_2^+ + 3H_2Y_2 \rightleftharpoons Ct(C_2H_3O_2)HY_2.4HY + H^+$$

H_2Y_2 denotes the molecule of di-(2-ethylhexyl) phosphate, which exists as a dimer.

In another paper the same authors[146] studied the effect of the concentration of the di-(2-ethylhexyl) phosphate in the paper. This may be denoted $[H_2Y_2]_p$ and may be expressed in $\mu g/cm^2$. They observed that the concentration in the paper is not proportional to the concentration in the benzene solution used to impregnate the paper. A derivation similar to that of eqn. (10.12) indicates that a plot of R_m vs. the logarithm of $[H_2Y_2]_p$ should be linear with a slope equal to the number of molecules of H_2Y_2 that react with one cation. The authors confirmed this relationship experimentally in the chromatography of calcium and magnesium with acetic acid of constant concentration as eluent but with varying values of $[H_2Y_2]_p$.

G.VI. COMPARISON OF PAPER CHROMATOGRAPHY WITH COLUMN CHROMATOGRAPHY

Any ion-exchange material that can be used in a column can also be incorporated either mechanically or chemically in paper. The advantages of paper chromatography are that it uses much smaller samples, eliminates the need of fraction collectors and monitors, and requires much less time. Also, in the separation of radioactive constituents, absorption of the rays by the paper is much less troublesome than absorption by a column.

On the other hand, the precision and accuracy of quantitative analysis by column chromatography is much better than by paper chromatography because of the large relative errors inherent in the analysis of very small samples. In fact, the vast majority of the articles on paper chromatography have dealt only with qualitative analysis. Even in the field of qualitative analysis, however, column chromatography has the advantage of being able to separate more complicated mixtures. Most of the methods of paper chromatography excepting, of course, two-dimensional chromatography, are capable of isolating only three constituents of a mixture whereas column chromatography has been used for the separation of each compound in a fifty-component mixture. The reason for the inferiority of paper chromatography in this respect is that the relative spreading of spots on paper is greater than that of zones in a column.

H. Ion-exchange Materials from Carbohydrates

After the successful use in paper chromatography of cellulose with ionogenic substituents (p. 240), it occurred to Sober and Peterson[147] that these substituted celluloses in the form of flocs or powder would probably be useful as the stationary phase in ion-exchange column chromatography. The very porous structure of cellulose permits rapid diffusion of even very large ions, such as the proteins, inside the cellulosic exchangers and hence, rapid exchange. In contrast, ions of this size are unable to penetrate the ordinary resinous or inorganic exchangers. In this paper, the authors describe the elution of 270 mg of dialyzed, lyophylized, kidney proteins through 5.0 g of diethylaminoethylcellulose. The eluent was 0.005 M phosphate at pH 7.0 gradiently changed to 0.1 M disodium phosphate in 0.5 M sodium chloride. The graph shows three well-defined peaks.

H.I. PREPARATION OF CELLULOSIC ION EXCHANGERS

The reactions that are used to introduce ionogenic groups into the cellulose of paper (p. 239) can also be used, of course, to prepare cellulosic exchangers in the form of powder or flocs.[148] The Bio-Rad products (distributed by Calbiochem, Los Angeles 63, California) are marketed as small cylinders, averaging 18 μ in diameter and 80 μ in length.

Cellulosic exchangers with specific capacities greater than those listed in Table 10.17 can be prepared; but they have the disadvantage that contact with water causes them to swell, become gelatinous, and impede chromatographic flow. This disadvantage can be overcome by treating the cellulose with a crosslinking agent such as formaldehyde or alkaline 1,3-dichloro-2-propanol. For example, Guthrie and Bullock[149] prepared diethylaminoethylcellulose by treating cotton three times with sodium hydroxide and diethylamino-2-chloroethane. Each treatment caused additional substitution; and the successive products had specific capacities of 0.75, 1.69, and 2.57. The last product was so gelatinous that it did not permit any flow through the column, but a product made by similar treatment of cellulose previously crosslinked by formaldehyde had a specific capacity of 2.92 and good chromatographic properties.

In a later paper, Guthrie et al.[150] raised serious doubts about the composition of products that had previously been regarded as strong-base cellulosic anion exchangers. They believe that diethylaminoethylcellulose can be quaternized with methyl iodide or ethyl bromide only under strictly anhydrous conditions that had not been employed in previous attempts to prepare this quaternary exchanger. In support of their contention that a commercially available "triethylaminoethylcellulose" contained mostly the diethylaminoethyl groups "with only slight conversion to a quaternary", they cite an experiment in which 1 g of the chloride-form exchanger was completely converted to the hydroxide form by batchwise equilibration with 100 ml of 0.01 M aqueous sodium hydroxide. These authors also present data of conductometric and potentiometric titrations to indicate that they succeeded in synthesizing the strong-base, quaternary exchangers.

Veder[151] prepared numerous batches of Ecteola-cellulose, varying the ratio of epichlorohydrin and of triethanolamine. In most experiments he used 30 g of cellulose and a total volume of 55 ml of epichlorohydrin and triethanolamine. The specific capacities of the products varied from 0.05 to 0.60, the maximum corresponding to 35 ml of epichlorohydrin and 20 ml of triethanolamine. An even larger Q, 0.78, was obtained by a repetition of the method of Peterson and Sober.[148]

Amphiprotic cellulosic exchangers have been prepared by the reaction between cellulose and methyl or ethyl esters of ethylenimino-N-acetic acid, β-ethylenimino-N-propionic acid, or β-ethylenimino-N-ethylphosphonic acid and subsequent hydrolysis of the ester groups.[152]

E. and K. Randerath[153] impregnated 20 g of cellulose with polyethylenimine by treating it with 120 ml of a 3% solution. After being washed, the product may be used as an anion exchanger. Or it may be treated with Graham salt and washed. Thus some of the long-chain phosphate ions are bound to the imine groups by a fraction of their charges, leaving other —PO_4^--units of the chain free to function as cation exchangers.

Since starch is a carbohydrate with a structure very similar to that of cellulose, it is not surprising that it has also been converted to an ion-exchange material. Wettstein et al.[154] treated starch with phosphorus oxychloride and pyridine. This caused the substitution of —$OPO(OH)_2$ for some of the hydrogen atoms of the starch. Also, some crosslinks are formed by —$OPO(OH)O$— groups.

Sephadex† is a crosslinked dextran gel. It is manufactured in various degrees of crosslinking and is used principally to separate mixtures into fractions of varying molecular size. Small solute molecules can enter the pores of the gel more easily than large solute molecules.

† Pharmacia Fine Chemicals Inc., 800 Centennial Avenue, Piscataway, N.J., 08854.

Therefore, elution of the mixture with water through a Sephadex column causes the larger molecules to emerge before the smaller ones. The ionogenic groups, $-C_2H_4-NEt_2$, $-CH_2COOH$, and $-C_2H_4SO_3H$, have been substituted in Sephadex to yield ion-exchange materials.

H.II. APPLICATIONS OF CELLULOSIC ION EXCHANGERS

The chief advantage of the exchangers prepared from carbohydrates in comparison with the resinous and inorganic exchangers lies in their very porous structure which allows very large ions to diffuse readily inside of the exchanger. Of course, resins of sufficiently low DVB content may also have a very porous structure; but such resins are too soft to permit satisfactory flow through the column and, in addition, undergo too much swelling and shrinking with changes in the total concentration of salt. Only a few representative examples of the many applications of cellulosic exchangers can be mentioned in this book.

H.II.a. *Proteins*

Sober et al.[155] studied the behavior of blood sera from the horse and from man on columns of dimethylaminoethylcellulose, 44 cm $\times$ 4.5 cm². The eluents were 0.005 M sodium phosphate of pH 7.0 at the beginning of each elution. This was gradiently changed to lower pH and greater concentrations of both sodium phosphate and chloride. Flow rates were about 0.041 cm/min, maintained by an eluent head of about 100 cm. Samples contained about 140 mg of protein nitrogen. A complete elution required about 8 days. The elution graphs indicated the partial separation of 10–20 constituents, indicated by peaks or shoulders.

Fahey and Horbett[156] separated human γ-globulins into five fractions on a 30-cm column of the same exchanger. The fractions differed from each other in their immunologic properties.

An example of the great fractionating ability of diethylaminoethylcellulose is found in the work of Finlayson and Mosesson.[157] They used the method of Blombäck[158] to separate a fraction of human blood plasma and then subjected this sample to chromatography. They found two incompletely separated peaks. The first contained 77% of the protein, but rechromatography of the second peak revealed that it contained 34% of the first subfraction. Thus the Blomback fraction contained 85% of the first subfraction and 15% of the second. These two subfractions could not be distinguished by their behavior in the ultracentrifuge, in immunologic tests, in their solubility in aqueous ethanol, in clotting times, in ultraviolet spectra, or in their terminal amino acids. They did differ slightly in electrophoretic behavior, the major subfraction being slightly less negative at pH 5.5 and 8.6.

Robinson et al.[159] used chromatography on the extract of hog renal cortex to isolate two enzymes. Both of them catalyze the oxidation of (—)-a-hydroxyacids according to the reaction

$$2C_nH_{2n+1}CHOHCOOH + O_2 \rightarrow C_nH_{2n+1}COCOOH + H_2O$$

but one acts upon short-chain acids while the other acts upon long-chain acids.

Ainbender et al.[160] separated poliovirus from SV_{40} virus using a 10-cm column of diethylaminoethylcellulose.

Maroux[161] et al. used a column of carboxymethylcellulose to purify beef trypsin and trypsinogen.

H.II.b. *Nucleotides*

Staehelin *et al.*[162] subjected a sample of yeast ribonucleic acid to partial depolymerization by pancreatic ribonuclease. Then they eluted it through a 30-cm column of diethylaminoethylcellulose with ammonium bicarbonate containing gradiently increased concentrations of ammonia. The elution graph had twenty-one peaks or shoulders. They identified the first thirteen of these as due to C, U, AC, AU, GC, GU + AAC, AAU, GAC, AGC, GAU, AGU, GGC, and GGU where A, C, G, and U denote respectively the adenylic, cytodylic, guanylic, and uridylic nucleotides. The last eight peaks of the chromatogram were probably due to still unidentified tetranucleotides.

H.II.c. *Carbohydrates*

Deuel and his colleagues[163] subjected wheat starch to chromatography through a column of diethylaminoethylcellulose with eluents of water and aqueous sodium hydroxide of discontinuously increasing concentration. The graph showed four very distinct peaks representing starches of different degrees of polymerization, the largest molecules emerging last from the column. The same group[164, 165] studied the structure of pectin by subjecting it to three different methods of partial hydrolysis (acid, base, and enzyme) and then chromatographing the various hydrolyzates. They determined both the ester groups and the polygalacturonic acid in each of the five fractions of each chromatographic separation. These data enabled them to draw conclusions not only about the structure of the pectin but also about the differences in the location of the ester groups hydrolyzed by the three different catalysts.

H.II.d. *Miscellaneous Organic Separations*

Eshelman *et al.*[166] recommend the following method for the determination of trisaturated glycerides in fats:

Treat the sample with mercaptoacetic acid. This adds H and —SHCH$_2$COOH at the double bonds, thus converting the unsaturated fats into acids. Shake with ethanol, petroleum ether, water, and ammonia, and discard the aqueous phase, which contains the excess mercaptoacetic acid and some mercaptoacetic-glycerides. Pass the organic phase through a column of diethylaminoethylcellulose to remove the rest of the mercaptoacetic-glycerides. Evaporate the effluent to dryness and weigh the residue of unsaturated fat. Unfortunately, the residue also contains a small amount of mono-olefinic-disaturated glyceride that failed to react with the mercaptoacetic acid. Correct for this error by either of the following methods: (1) Determine the iodine number of the residue and calculate the weight of the unsaturated constituent on the assumption that it is the glyceride of one molecule of oleic acid with two molecules of the most abundant saturated fatty acid in the sample. (2) Repeat the entire procedure until the amount of unsaturated compound in the residue is negligible, as indicated by the iodine number.

The data obtained on a sample of lard indicate how serious may be the contamination of the residue with unsaturated fat. The first residue amounted to 6.72% of the sample and had an iodine number of 19.6. Each repetition of the procedure gave a smaller residue with a smaller iodine number. After three repetitions, the residue was 2.72% of the sample and had an iodine number of 2.4.

Quaternary ammonium compounds are sometimes added to animal feeds in very small amounts. The colorimetric determination of these compounds is subject to interference from fats and colored compounds in the sample. Metcalfe[167] recommended the following method.

Pass a chloroform extract of the sample through a column of carboxymethylcellulose. The quaternary nitrogen compounds are retained by the exchanger. Wash the column with ethanol and water. Then pass 1 M hydrochloric acid in ethanol through the column followed by water. The quaternary compounds are now in the eluate. Add bromophenol blue and shake with chloroform. The quaternary nitrogen ions accompanied by an equivalent amount of indicator anions enter the organic phase. Finally, determine photometrically the amount of bromophenol blue in the chloro- form. The author analyzed eight samples of feeds to which known amounts (4–120 ppm) had been added. The standard relative error was 4.0%.

Hendrickson and Ballou[168] used a column of diethylaminoethylcellulose to isolate three inositides (derivatives of cyclohexanehexol) from human and beef brains. They subjected the Folch fraction of the brain tissue to gradient elution, using as eluent a mixture of chloro- form, methanol, and water containing 0.0–6.0 M ammonium acetate. The first compound to be eluted was the calcium chelate of phosphoinositide, giving a nearly Gaussian curve well separated from the other constituents. Two badly overlapping peaks followed, representing two unidentified lipoids. The fourth and fifth curves were almost quantitatively separated and nearly Gaussian in shape; they represented di- and triphosphinositides.

H.II.e. *Inorganic Ions*

The following two applications of cellulose phosphate depend on its ability to absorb cerium, thorium, uranium(IV), uranium(VI), titanium, zirconium, and iron(III) very strongly even from 5 M strong acids.

Head *et al.*[169] used this exchanger for the large-scale separation of thorium from most of the other constituents of monazite.

They treated 1 kg of the ore with 1.1 l. of concentrated sulphuric acid. They diluted this to 5 l., filtered and then diluted to 15 l. They passed this solution through a column, 85 cm × 3.8 cm² of cellulose phosphate until thorium broke through. Then they eluted most of the thorium from the column with 1.0 M ammonium carbonate. A typical ore contained 6.8% thorium oxide and 50% rare-earth oxides, i.e. a weight ratio of 0.136. After the solution of the ore had been passed through the column, the weight ratio of thorium oxide to rare-earth oxide on the column was 67. Since carbonate ion complexes with thorium but not with the rare earths, the elution step increased this ratio still further to 930.

Traces of copper, lead, cadmium, and zinc can be determined very satisfactorily by square-wave polarography without interference from each other. This method is applicable to the determination of traces of these metals in uranium provided that they are first separated from the uranium.

Goode and Campbell[170] used cellulose phosphate for this purpose. They dissolved a 0.5-g sample in hydrochloric acid with the aid of hydrogen peroxide. They evaporated this solution to dryness, dissolved the residue in 2 ml of 1 M hydrochloric acid, transferred it to a tube containing 3 g of cellulose phosphate and passed 2 M hydrochloric acid through the column. This removed the copper, lead, cadmium, and zinc from the column, leaving uranium and iron behind. Before the polaro- graphic determination it was necessary to destroy traces of organic matter in the eluate by evapora- tion with nitric and perchloric acids. It was possible to remove the uranium from the column with sodium carbonate, but this treatment made the exchanger gelatinous so that satisfactory flow rates could not be maintained. Therefore, a fresh column was used for each separation.

I. Thin-layer Chromatography

In thin-layer chromatography[171] the stationary phase, usually an inorganic adsorbent, is caused to adhere, usually with the aid of a binding agent, as a uniform, thin layer to some smooth, flat surface such as a plate of glass or plastic. The technique is very similar to that of paper chromatography. Sample spots are placed near one edge; this edge is submerged in a shallow layer of eluent. The eluent travels upward through the stationary phase by capillary action carrying the spots with it at a slower rate. Two-dimensional migration may also be applied to thin-layer chromatography. The chief advantage of thin-layer chromatography over paper chromatography lies in the more rapid migration and shorter time required for separations. In ion-exchange thin-layer chromatography, any kind of ion-exchange material may be used as the stationary phase.

I.I. CELLULOSIC EXCHANGERS

At first Randerath[172] used Ecteola-cellulose in the separation of nucleotides, but he later used poly(ethyleneimine)-cellulose[173] for similar purposes. A typical example of his work[174] is the separation of the triphosphates of cytosine, adenine, guanine, uracil, hypoxanthine, desoxycytosine, desoxyadenine, desoxyguanine, and desoxythymine, from each other and from the corresponding diphosphates by two-dimensional chromatography. The first eluent was 1.0 M lithium chloride saturated with boric acid. The latter compound formed complex borate anions with the unreduced compounds but not with the desoxy-compounds; thus, the desoxy-compounds had smaller negative charges and were eluted more rapidly. After drying the plate and washing away the boric acid and lithium chloride with methanol, they passed the second eluent, 0.8 M lithium chloride in 1 M acetic acid, in the same direction. This served to separate compounds according to their ionization constants. They dried the plate and washed it again with methanol. Then they eluted it at right angles with ammonium sulphate.

R_f values of 20 amino acids on diethylaminoethylcellulose with nine different eluents have been reported.[175] Wieland and Determann[176] separated two lactic-acid-dehydrogenases with diethylaminoethyl-Sephadex. The eluent was 0.020 M phosphate buffer at pH 7.2 with gradiently increasing concentration of sodium chloride.

I.II. ION-EXCHANGE RESINS

Hüttenrausch et al.[177] were the first to use ion-exchange resins for thin-layer chromatography. Although they had difficulty preparing a layer of sufficient mechanical stability, they succeeded in separating most of the members of the vitamin-B complex on a thin layer of a carboxylic resin with a little collodion as binder. Berger et al.[178] used cellulose as a binder with Dowex 1-X10 to separate radioactive chloride, bromide, and iodide ions. A plot of radioactivity vs. the distance from the origin resembled an elution graph. None of the three peaks was completely isolated. Sherma[179] used starch as binder with Dowex 1 and Dowex 50 to separate metals. His eluents were 0.50 M hydrochloric acid in 74% acetone for the Dowex 50 and 0.6 M hydrochloric acid in 90% methanol for the Dowex 1. He mentions seven pairs of metals that can be separated but apparently did not succeed in the complete separation of a three-component sample. It is likely that the slow diffusion of the ions through the resin in the nearly nonaqueous medium accounts for the rather poor separations.

R_f values of eight metals have been reported[180] on thin layers of hydrous zirconium oxide and zirconium phosphate with eluents of hydrochloric acid, ammonium chloride, and ammonium nitrate. Holzappel et al.[181] have studied the behavior of the rare earths on thin layers of diatomaceous earth impregnated with di(2-ethylhexyl) phosphate, using eluents of hydrochloric or nitric acid. They report the various R_f values and show chromatograms of nine mixtures of three or four metals each that yielded complete separations. Many of these are mixtures of adjacent rare earths.

J. Chromatographic Resolution of Racemic Substances

Many attempts have been made to resolve racemic substances chromatographically. Some experiments in this field were based on adsorption chromatography with optically active natural products such as lactose[182, 183] or starch[184–186] as the stationary phases (Table 10.20).

Cellulose has been used as the stationary phase in the resolution of some amino acids by paper chromatography.[187, 188] Some of these acids are easily resolvable; others have yielded only negative results. Dahlgleish[188] tried to correlate the incidence of successes and failures with the structure of the molecules. He postulates that an amino acid in order to be resolved must establish a "three-point contact" with the cellulose. Van der Waals forces between a benzene ring and the dextrose ring of the cellulose may cause one point of contact. Hydrogen bonds between the hydroxyl groups of the cellulose and carboxyl and amino groups of the acid form the other two points of contact. In addition, the three contact points must be near the asymmetric carbon atom of the amino acid. Most of the experimental observations are in accord with Dahlgleish's postulates. Although each of several racemic amino acids yields a pair of widely separated spots, the quantities of the acids resolved are always very small in ordinary paper chromatography. Kotake et al.[187] avoided this difficulty by using a pile of filter papers, 9 cm in diameter and 15 cm high. Thus they resolved completely a sample of 250 mg of tyrosin-3-sulphonic acid.

Leitch et al.[190] obtained a partial resolution of mandelic acid by elution through Sephadex G-25, a dextran gel, with 3.0 M aqueous sodium chloride as eluent.

The first successful application of optically active ion-exchange resins to chromatographic resolutions was achieved by Grubhofer and Schlieth.[191] They treated Amberlite XE64, a carboxylic resin, with sulphuryl chloride to convert the —COOH groups to —COCl. Then they treated the resin with quinine. They believe that the carbonyl group reacted with the hydroxyl group of the quinine to give a polymeric ester that would be a weak-base resin by virtue of the amino groups in the quinine. Less than 16% of the original carboxyl groups were thus converted to quinine ester groups. The salient features of their resolution with this resin are recorded in Table 10.20. Other investigators treated chloromethylated polystyrene resins with optically active tertiary amines to synthesize optically active strong-base anion-exchange resins. Suda and Oda[192] used brucine for this purpose while Lott and Rieman[193] used (—)-N,N-dimethyl-α-phenethylamine.

TABLE 10.20. SOME CHROMATOGRAPHIC RESOLUTIONS

Expt. No.	Stationary phase	Column H cm	Column A cm²	Racemic substance	Amount (mmol)	Mobile phase	Time (hr)	Recovery Amount (mmol)	Recovery Purity (%)	ε	Ref.
1.	Lactose	130	79	p-Phenylenebisiminocamphor	0.075	Light petroleum + benzene (8/1)	60	0.0068	7.7		182
2.	Lactose	84	44	Troger's base	24	Petroleum ether	120	0.080	26		183
3.	Starch	45	2.8	Mandelic acid	13	Water		0.67	28		184
4.	Starch	150	2.0	Mandelic acid	1.6	Water	21	0.028	100	0.06	185, 186
5.	Sephadex G-25	310	0.78	Mandelic acid	0.71	0.3 M NaCl		0.099	17		190
6.	Quinine ester of Amberlite IRC 50	100	1.1	Mandelic acid	88	$CHCl_3$	38	0.0055	100		191
7.	Brucine resin	120	1.1	Mandelic acid	30	Water	9.3	0.063	27		192
8.	Resinwith—CH_2—NMe_2—CHMePhCl	490	0.27	Sodium mandelate	5.0	Water	200	0.35	8.0	0.005	193
9.	Optically active liq. anion exchanger			Sodium mandelate						0.42	194
10.	Amberlite CG-50, (−)—$Co(en)_3R_3$	16	1.3	KCoEDTA	1.0	0.2 M KCl		1.0	100		196
11.	Dowex 1-X4, (+)—tartrate form	55	2.3	$Co(en)_3Br_3$	0.80	Water	1.2	0.0054	40		196
12.	Dowex 50-W-X2, NaR	1000	1.3	Butanol-2 esterfied with (−)—mandelic acid	32	Water	650	7.2	100	0.082	197
13.	Dowex 50-W-X2, NaR	320	0.78	3-Methylbutanol-2 esterfied with (+)—lactic acid	26	0.50 M Na_2SO	200	2.7	100	0.10	198

Romano[194] *et al.* synthesized an optically active liquid anion exchanger, (−)-*N*-(1-naphthyl)methyl-α-methylbenzylamine, $C_{10}H_7\text{-}CH_2\text{-}NH\text{-}CH(C_6H_5)CH_3$, to which they gave the trivial name S-amine. Batchwise equilibration experiments between a solution of the hydrochloride of this amine in chloroform and an aqueous solution of sodium (±)-mandelate indicated that the selectivity coefficient of the (+)-mandelate relative to the (−)-mandelate was 1.22 to 1.42, depending on the concentrations of the solutions. Attempts to use this liquid anion exchanger in chromatographic resolutions were unsuccessful because the investigators could not find a satisfactory support.

Then the investigators used a Craig countercurrent apparatus consisting of 200 tubes of 80-ml capacity each. With 40.0 ml of 0.151 M (nearly saturated) S-ammonium chloride in each tube as the stationary phase, they "eluted" a sample of 12.1 mmol of sodium (±)-mandelate through the apparatus with 5.00 M aqueous sodium chloride. Although the enantiomers were not quantitatively separated, 3.0 mmol of each was obtained in an optical purity of 99.96%. Then they used the countercurrent apparatus in an experiment analogous to displacement chromatography. The stationary phase was 0.100 M S-ammonium chloride in chloroform, 40.0 ml in each tube. Sixty-seven 40.0-ml portions of 0.100 M aqueous sodium (±)-mandelate were fed into the apparatus, followed by 0.100 M aqueous sodium hydroxide as the displacing agent. Thirty-six mmol of the (−)-enantiomer of mandelate was obtained with an optical purity of 99% or better. An equal amount of the (+)-isomer should have been obtained with the same purity; but the displacing agent, sodium hydroxide, not only entered into the ion-exchange reaction but also reacted with the stationary solvent, probably thus

$$CHCl_3 + 4OH^- \rightarrow 3Cl^- + HCOO^- + 2H_2O$$

Chloroform is the only solvent in which S-ammonium chloride is sufficiently soluble to serve as a satisfactory liquid ion exchanger. However, solutions of S-ammonium benzene-sulphonate in nitrobenzene seem to offer several advantages over the solutions of S-ammonium chloride in chloroform:[194a] (1) The solubility of the benzenesulphonate in nitrobenzene is greater than that of the chloride in chloroform. (2) The solubility of the benzenesulphonate in water is less than that of the chloride. (3) It is likely that aqueous sodium hydroxide will be a satisfactory displacing agent if the exchanger is dissolved in nitrobenzene. (4) The selectivity coefficient of (+)-mandelate vs. (−)-mandelate is slightly greater with S-ammonium benzenesulphonate in nitrobenzene than with the hydrochloride of the amine in chloroform.

J.II. OPTICALLY INACTIVE RESINS
WITH OPTICALLY ACTIVE COUNTERIONS

When a solution of a racemic nonelectrolyte is passed through a column of an ion-exchange resin, it seems reasonable to expect that partial resolution will occur if the matrix, the fixed ionogenic group, or the counterions are optically active. Since the synthetic problem is easiest if the asymmetry of the stationary phase is located in the counterions, Leitch[195] investigated the resolution of nonelectrolytes by chromatography through Dowex 50-X2 with optically active counterions. He used only quaternary ammonium ions for this purpose to avoid loss of primary, secondary, or tertiary ammonium counterions by reactions such as

$$C_6H_5CH(CH_3)NH_3R \rightleftharpoons HR + C_6H_5CH(CH_3)NH_2.$$

He obtained only negative results with *N,N,N*-trimethyl-α-phenethylammonium ion as counterion and α-phenethanol or methyl mandelate as racemates. He obtained a slight but definite resolution of α-phenethanol with the resin in the (—)-methyl-brucinium form, but the instability of this counterion was a serious handicap.

Yoshino et al.[196] also failed in the resolution of mandelic acid and anionic complexes of cobalt(III) using the (—)-cobalt(III) trisethylene-diamine form of Dowex 50W-X4. However, with optically active $Co(H_2NCH_2CH_2NH_2)_3^{3+}$ as the counterion of Amberlite CG-50 (a carboxylic resin), they obtained an apparently quantitative resolution of the CoEDTA⁻ anion as potassium salt. Because of the ion-exchange reaction

$$Co(en)_3R_3 + 3K^+ \rightleftharpoons 3KR + Co(en)_3^{3+}$$

some of the cationic cobalt complex accompanies the anionic complex into the effluent· They avoided this difficulty by passing the effluent through a shorter column of Dowex 50W-X4 in the potassium form.

They also obtained a partial resolution of cobalt(III) trisethylenediamine bromide by eluting it through a column of Dowex 1-X4 in the (+)-tartrate form. They passed the effluent from this column through a column of hydroxide-form Dowex 1 to prevent tartrate ion from contaminating the resolved cobalt cations.

J.III. CHROMATOGRAPHIC SEPARATION OF DIASTEREOISOMERS WITH AN OPTICALLY INACTIVE STATIONARY PHASE

The reaction between a racemic alcohol such as butanol-2 and an optically active acid such as (—)-mandelic acid yields a pair of diastereoisomers. These differ from each other in physical properties and therefore should be separable by elution through an optically inactive ion-exchange resin. Then the hydrolysis of the separated diastereoisomeric esters would complete the resolution of the alcohol. Spitz et al.[197] used this method to resolve butanol-2, obtaining 2.2% of the sample in practically 100% optical purity.

Leitch et al.[195, 198] found (+)-lactic acid to be better for this purpose than mandelic acid because the lactate ester of any given alcohol is more soluble in water than the corresponding mandelate ester. The greater solubility permits the use of more concentrated solutions of the ester and hence larger samples with any given column. Furthermore, the more hydrophylic nature of the lactate esters causes smaller C values and hence permits separations with smaller volumes of eluent. In addition, salting-out chromatography with the subsequent increase in the ratio of C_2/C_1 of the diastereoisomeric esters is applicable to some lactates whereas the corresponding mandelates would be too slightly soluble in the salt solution to allow efficient separations. Leitch studied most thoroughly the resolution of 3-methyl-2-butanol by this method (Table 10.20). He also obtained less satisfactory partial resolutions of pentanol-2 and hexanol-2.

Diastereoisomeric esters have also been separated by gas–liquid chromatography.[199]

Table 10.20 summarizes the data of several chromatographic resolutions. The sixth column gives the amount of racemic (or diastereoisomeric) substance put into the column. The eighth column shows the time required by the chromatographic process, exclusive of the time spent in the preparation of the column. The ninth column reveals the quantity of one enantiomer (or diastereoisomer) recovered in the degree of optical purity indicated in the tenth column. In general, the "recovery" consisted in a determination of the amount and optical purity of the enantiomer (or diastereoisomer) in the eluate fraction, not in separating it from the large volume of solvent and perhaps also from other solutes. The parameter ϵ (eleventh column) is defined by the equation

$$\epsilon = 1 - (C_2/C_1),$$

where C_2 and C_1 are the distribution ratios of the two enantiomers or diastereoisomers. Obviously, large values of ϵ favor efficient separation.

The amounts and/or the optical purities of the enantiomers obtained by chromatographic resolution are not sufficiently great to enable this method to compete with the classical method of fractional crystallization of diastereoisomeric salts. Nevertheless, the data of Table 10.20 prove that chromatographic resolutions are possible and support the hope that continued research in this field will produce chromatographic methods of practical importance.

References

1. R. KUNIN, E. MEITZNER and N. BORTNICK, *J. Am. Chem. Soc.*, **84**, 305 (1962).
2. K. A. KUN and R. KUNIN, *J. Polymer Sci.*, **B2**, 587 (1964).
3. J. R. MILLAR, D. G. SMITH, W. E. MARR and T. R. E. KRESSMAN, *J. Chem. Soc.*, 218 (1963).
4. J. R. MILLAR, D. G. SMITH, W. E. MARR and T. R. E. KRESSMAN, *J. Chem. Soc.*, 2779 (1963).
5. J. R. MILLAR, D. G. SMITH, W. E. MARR and T. R. E. KRESSMAN, *J. Chem. Soc*, 2740 (1964).
6. R. KUNIN, E. F. MEITZNER, J. A. OLINE, S. A. FISHER and N. W. FRISCH, *Ind. Eng. Chem.*, *Product Research and Development*, **1**, 140 (1962).
7. J. DAHLBERG and O. SAMUELSON, *Svensk Chem. Tidskr.*, **75**, 178 (1963).
8. J. S. FRITZ and H. WAKI, *Anal. Chem.*, **35**, 1079 (1963).
9. J. S FRITZ and R. G. GREENE, *Anal. Chem.*, **36**, 1095 (1964).
10. E. A. EMKEN, C. R. SCHOLFIELD and H. L. DUTTON, *J. Am. Oil Chemists' Soc.*, **41**, 388 (1964).
11. J. E. CASSIDY and C. A. STREULI, *Anal. Chim. Acta*, **31**, 86 (1964).
12. J. R. MILLAR, *J. Chem. Soc.*, 1311 (1960).
13. J. R. MILLAR, D. G. SMITH and W. E. MARR, *J. Chem. Soc.*, 1789 (1962).
14. M. J. HATCH, J. A. DILLON and H. B SMITH, *Ind. Eng. Chem.*, **49**, 1812 (1957).
15. M. J. HATCH and H. B. SMITH, *J. Am. Oil Chemists' Soc.*, **38**, 470 (1961).
16. C. ROLLINS, L. JENSEN and A. N. SCHWARTZ, *Anal. Chem.*, **34**, 711 (1962).
17. K. H. MEYER and F. J. SIEVERS, *Helv. Chim. Acta*, **19**, 665 (1963).
18. T. TEORELL, *Discussions Faraday Soc.*, **21**, 9 (1956).
19. F. HELFFERICH, *Ion Exchange*, McGraw-Hill, New York, 1962, pp. 368 ff.
20. F. HELFFERICH, *Ion Exchange*, McGraw-Hill, New York, 1962, p. 385.
21. C. E. MARSHALL, *J. Phys. Chem.*, **43**, 1155 (1939); **48**, 67 (1944).
22. A. S. BASU, *J. Ind. Chem. Soc.*, **39**, 619 (1962).
23. U. SCHINDEWOLF and K. BONHOEFFER, *Z. Elektrochem.*, **57**, 216 (1953).
24. C. W. CARR, *Arch. Biochem. and Biophys.*, **62**, 476 (1956).
25. S. K SINHA, *J. Ind. Chem. Soc.*, **32**, 35 (1955).
26. H. G. SPENCER and F. LINDSTROM, *Anal. Chim. Acta*, **27**, 573 (1962).
27. H. L. ROTHBART, private communication.
28. Y. HARA, *Bull. Chem. Soc. Japan*, **36**, 1373 (1963).
29. S. W. FELDBERG and C. E. BRICKER, *Anal. Chem.*, **31**, 1852 (1959).
30. P. P. L. HO and M. M. MARSH, *Anal. Chem.*, **35**, 610 (1963).
31. R. B. HANSELMAN and L. B. ROGERS, *Anal. Chem.*, **32**, 1240 (1960).
32. S. R. CAPLAN, *J. Electrochem. Soc.*, **108**, 577 (1961).
33. W. T. GRUBB and P. D. ZEMANY, *Nature*, **176**, 221 (1955).
34. P. D. ZEMANY, W. W. WELBON and G. L. GAINER, *Anal. Chem.*, **30**, 299 (1958).
35. C. B. AMPHLETT, *Inorganic Ion Exchangers*, Elsevier, New York, 1964.
36. K. A. KRAUS, H. O. PHILLIPS, T. A. CARLSON and J. S. JOHNSON, *Proc. 2nd Intern. Conf. on Peaceful Uses of Atomic Energy*, Geneva, **28**, 3 (1958).
37. W. J. MAECK, M. E. KUSSY and J. E. REIN, *Anal. Chem.*, **35**, 2086 (1963).
38. S. AHRLAND, I. GRENTHE and B. NOREN, *Acta Chem. Scand.*, **14**, 1059 (1960).
39. S. AHRLAND, J. GRENTHE and B. NOREN, *Acta Chem. Scand.*, **14**, 1077 (1960).
40. F. NYDAHL and L. A. GUSTAFSSON, *Acta Chem. Scand.*, **7**, 143 (1953).
41. F. NYADHL, *Anal. Chem.*, **26**, 580 (1954).
42. L. BAETSLÉ and J. PELSMAEKERS, *J. Inorg. Nucl. Chem.*, **21**, 124 (1961).
43. A. CLEARFIELD and J. A. STYNES, *J. Inorg. Nucl. Chem.*, **26**, 117 (1964).
44. G. ALBERTI and A. CONTE, *J. Chromatog.*, **5**, 244 (1961).
45. S. AHRLAND, J. ALBERTSON, L. JOHANSSON, B. NIHLGARD and L. NILSSON, *Acta Chem. Scand.*, **18**, 1357 (1964).

46. L. BAETSLÉ, *J. Inorg. Nucl. Chem.*, **25**, 271 (1963).
47. O. D. BONNER, *J. Phys. Chem.*, **59**, 719 (1955).
48. O. OSTERRID, *Z. anal. Chem.*, **199**, 260 (1963).
49. I. GAL and A. RUVARAC, *J. Chromatog.*, **13**, 549 (1964).
50. H. O. PHILLIPS and K. A. KRAUS, *J. Am. Chem. Soc.*, **84**, 2267 (1962).
51. M. H. CAMPBELL, *Anal. Chem.*, **37**, 252 (1965).
52. Y. INOUE, S. SUZUKI, H. GOTO, *Bull. Chem. Soc. Japan*, **37**, 1547 (1964).
53. J. KRTIL, *J. Inorg. Nucl. Chem.*, **24**, 1139 (1962).
54. R. W. C. BROADBANK, S. DHABANANDANA and R. D. HARDING, *J. Inorg. Nucl. Chem.*, **23**, 311 (1961).
55. J. VAN R. SMIT, J. J. JACOBS and W. ROBB, *J. Inorg. Nucl. Chem.*, **12**, 95 (1959).
56. J. KRTIL and I. KRIVY, *J. Inorg. Nucl. Chem.*, **25**, 1191 (1963).
57. J. VAN R. SMIT, *J. Inorg. Nucl. Chem.*, **27**, 227 (1965).
58. J. VAN R. SMIT, W. ROBB and J. J. JACOBS, *J. Inorg. Nucl. Chem.*, **12**, 104 (1959).
59. H. L. CARON and T. T. SUGIHARA, *J. Inorg. Nucl. Chem.*, **34**, 1082 (1962).
60. J. KRTIL, *J. Inorg. Nucl. Chem.*, **19**, 298 (1961).
61. J. VAN R. SMIT and W. ROBB, *J. Inorg. Nucl. Chem.*, **26**, 509 (1964).
62. R. W. C. BROADBANK, S. DHABANANDANA and R. D. HARDING, *Analyst*, **85**, 365 (1960).
63. T. L. THOMAS and R. L. MAYS, Separations with molecular sieves, in W. G. BERL's *Physical Methods in Chemical Analysis*, Academic Press, vol. 4, New York, 1961, p. 55.
64. G. T. KERR and G. T. KOKTAILO, *J. Am. Chem. Soc.*, **83**, 4675 (1961).
65. R. M. BARRER and D. C. SAMMON, *J. Chem. Soc.*, 675 (1956).
66. V. KOUŘÍM, J. RAIS and B. MILLION, *J. Inorg. Nucl. Chem.*, **26**, 1111 (1964).
67. J. KRTIL, *J. Inorg. Nucl. Chem.*, **27**, 233 (1965).
68. W. E. PRONT, E. R. RUSSELL, H. J. GROH, *J. Inorg. Nucl. Chem.*, **27**, 473 (1965).
69. D. HUYS and L. H. BAETSLÉ, *J. Inorg. Nucl. Chem.*, **26**, 1329 (1964).
70. L. H. BAETSLÉ, D. VAN DEYCK and D. HUYS, *J. Inorg. Nucl. Chem.*, **27**, 683 (1965).
71. H. O. PHILLIPS and K. A. KRAUS, *J. Am. Chem. Soc.*, **85**, 486 (1963).
72. K. H. LIESER and W. HILD, *Naturwissenschaften*, **47**, 494 (1960).
73. E. L. SMITH and J. E. PAGE, *J. Soc. Chem. Ind.* (London), **67**, 48 (1948).
74. R. KUNIN and A. G. WINGER, *Angew. Chem., Intern. Ed.*, **1**, 149 (1962).
75. K. SOLLNER and G. M. SHEAN, *J. Am. Chem. Soc.*, **86**, 1901 (1964).
76. O. D. BONNER and D. C. LUNNEY, *J. Phys. Chem.*, **70**, 1140 (1966).
77. K. B. BROWN, C. F. COLEMAN, D. J. CROUSE, C. A. BLAKE and A. D. RYAN, *Proc. 2nd Intern. Conf. on Peaceful Uses of Atomic Energy*, Geneva, **3**, 472 (1958).
78. R. J. SOCHACKA and S. SIEKERSKI, *J. Chromatog.*, **16**, 376 (1964).
79. A. S. KERTES and I. T. PLATZNER, *J. Inorg. Nucl. Chem.*, **24**, 1417 (1962).
80. T. F. YOUNG, L. F. MARANVILLE and H. M. SMITH, Raman spectral investigations of ionic equilibria in solutions of strong electrolytes, in W. J. HAMER, *The Structure of Electrolytic Solutions*, John Wiley, New York, 1959, p. 42.
81. F. BARONCELLI, G. SCIBONA and M. ZIFFERERO, *J. Inorg. Nucl. Chem.*, **24**, 405 (1962).
82. W. E. KEDER, J. C. SHEPPARD and A. S. WILSON, *J. Inorg. Nucl. Chem.*, **12**, 327 (1960).
83. A. D. NELSON, J. L. FASHING and R. A. McDONALD, *J. Inorg. Nucl. Chem.*, **27**, 439 (1965).
84. S. LINDENBAUM and G. E. BOYD, *J. Phys. Chem.*, **67**, 1238 (1963).
85. C. F. BAES, JR., *J. Inorg. Nucl. Chem.*, **24**, 707 (1962).
86. C. F. BAES, JR., R. A. ZINGARO and C. F. COLEMAN, *J. Phys. Chem.*, **62**, 129 (1958).
87. F. E. BUTLER, *Anal. Chem.*, **37**, 340 (1965).
88. R. J. KNAPP, R. E. VAN AMAN and J. H. KANZELMEYER, *Anal. Chem.*, **34**, 1374 (1962).
89. G. NAKAGAWA, *Bunseki Kagaku*, **9**, 721 (1960); *Chem. Abs.*, **55**, 26857g (1961).
90. H. M. N. H. IRVING and A. D. DAMODARAN, *Analyst*, **90**, 443 (1965).
90a. H. FREISER, *Anal. Chem.*, **40**, 522R (1968).
91. E. CERRAI and C. TESTA, *J. Inorg. Nucl. Chem.*, **25**, 1045 (1963).
92. J. W. WINCHESTER, *J. Chromatog.*, **10**, 502 (1963).
93. R. J. SOCHACKA and S. SIEKERSKI, *J. Chromatog.*, **16**, 376 (1964).
94. S. SIEKERSKI and R. J. SOCHACKA, *J. Chromatog.*, **16**, 385 (1964).
95. M. S. FRANT and J. W. ROSS, JR., *Science*, **154**, 1553 (1966).
96. J. W. ROSS, JR., *Science*, **156**, 1378 (1967).
97. J. W. ROSS, JR., private communication.
98. M. LEDERER and S. KERTES, *Anal. Chim. Acta*, **15**, 226 (1956).
99. G. ALBERTI, F. DOBICI and G. GRASSINI, *J. Chromatog.*, **8**, 103 (1962).
100. E. C. YACKEL and O. KENYON, *J. Am. Chem. Soc.*, **64**, 121 (1942).
101. W. LANTSCH, G. MANECKE and W. BROSER, *Z. Naturforsch.*, **8b**, 232 (1953).
102. F. C. McINTIRE and J. R. SCHENK, *J. Am. Chem. Soc.*, **70**, 1193 (1948).

103. R. W. LITTLE (ed.), *Flameproofing Textile Fabrics*, A.C.S. Monograph 104, Reinhold Publishing Corp. New York, 1947, p. 179.
104. N. F. KEMBER and R. A. WELLS, *Nature*, **175**, 512 (1955).
105. T. TIMELL, *Svensk Papperstidn.*, **51**, 254 (1948); *Chem. Abs.*, **43**, 396d (1949).
106. C. L. HOFFAUIR and J. D. GUTHRIE, *Text. Res. J.*, **20**, 617 (1950).
107. T. WIELAND and A. BERG, *Angew. Chem.*, **64**, 418 (1952).
108. C. S. KNIGHT, *Nature*, **183**, 165 (1959).
109. J. B. MOORE, *Anal. Chem.*, **34**, 1506 (1962).
110. E. CERRAI and C. TRIULZI, *J. Chromatog.*, **16**, 365 (1964).
111. C. PAOLINI and G. SERLUPI-CRESCENZI, *Gazz. Chim. Ital.*, **94**, 181 (1964).
112. J. HARTEL and A. J. G. PLEUMEEKERS, *Anal. Chem.*, **36**, 1021 (1964).
113. M. LEDERER, *Anal. Chim. Acta*, **12**, 142 (1955).
114. H. T. PETERSON, *Anal. Chem.*, **31**, 1279 (1959).
115. J. SHERMA, *Talanta*, **9**, 775 (1962).
116. J. SHERMA and C. W. CLINE, *Talanta*, **10**, 787 (1963).
117. J. SHERMA, *Anal. Chem.*, **36**, 690 (1964).
118. J. SHERMA, *Talanta*, **11**, 1373 (1964).
119. J. SHERMA and W. CLINE, *Anal. Chim. Acta*, **30**, 139 (1964).
120. J. SHERMA and K. M. RICH, *J. Chromatog.*, **26**, 327 (1967).
121. M. GRIMALDI, A. LIBERTI and M. VICEDOMINI, *J. Chromatog.*, **11**, 101 (1963).
122. J. SHERMA, *Anal. Chim. Acta*, **36**, 138 (1966).
123. C. HEININGER and F. M. LANZAFAMA, *Anal. Chim. Acta*, **30**, 148 (1964).
124. L. OSSICINI, *J. Chromatog.*, **9**, 114 (1962).
125. R. HÜTTENRAUCH and L. KLOTZ, *Experientia*, **19**, 95 (1963).
126. D. LOCKE and J. SHERMA, *Anal. Chim. Acta*, **25**, 312 (1961).
127. J. SHERMA and D. E. THOMPSON, JR., *Anal. Chim. Acta*, **32**, 181 (1966).
128. J. SHERMA and L. H. PIGNOLET, *Anal. Chim. Acta*, **34**, 185 (1966).
129. G. P. McNICOL, A. P. FLETCHER, N. ALKJAERSIG and S. SHERRY, *J. Lab. Clin. Med.*, **59**, 7 (1962); *Chem. Abs.*, **56**, 11944b (1962).
130. A. LEWANDOWSKI and A. JARCZEWSKI, *Talanta*, **4**, 174 (1960).
131. G. ALBERTI and G. GRASSINI, *J. Chromatog.*, **4**, 83 (1960).
132. K. SAKORDINSKY and M. LEDERER, *J. Chromatog.*, **20**, 358 (1965).
133. J. M. P. CABRAL, *J. Chromatog.*, **4**, 86 (1961).
134. J. P. ADLOFF, *J. Chromatog.*, **5**, 366 (1961).
135. G. ALBERTI and G. GRASSINI, *J. Chromatog.*, **4**, 423 (1960).
136. M. N. SASTRI and A. P. RAO, *J. Chromatog.*, **9**, 250 (1962).
137. P. CATELLI, *J. Chromatog.*, **9**, 534 (1962).
138. I. D. COUSSIO, G. B. MARINI-BETTOLO and V. MOSCATELLI, *J. Chromatog.*, **11**, 238 (1963).
139. C. TESTA, *J. Chromatog.*, **5**, 236 (1961).
140. E. CERRAI and C. TESTA, *J. Chromatog.*, **7**, 112 (1962).
141. K. RANDERATH, *J. Chromatog.*, **10**, 235 (1963).
142. E. CERRAI and C. TESTA, *J. Chromatog.*, **8**, 232 (1962).
143. M. N. SASTRI and A. P. RAO, *Z. Anal. Chem.*, **196**, 166 (1963).
144. C. TESTA, *Anal. Chem.*, **34**, 1556 (1962).
145. E. CERRAI and G. GHERSINI, *J. Chromatog.*, **13**, 211 (1964).
146. E. CERRAI and G. GHERSINI, *J. Chromatog.*, **15**, 236 (1964).
147. H. A. SOBER and E. A. PETERSON, *J. Am. Chem. Soc.*, **76**, 1711 (1954).
148. E. A. PETERSON and H. A. SOBER, *J. Am. Chem. Soc.*, **78**, 751 (1956).
149. J. D. GUTHRIE and A. L. BULLOCK, *Ind. Eng. Chem.*, **52**, 935 (1960).
150. R. R. BENERITO, B. B. WOODWARD and J. D. GUTHRIE, *Anal. Chem.*, **37**, 1693 (1965).
151. H. A. VEDER, *J. Chromatog.*, **10**, 507 (1963).
152. G. MANECKE and P. GERGS, *Naturwissenschaften*, **50**, 329 (1963).
153. E. RANDERATH and K. RANDERATH, *J. Chromatog.*, **10**, 509 (1963).
154. F. WETTSTEIN, H. NEUKOM and H. DEUEL, *Helv. Chim. Acta*, **44**, 1949 (1961).
155. H. A. SOBER, J. F. GUTTER, M. M. WYCKOFF and E. A. PETERSON, *J. Am. Chem. Soc.*, **78**, 756 (1956).
156. J. L. FAHEY and A. P. HORBETT, *J. Biol. Chem.*, **234**, 2645 (1959).
157. J. S. FINLAYSON and M. W. MOSESSON, *Biochem.*, **2**, 42 (1963).
158. B. BLOMBÄCK and M. BLOMBÄCK, *Ark. Kemi*, **21**, 299 (1956).
159. J. C. ROBINSON, L. KEAY, R. MOLINARI and I. W. SIZER, *J. Biol. Chem.*, **237**, 2001 (1962).
160. E. AINBENDER, H. D. ZEPP and H. L. HODER, *Proc. Soc. Exptl. Biol. Med.*, **110**, 271 (1962).
161. S. MAROUX, M. ROVERY and P. DESNUELLE, *Biochim. Biophys. Acta*, **56**, 202 (1962).
162. M. STAEHELIN, E. A. PETERSON and H. A. SOBER, *Arch. Biochem., Biophys.*, **85**, 289 (1959).

163. H. NEUKOM, H. DEUEL, W. J. HERI and W. KUNDIG, *Helv. Chim. Acta*, **43**, 64 (1960).
164. W. HERI, H. NEUKOM and H. DEUEL, *Helv. Chim. Acta*, **44**, 1939 (1961).
165. W. HERI, H. NEUKOM and H. DEUEL, *Helv. Chim. Acta*, **44**, 1945 (1961).
166. L. R. ESHELMAN, E. Y. MANZO, S. J. MARCUS, A. E. DECOUTEAU and E. G. HAMMOND, *Anal. Chem.*, **32**, 844 (1960).
167. L. D. METCALFE, *Anal. Chem.*, **32**, 70 (1960).
168. H. S. HENDRICKSON and C. E. BALLOU, *J. Biol. Chem.*, **239**, 1369 (1964).
169. A. J. HEAD, W. F. KEMBER, R. P. MILLER and R. A. WELK, *J. Appl. Chem.*, **9**, 599 (1959).
170. G. C. GOODE and M. C. CAMPBELL, *Anal. Chim. Acta*, **27**, 422 (1962).
171. K. RANDERATH, *Dünnschicht-Chromatographie*, Verlag Chemie, Weinheim, 1962.
172. K. RANDERATH, *Angew. Chem.*, **73**, 436 (1961).
173. K. RANDERATH, *Biochem. Biophys. Acta*, **76**, 622 (1963).
174. J. NEUHARD, E. RANDERATH and K. RANDERATH, *Anal. Biochem.*, **13**, 211 (1965).
175. P. DE LA LLOLSA, C. TERTRIN and M. JUTISZ, *J. Chromatog.*, **14**, 136 (1964).
176. T. WIELAND and H. DETERMANN, *Experientia*, **18**, 431 (1962).
177. R. HÜTTENRAUCH, L. KLOTZ and W. MULLER, *Z. Chem.*, **3**, 193 (1963).
178. J. BERGER, G. MEYNIEL, J. PETIT and P. BLANQUET, *Bull. Soc. Chim. France*, 2662 (1963).
179. J. SHERMA, *Chemist-Analyst*, **55**, 86 (1966).
180. B. A. ZABIN and C. B. ROBBINS, *J. Chromatog.*, **14**, 534 (1964).
181. H. HOLZAPPEL, L. V. LAN and G. WERNER, *J. Chromatog.*, **20**, 580 (1965).
182. G. M. HENDERSON and H. G. RULE, *J. Chem. Soc.*, 1568 (1939).
183. V. PRELOG and P. WIELAND, *Helv. Chim. Acta*, **27**, 1127 (1954).
184. H. KREBS, J. A. WAGNER and J. DIEWALD, *Chem. Ber.*, **89**, 1875 (1956).
185. M. OHARA, I. FUGITA and T. KWAN, *Bull. Chem. Soc. Japan*, **35**, 2049 (1962)
186. M. OHARA, C. CHEN and T. KWAN, *Bull. Chem. Soc. Japan*, **39**, 137 (1966).
187. M. KOTAKE, T. SAKAN, N. NAKAMURA and S. SENOH, *J. Am. Chem. Soc.*, **73**, 2973 (1951).
188. C. E. DALGLEISH, *J. Chem. Soc.*, 3940 (1952).
189. J. P. LAMBOOY, *J. Am. Chem. Soc.*, **76**, 133 (1954).
190. R. E. LEITCH, H. L. ROTHBART and W. RIEMAN, *J. Chromatog.*, **28**, 132 (1967).
191. N. GRUBHOFER and L. SCHLIETH, *Z. Physiol. Chem.*, **296**, 262 (1954).
192. H. SUDA and R. ODA, *Kanazawa Daikagu Kogakubu Kiyo*, **2**, 215 (1960).
193. J. A. LOTT and W. RIEMAN, *J. Org. Chem.*, **31**, 561 (1966).
194. S. ROMANO, K. H. WELLS, H. L. ROTHBART and W. RIEMAN, *Talanta*, **16**, 581 (1969).
194a. W. RIEMAN, unpublished work.
195. R. E. LEITCH, Thesis, Rutgers, The State University, 1967.
196. Y. YOSHINO, H. SUGIYAMA, S. NOGAITO and H. KINOSHITA, *Scientific Papers of the College of General Education, University of Tokyo*, **16**, 57 (1966).
197. H. D. SPITZ, H. L. ROTHBART and W. RIEMAN, *J. Chromatog.*, **29**, 94 (1967).
198. R. E. LEITCH, H. L. ROTHBART and W. RIEMAN, *Talanta*, **15**, 231 (1968).
199. H. C. ROSE, R. L. STERN and B. L. KARGER, *Anal. Chem.*, **38**, 469 (1966).

CHAPTER 11

STUDY OF COMPLEX IONS

COMPLEX ions are all-important in ion exchange as it is applied to inorganic analysis. In the separation of metals by cation exchange, complexes formed in the solution tend to keep the metal out of the exchanger; in separations by anion exchange, complexes formed in the solution, if they carry a negative charge, act to fix the metal on the exchanger. Finally, cationic complexes are absorbed to a greater or lesser extent by cation exchangers, leading to the possibility of exchanging the ligands as well as the cations themselves.

The nature of the metal complexes formed within the exchanger is interesting, but a matter of more general interest is the use of ion-exchange distribution studies to determine the nature and stability of complexes existing in the solution. We shall discuss this aspect first.

A. Separation of Kinetically Stable Complexes

Complexes of chromium(III) and cobalt(III) are very slow to react, and well-known theories have been advanced to account for the fact. The stability is due to slowness of reaction, that is to an activation energy barrier, and not to a low free-energy state. Thus a mixture of the salt $Co(NH_3)_6Cl_3$ with 6 M hydrochloric acid can be heated under reflux for several hours without apparent change, even though the stable products of the reaction are $CoCl_2$, ammonium chloride, and chlorine.

The salt $[Co(NH_3)_5Cl]Cl_2$, prepared in the laboratory from the salt $[Co(NH_3)_4CO_3]NO_3$ by way of $[Co(NH_3)_5OH_2]Cl_3$, nearly always contains some $[Co(NH_3)_6]Cl_3$ as an impurity. The ions $[Co(NH_3)_6]^{3+}$ (yellow) and $[Co(NH_3)_5Cl]^{2+}$ (red) can readily be separated on a column of a cation-exchange resin such as Dowex 50W. The mixture is washed down the column by a solution of sodium chloride, some 2–4 M. Soon the red band of divalent ions sweeps ahead of the more strongly absorbed, yellow, trivalent ions. This is a typical separation based on charge; the more highly charged ions are held more strongly by the exchanger.

Many examples of this type of separation may be found in the current literature. Thus the salt $[Cr(OH_2)_5SO_4]Cl.0.5H_2O$ has been prepared by reducing chromic acid with sulphur dioxide at $-10°$ in the presence of perchloric and hydrochloric acids; the product is placed on a cation-exchange resin column and washed with dilute hydrochloric acid, a violet band of the trivalent $[Cr(OH_2)_6]^{3+}$ stays at the top of the column, while the green, singly charged cation $[Cr(OH_2)_5SO_4]^+$ passes down the column and is eluted.[1] Another example is the ion $[Cr(OH_2)_5I]^{2+}$, which is prepared from chromium(II) perchlorate by oxidation with iodine; it is freed from small amounts of accompanying $[Cr(OH_2)_6]^{3+}$ by cation exchange,[2, 3] again taking advantage of the effect of ionic charge on selectivity. Because the iodopentaquo-chromium(III) ion is hydrolyzed at a significant rate at room temperature, the preparation and purification are performed at 0–1°. The chloropentaquochromium(III) complex is prepared and purified in a similar way.[4]

The optical isomers of a kinetically stable, positively charged, trinuclear cobalt(III) complex were separated on a column of carboxymethylcellulose by Brubaker *et al.*[4a]

Of course ion exchange can be used to separate ions of opposite charge, that is, to exchange ionic partners and prepare two separate salts. Chromium(III) fluoride solution, treated with ethylenediamine, gives the ions [Cr en$_2$F$_2$]$^+$ (red) and [Cr en F$_4$]$^-$ (blue); the former is retained on a cation-exchange resin, the latter on an anion-exchanging cellulose.[5] Likewise, when a mixture of iridium(IV) and palladium(II) is placed in a solution of ammonia and ammonium chloride, the former forms the anion IrCl$_6^{2-}$ and the latter the cation Pd(NH$_3$)$_4^{2+}$, which can be absorbed on a cation exchanger.[6]

The annoying occurrence of "double peaks", where one metal gives rise to two or more maxima in chromatographic elution curves is due to the formation of two or more complex ions which are sufficiently long-lived that they do not come to equilibrium with their environment in the process of ion-exchange chromatography. This situation arises sometimes with titanium and other transition metals. The speed of reaction depends on the ligand as well as the central ion.

B. Measurement of Stability Constants by Ion Exchange

The use of ion-exchange equilibria to measure the stability of "labile" complexes (which are the commonest kind of complex ions) in aqueous solutions is now a standard technique which supplements the more general and more rapid methods of potentiometric titration or photometric measurement. Most determinations are made by equilibrating the solutions with a bead-type cation-exchange resin; some have been made with anion-exchange resins, and recently resin-impregnated papers and ion-exchange resin membranes have been used.

B.I. EQUILIBRIUM STUDIES WITH CATION-EXCHANGE RESINS

In this section we shall use the following symbols, in accordance with Fronaeus;[10]

Partition coefficients:

$$\phi = \text{gross partition coefficient} = \frac{\text{total metal in resin}}{\text{total metal in solution}}.$$

$$l_0 = \frac{\text{uncomplexed metal in resin}}{\text{uncomplexed metal in solution}}.$$

$$l_1 = \frac{\text{concentration of MA}^+ \text{ in resin}}{\text{concentration of MA}^+ \text{ in solution}}.$$

$$l = l_1\beta_1/l_0$$

Equilibrium constants

β_1, β_2, etc. = cumulative formation constants.

K_1, K_2, etc. = successive formation constants; $K_1 = \beta_1$, $K_1 K_2 = \beta_2$, etc.

$\displaystyle X = \sum_{j}^{N} (1 + \beta_j \,[A^-]^j)$; $[A^-]$ = ligand concentration, N = maximum coordination number.

B.I.a. *No Complexes Absorbed*

Some of the earliest of these studies were made by Schubert,[7, 8] and applied to complexes such as that between strontium and citrate, a type of complex that is important in chromatographic separations (see Chapter 8) and whose stability is hard to determine by other methods.

The equilibrium between strontium and citrate ions can be written simply as:

$$Sr^{2+} + Cit^{3-} \rightleftharpoons SrCit^-$$

There is a reasonable presumption that, at any rate in the higher pH range, only the one complex is formed, and one may also assume that the only form in which strontium enters the resin is as the simple cation Sr^{2+}. Thus the resin acts as a selective "solvent" for the ion Sr^{2+}, and does not absorb the complex. If, in addition, the strontium concentration is kept very low (Schubert worked at the tracer level) and the solution contains a constant concentration of another positive ion such as sodium, one may consider that, for the distribution of strontium between a sodium-loaded resin and a sodium salt solution,

$$\frac{\text{Concentration of } Sr^{2+} \text{ in resin}}{\text{Concentration of uncomplexed } Sr^{2+} \text{ in solution}} = \text{constant} = l_0.$$

(It is considered that activity coefficients remain constant.) If citrate is added to the solution, the distribution will shift. Strontium will be brought out of the resin, and the overall distribution coefficient,

$$\frac{\text{Concentration of } Sr^{2+} \text{ in resin}}{\text{Total } Sr^{2+} \text{ concentration in solution, free and complexed}} = \phi$$

will diminish as the citrate concentration rises. Experiments are set up as follows: equal weights of air-dried sodium resin are placed in several tubes; to the first tube, a known volume of sodium perchlorate solution of known concentration is added; to the other tubes, mixed solutions of sodium perchlorate and sodium citrate are added, keeping the total sodium-ion concentration the same, as well as the total volume. Then small, equal amounts of radioactive strontium salt are added to the tubes, and the tubes are shaken for several hours until equilibrium is reached. Samples of the solution are "counted" and the distribution ratios ϕ are calculated.

In this example it is impossible to keep the ionic strength constant as well as the sodium ion concentration, and if desired, a correction may be made for the changing activity coefficients, with the second-approximation Debye–Hückel equation. If this effect is neglected,

$$\phi = \frac{[\overline{Sr^{2+}}]}{[Sr^{2+}] + K\,[Sr^{2+}]\,[Cit^{3-}]} \tag{11.1}$$

(barred symbols indicate the resin phase; K is the formation constant of the complex $SrCit^-$), and

$$\frac{l_0}{\phi} = 1 + K\,[Cit^{3-}]. \tag{11.2}$$

If the complex is very stable, that is, K is very large, it may be difficult to arrange the experiments to give a reliable value of l_0 by direct measurement. Thus, a quantity of tracer which gives convenient counting rates in presence of citrate (or other complexing agent) may give an inconveniently low counting rate in its absence. This situation is more serious when nonradioactive methods, such as photometry, are used to measure the metal concentrations. In this case it is best to determine the distribution coefficient l_0 by plotting $(1/\phi)$ against the concentration of the complexing agent and extrapolating to zero; the intercept is $(1/l_0)$.

For a more general case in which two or more complexes are formed in the solution, but still restricted to the case in which only the noncomplexed metal ions are absorbed by the resin:

$$M^{2+} + A^{2-} \rightleftharpoons MA \text{ (constant } \beta_1\text{)},$$

$$M^{2+} + 2A^{2-} \rightleftharpoons MA_2^{2-} \text{ (constant } \beta_2\text{), etc.,}$$

we see that

$$\frac{l_0}{\phi} = 1 + \beta_1 [A^{2-}] + \beta_2 [A^{2-}]^2 + \cdots \tag{11.3}$$

(The notation of Bjerrum *et al.* is used; the β values mean cumulative formation constants.)[9] If β_1 and β_2 differ sufficiently, it is possible to determine β_1 reasonably accurately, and β_2 less accurately, by analysis of the graph of l_0/ϕ (or of $1/\phi$) against $[A^{2-}]$. An example in which this can be done is the copper(II)-tartrate association, where $\log \beta_1 = 3.2$, $\log \beta_2 = 5.1$ (at ionic strength 1.0).[9] The copper in the solutions can be determined at very low concentrations by a photometric method, such as the reaction with sodium diethyldithiocarbamate or 2,2'-biquinoline, and in this case it is definitely preferable to determine l_0 by extrapolation. Another caution is necessary; cupric ions hydrolyze significantly above pH 7 ($\log \beta_1$ for the formation of $CuOH^+$ is 6.0) and tartrate ions, T^{2-}, form HT^- to a significant extent below pH 5.5 (pK_2 for tartaric acid is 4.34), so that careful pH control is needed. Nevertheless, reasonable values for the first formation constant of copper(II) tartrate have been obtained in student laboratories (in the experience of one of the authors).

B.I.b. *One Cationic Complex Absorbed*

A more general type of association is that in which a di- or trivalent cation combines by stages with a singly charged anion; with a cation M^{2+} and an anion A^-, these equilibria must be considered:

$$M^{2+} + A^- \rightleftharpoons MA^+; \beta_1 = [MA^+]/[M^{2+}][A^-],$$

$$M^{2+} + 2A^- \rightleftharpoons MA_2; \beta_2 = [MA_2]/[M^{2+}][A^-]^2,$$

$$M^{2+} + 3A^- \rightleftharpoons MA_3^-; \beta_3 = [MA_3^-]/[M^{2+}][A^-]^3, \text{ etc.}$$

We can reasonably suppose that the complexes MA_2, MA_3^- and higher complexes will not be absorbed by the cation-exchange resin, unless the electrolyte concentration is of the order of 2 M and above, but we cannot assume that MA^+ is not absorbed. It will be absorbed, and the observed distribution coefficient ϕ will depend, not only on the formation constant for MA^+, but also on its ion-exchange distribution coefficient, $l_1 = \overline{[MA^+]}/[MA^+]$. This cannot be determined directly, but must be found by solving some rather complicated equations.

The first treatment of this system was given by Fronaeus.[10] His experiments and mathematical treatment were more complicated than would be necessary today, because he used a cation-exchange resin of the sulphonated phenolic type whose exchange capacity depended on the pH and for which the distribution coefficients l_0 and ϕ depended on the amount of metal in the resin. He started by studying the cupric ion-acetate association. He determined l_0 by extrapolation at constant pH, and to overcome the marked dependence of ϕ on metal-ion concentration, he made experiments at several different metal-ion loadings and reduced his data to a single curve of ϕ against the concentration of free acetate ions for a constant total metal concentration in the resin, $[\overline{M^{2+}}] + [\overline{MA^+}]$. From this curve the formation constants β could be derived by the following procedure:

The overall distribution coefficient ϕ is given by:

$$\phi = \frac{[\overline{M^{2+}}]}{[M^{2+}]} \frac{(1 + l_1\,\beta_1\,[A^-]/l_0)}{(1 + \beta_1\,[A^-] + \beta_2\,[A^-]^2 + \cdots)}$$

$$= l_0 \frac{1 + l\,[A^-]}{X}, \tag{11.4}$$

where
$$l = l_1\beta_1/l_0; \quad X = 1 + \sum_{j=1}^{N}\beta_j\,[A^-]^j.$$

Rearranging,

$$\phi X = l_0\,(1 + l\,[A^-]).$$

Differentiating twice by $[A^-]$, and representing

$d\phi/d[A^-]$ by ϕ', $d^2\phi/d[A^-]^2$ by ϕ'', etc.,
$$\phi''\,X + 2\phi'\,X' + \phi X'' = 0. \tag{11.5}$$

This eliminates the troublesome unknown constant l. (If *two* cationic complexes were formed, such as MA^{2+} and MA_2^+ by a trivalent metal, it would be necessary to differentiate three times to make the derivative function equal to zero.)

Now the quantity X and its derivatives X' and X'' are known functions of $[A^-]$ (known) and the β (unknown). If we measure ϕ, ϕ', and ϕ'' at different concentrations of A, as many concentrations as there are complexes formed (i.e., at N concentrations), we will obtain a set of N simultaneous equations which can be solved to get $\beta_1, \beta_2, \ldots, \beta_N$. Of course, this procedure requires that the experimental data are very accurate indeed, more accurate than they ever are. Yet it is not hard to get a good value for β_1 from the measurement at low $[A^-]$. Here, as a general rule, only the first complex, MA^+, is formed in appreciable amount; i.e. the quantities $\phi''\beta_1[A^-]$ and $\phi'\beta_1$ are much larger than the terms in β_2. Equation (11.5) now becomes, in the limit of low $[A^-]$,

$$\phi''\,(1 + \beta_1\,[A^-]) + 4\,\phi'\,\beta_1 = 0 \tag{11.5a}$$

and can be solved for β_1.

Calculation of β_2 and successive formation constants from eqn. (11.5) is less exact, and an alternative way to calculate these was suggested by Fronaeus. Knowing β_1, l can be evaluated as follows. A function

$$\phi_1 = \left(\frac{1}{\phi} - \frac{1}{l_0}\right)\frac{1}{[A^-]}$$

is defined. Inserting the value of ϕ from eqn. (11.4), we find that, in the limit of low $[A^-]$,

$$\lim_{[A^-]\to 0} \phi_1 = \frac{\beta_1 - l}{l_0}. \tag{11.6}$$

Thus the constant l is found and inserted into eqn. (11.4). The polynomial X can now be calculated from any desired values of $[A^-]$ and ϕ, and from it the formation constants β_2, β_3, etc., can be found with greater precision.

Fronaeus later[11] offered another calculation method which is better for weaker complexes. He introduced a new function f defined as follows:

$$f = \frac{1}{[A^-]^2} \left\{ \frac{l_0}{\phi} [(\beta_1 - l) [A^-] - 1] + 1 \right\}. \tag{11.7}$$

By writing the functions ϕ and ϕ_1 in terms of X one can show that

$$f = \beta_1 l_0 \phi - \sum_{J=2}^{N} \beta_J [A^-]^{J-2}. \tag{11.8}$$

In the limit of low $[A^-]$ this reduces to

$$f^0 = \beta_1(\beta_1 - l) - \beta_2. \tag{11.8a}$$

It is found experimentally that for complexes which are not excessively stable the functions f and ϕ are nearly linear in $[A^-]$ at low concentrations. The ratio of the slopes of these two curves gives β_1, for, if $[A^-]$ is sufficiently low that only the first *three* complexes need be considered, eqn. (11.8) becomes

$$\Delta f = \beta_1 l_0 \Delta \phi - \beta_2 \Delta [A^-], \tag{11.8b}$$

where $\Delta f = f - f^0$, etc. The constants β_1 and β_3 can be determined by this relationship, β_2 by eqn. (11.8a), and the higher constants by the expansion of eqn. (11.8).

Results obtained by this method agree satisfactorily with those obtained potentiometrically, and are more reliable[11] for weak complexes such as $Ni(OAc)_2$ and $Ni(OAc)_3^-$. In the cupric ion-acetate system the two methods gave these results in 1 M sodium perchlorate solutions:[10]

	Ion exchange	Potentiometric
β_1	45 ± 2	47 ± 1
β_2	440 ± 60	450 ± 50
β_3	1000 ± 300	1150 ± 150
β_4	—	750 ± 200

A comparison of l_0 and l_1, the ion-exchange distribution coefficients for Cu^{2+} and $CuOAc^+$, is interesting; they were 0.14 and 0.055 l. per gram of dry sodium form resin respectively. The absorption of the positively charged complex $CuOAc^+$ is certainly not negligible. In more dilute sodium salt solutions, however, the ion-exchange absorption of Cu^{2+} would naturally be greater in proportion to that of $CuOAc^+$.

Fronaeus discusses the possible formation of polynuclear complexes, and points out that if the proportion of polynuclear complex is small, the terms in eqn. (11.4) will cancel. Of course, the lower the metal-ion concentration the less will be the proportion of polynuclear complexes, and at tracer levels these can usually be disregarded completely.

B.II. STABILITY CONSTANTS BY PAPER CHROMATOGRAPHY

An interesting application of the Fronaeus method is to chromatography on resin-impregnated paper.[12, 13] The distribution coefficient ϕ of the metal between the resin and the solution is simply related to the R_f value:

$$\phi = \frac{1 - R_f}{R_f} \frac{v}{w}, \tag{11.9}$$

where v and w are the volume of solvent and weight of resin, respectively, in a unit cross-section of the paper. The complexing agent causes more of the metal to be in the solution at any given time and therefore increases the R_f value. The *ratio* of the distribution coefficient ϕ to the distribution coefficient in the absence of ligand, l_0 (or ϕ_0) is what gives the stability constants, not the actual distribution coefficients themselves, and it would not be necessary to measure the ratio v/w were it not that this is affected slightly by the presence of the ligand. Thus this ratio must be measured. It is also necessary to avoid frontal disturbances where the solution meets the dry paper; therefore, the spot of metal salt solution is applied *behind* the solution front. (If the spots are applied ahead of the front, in the normal manner of paper chromatography, bad tailing will result.) Experiments are run with a series of developing solutions of constant displacing-ion concentration and varying amounts of complexing anion; thus, sodium perchlorate solutions containing varying proportions of sodium acetate may be used. The values of the distribution coefficient ϕ are extrapolated to zero ligand concentration, as before, to obtain l_0.

This technique is not as accurate as the batch equilibration method, and it fails if the R_f values are very low (below 0.05) or very high (above 0.9). Its validity depends on a "linear isotherm", or constancy of distribution coefficients at different metal loadings of the resin, but this condition is found to hold on paper impregnated with sulphonated polystyrene resin; the spots remain circular and the R_f values are independent of metal loading within the limits used. The advantages of the method are, first, that it does not require an accurate quantitative determination of trace-metal concentrations, but only a spray reagent that will reveal the spot; and second, that it is fast and simple. It has been used to study the effects of nonaqueous solvents on complex-ion formation.[13] The mathematical treatment is exactly like that described above.

B.III. RESIN MEMBRANES

A new method for studying complex-ion equilibria uses ion-exchange membranes.[14] A cation-exchange resin membrane separates two solutions—one containing the complexing anion, the other a noncomplexing salt. The metal to be studied is added as a radioactive tracer in low concentration, and is allowed to distribute itself across the membrane between the solutions on the two sides. At equilibrium there will be more of the metal on the side containing the complexing agent.

This method depends on the ability of a cation-exchange membrane to permit the passage of positive ions while obstructing the passage of negative ions. This ability in turn depends on the exclusion of salt from the membrane by the Donnan equilibrium (Chapter 2). The salt exclusion falls rapidly at high concentrations, and at the concentrations customarily used in complex-ion studies, 1 M and more, the salt exclusion and therefore the selective permeability of the membrane is greatly reduced. This limits the application of the method to low salt concentrations (0.1–0.2 M). On the other hand, the method has the great advantage that the distribution coefficients of the various cationic species between the resin and the solution need not be known. This greatly simplifies the evaluation of formation constants from the experimental data.

In the paper mentioned[14] Wallace measured the formation constants of the uranyl ion-sulphate complexes, using uranium-233 as a tracer. He also added the tracer sodium-22, and measured the distribution of both tracer ions simultaneously. The reason for the two measurements is this. Equilibrium across the membrane does not mean that the concentrations of the mobile cations must be the same on both sides of the membrane, but rather that the activity of a particular salt must be the same on both sides. For the sodium salt of an anion X, for example, the membrane equilibrium may be written:

$$[Na^+]_1 [X^-]_1 = [Na^+]_2 [X^-]_2 \cdot \gamma_2^2/\gamma_1^2,$$

where the subscripts 1 and 2 refer to the two sides of the membrane, and γ_1 and γ_2 are mean activity coefficients ($\gamma_\pm$).

A similar relation holds for a divalent ionic species M^{2+}:

$$[M^{2+}]_1 [X^-]_1^2 = [M^{2+}]_2 [X^-]_2^2 \cdot \gamma_2^3/\gamma_1^3$$

Combining the two, we get

$$\frac{[M^{2+}]_2 [Na^+]_1^2}{[M^{2+}]_1 [Na^+]_2^2} = E, \tag{11.10}$$

where the quotient E depends on the ionic strengths of the two solutions. It is nearly unity for two dilute solutions of equal ionic strengths. If, therefore, side 2 contains a complexing anion and side 1 does not, the measured total concentration of M, divided by the concentration $[M^{2+}]$ calculated from eqn. (11.10), equals the quantity X in eqn. (11.4). This is the series containing the formation constants. These may be evaluated from a series of measurements. The exact computation method chosen will depend, of course, on the orders of magnitude of the different constants.

For the uranyl ion-sulphate association Wallace found $\beta_1 = 1390 \pm 46$, $\beta_2 = 16,100 \pm 1600$ ("thermodynamic" values corrected to zero ionic strength) at 25°. The first constant compares with 910 found spectrophotometrically and 1700 found by conductivity. Clearly the membrane method will be valuable for investigating metal complexes. A couple of experimental points should be noted; first, it is desirable to keep the ionic strength the same on both sides of the membrane, and it is also desirable to keep the osmotic pressures the same to avoid water transport; one cannot do both at once with a divalent complexing anion and a univalent supporting anion. Second, at low salt concentrations much of the divalent tracer goes into the membrane. Wallace loaded the membrane with the tracers before placing the salt solutions on the two sides.

B.IV. ANION-EXCHANGE RESINS

In principle anion-exchange resins can be used to measure stability constants in the same way as cation-exchange resins. It is necessary to know the ionic species taken up by the exchanger, and fortunately there is good reason to suppose that this is the coordinatively saturated complex anion. The concentration of the ligand anions inside a strong-base anion-exchange resin is very large. The very thorough studies of Horne et al.[15] on the absorption

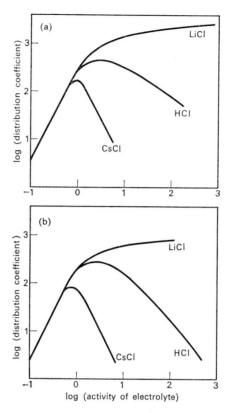

FIG. 11.1. Distribution of tracer zinc(II) between (a) a quaternary-ammonium anion-exchange resin, (b) a 10% solution of methyldioctylamine hydrochloride in trichloro-ethylene and aqueous solutions. (Reproduced by permission from U. Schindewolf, *Zeitschrift für Elektrochemie*, **62**, 335 (1958).)

of zinc(II) by anion-exchange resins showed that when zinc is absorbed from chloride solutions by a chloride-form resin, two chloride ions enter the resin with every zinc ion, corresponding to the ion $ZnCl_4{}^{2-}$, and furthermore, that the maximum capacity of the resin for holding zinc is equivalent to its ion-exchange capacity. In this case at least there is little doubt that the metal is bound as its fully coordinated complex, $ZnCl_4{}^{2-}$. This is also deduced from the results of Kraus *et al.*[16] (see Fig. 11.1). At low concentrations of hydrochloric acid or alkali-metal chloride, below 0.5 M, the distribution coefficient of zinc and other divalent metals increases with the square of the chloride-ion concentration, indicating an equilibrium

$$M^{2+} + 2Cl^- + 2RCl \rightleftharpoons R_2MCl_4$$

where R^+ is the fixed ionic group of the resin. That this is the principal equilibrium is confirmed by the work of Schindewolf[17] on liquid anion-exchangers. By using solutions of the liquid ion-exchanger methyldioctylamine hydrochloride in an inert solvent he varied the concentration of the functional groups R^+, and showed that the distribution coefficient for zinc depended on the square of the concentration of functional groups.

Thus one can write an equation for the distribution coefficient which is analogous to eqn. (11.4):

$$\phi = \frac{[\overline{ZnCl_4^{2-}}]}{[Zn^{2+}] \left(1 + \sum\limits_{j=1}^{4} \beta_j \, [Cl^-]^j\right)} \qquad (11.11)$$

or

$$\phi = \frac{K \, [\overline{RCl}][Cl^-]^2}{\left(1 + \sum\limits_{j=1} \beta_j \, [Cl^-]^j\right)}. \qquad (11.11a)$$

The constant K is the equilibrium constant for the reaction between M^{2+}, Cl^- and the resin chloride. It may be measured from the graph of ϕ against chloride-ion concentration at low concentrations. The calculation of the formation constants follows by methods similar to those discussed above. Several formation constants of chloride[18, 19] and bromide[15] complexes have been determined in this way, also those of sulphate complexes.[20, 21]

C. Anionic Chloride Complexes

Anion-exchange chromatography of metal ions in hydrochloric acid solutions was discussed in Chapter 8. It is a very powerful method of separation, and a number of studies have been made to try to understand how it works.

First, it is clear that the stability of the chloride complexes in the resin is of a different order of magnitude from their stability in aqueous solutions and bears little relation to the stability in solution. Iron (III) is absorbed from hydrochloric acid solutions by quaternary-base ion-exchange resins with distribution coefficients exceeding 10^4, yet the *first* stage of association between iron(III) and Cl^- in solution[22] has a formation constant β_1 of only 3. For the cobalt(II)-Cl^- association, $\beta_2 = 0.2$, yet the distribution coefficient rises to 100. For the complexes of zinc(II) and cadmium(II) with chloride, $\beta_3 = 1.6$ and 260 respectively, yet the maximum distribution coefficients are about the same, 500–1000. Quite obviously the complex ions in the resin, $FeCl_4^-$, $CoCl_4^{2-}$, and so on, are very closely associated with the fixed ions, presumably as ion pairs. Evidence for this view comes from a study of the distribution of metal ions between chloride solutions and a liquid anion exchanger. The curves of distribution coefficient against chloride-ion concentration closely parallel the curves obtained with a quaternary ammonium-type anion-exchange resin[17] (Fig. 11.1).

At high chloride concentrations the distribution coefficients fall again. This is due to a displacement of the complex metal-chloride ions by the chloride ions of the solution

$$R_2MCl_4 + 2Cl^- \rightleftharpoons 2RCl + MCl_4^{2-}$$

and, indeed, the curves for cesium chloride in Fig. 11.1 have the slope of -2 that one expects from this equation. The curves for lithium chloride and hydrochloric acid do not have this slope, however; the absorption from lithium chloride solutions, in fact, continues to rise, and very large distribution coefficients are obtained at high lithium chloride concentrations (Fig. 11.2).

Many studies have been made of anion-exchange distributions in concentrated solutions of the various alkali metal chlorides and hydrochloric acid. The very strong absorption from lithium chloride solutions was noted by Kraus.[24] In 12 M lithium chloride the distribution coefficients of several metals are over 100 times as large as they are in 12 M hydrochloric acid, and this effect can be used in chromatography; beryllium is absorbed by an anion-exchange resin from 12 M lithium chloride (distribution coefficient 8 ml/g) whereas its absorption from hydrochloric acid is negligible.

The "lithium chloride effect" is understood as soon as one realizes that this salt has a very high activity coefficent in concentrated solutions. This means that the Donnan equilibrium invasion of the resin by chloride ions is very much greater from lithium chloride solutions than from other alkali metal chloride solutions of equal concentration.[25] When activity coefficients (which are, of course, related to ionic hydration) are taken into account, with activities in the resin being found from measurements of electrolyte invasion, all alkali chlorides are

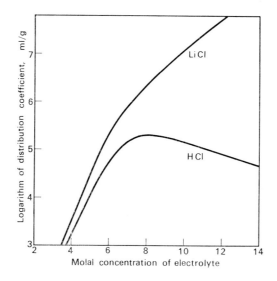

FIG. 11.2. Absorption of gallium(III) by a quaternary-ammonium anion-exchange resin. (Reproduced by permission from Volume 7, *Proceedings of the First International Conference on the Peaceful Uses of Atomic Energy*, United Nations, 1958, p. 136.)

seen to follow the same pattern.[26] It is *hydrochloric acid* that gives anomalous results, for the activity coefficients of hydrochloric acid and lithium chloride are very similar at high concentrations. One has to explain why the absorption of metals from hydrochloric acid solutions is not much larger, and a probable reason is the association of hydrogen and chloride ions to form molecular hydrochloric acid.

D. Metal-ammonia and Metal-amine Complexes

It is well known that resinous cation exchangers take up metal-ammonia complex ions. Ligand exchange (Chapter 8) depends on this fact. In crosslinked polystyrene-sulphonic acid exchangers the complexes of copper(II), nickel(II), and silver(I), and presumably other cations as well, are as stable within the resin as they are in the aqueous solution. That is to say, if a resin containing copper ions is brought to equilibrium with an aqueous solution of ammonia, with or without added ammonium salt, the ratio of bound ammonia to total copper (the $\bar{n}$ of J. Bjerrum) is the same inside the resin as it is in the solution. The formation curves of $\bar{n}$ against log [NH$_3$] coincide.[27]

Resins with weakly acidic functional groups behave differently. The metal-ammonia complexes are less stable in these resins than they are in water. In the case of carboxylic acid resins there seems to be a general destabilizing effect equivalent to about 1.5 to 2 logarithmic units, i.e. the curve of $\bar{n}$ against pNH$_3$ is displaced to lower pNH$_3$ values by 1.5–2 pNH$_3$ units. The maximum coordinating capacity does not seem to be reduced.[24] With iminodiacetate functional groups, however, the chelating iminodiacetate groups "block" three of the coordinative valences of nickel(II), rendering them unavailable for binding ammonia, and destabilizing the other three. The formation constants for the three ammonias that are bound are roughly equal to K_4, K_5, K_6 for the nickel ion-ammonia association in water.[28] A similar effect is noted with copper(II). The reduction in the power to bind ammonia makes chelating resins unattractive for ligand-exchange chromatography. The situation with phosphonic acid resins and with inorganic exchangers is even more extreme, and the sharp reduction in the power to bind ammonia shows that in these exchangers the association between the cations and the fixed ionic groups is a much stronger one than in sulphonic acid resins. This is in accordance with the well-known theory of Eisenman (Chapter 3).

The stability of metal-amine complexes within cation exchangers depends on several factors. In certain cases, such as the association of n-butylamine with silver ions, the complexes seem to be equally stable in the resin and in the solution.[27] Large amine molecules tend to be excluded from the resin; the molecules of aromatic amines, especially benzylamine, are preferentially absorbed by a polystyrene-type resin, making their metal complexes more stable within the resin. The case of 1,2-diamines is very interesting. The complexes of ethylenediamine and 1,2-propanediamine with silver, copper(II), and nickel(II) are decidedly more stable within the resin than they are in solution.[29, 30] The graph of ($\bar{n}$ in the resin) vs. log (concentration of diamine in solution), for silver(I) and ethylenediamine shows a stabilization of three powers of ten in the resin as compared to the solution; yet the silver ions in the resin coordinate only one molecule of ethylenediamine apiece,[30] whereas in the solution they coordinate two molecules. Possibly ethylenediamine acts as a bridging ligand in the resin to form long polynuclear complexes. Another case, where perhaps polynuclear complexes are formed, is the association of nickel(II) with hydrazine. Here the stabilization is no more than 1.5 logarithmic unit, but the nickel reaches its maximum saturation with four hydrazine molecules per nickel ion, instead of six as occurs in dilute aqueous solutions. This suggests six-coordinated nickel in a two-dimensional structure, with two hydrazines per nickel ion serving as "bridging ligands". An interesting fact is that the swelling and water content of sulphonic-acid resins saturated with nickel and hydrazine is far less than normal.

We have noted the extremely strong binding of 1,2-diamines by metal ions in sulphonic-acid resins (Chapter 8) and the fact that the binding of small quantities of hydrazine, while not nearly as strong as that of ethylenediamine, is nevertheless a good deal stronger than the binding of methylhydrazines, permitting their separation by ligand-exchange chromatography. The problem of ligand-exchange selectivity includes the problem of the stabilities of cationic complexes within the resins. Before this problem can be approached satisfactorily it will be necessary to know more about the stability of metal-amine complexes in solution. Many metal-amine complexes are so unstable that they cannot readily be investigated by pH titration, yet the amines are bound quite satisfactorily by ligand exchange on cation-exchange resins. Perhaps the resin environment stabilizes these complexes. A certain analogy with the anion-exchange of metal chloride complexes can be drawn. When the resin gets crowded with large ions and molecules and its water content drops, the resin interior ceases to resemble an aqueous solution, and interactions with the fixed ions and the polymer network become much more important than those with water molecules.

References

1. J. E. FINHOLT, R. W. ANDERSON, J. A. FYFE and K. G. COULTON, *Inorg. Chem.*, **4**, 43 (1965).
2. M. ARDON, *Inorg. Chem.*, **4**, 372 (1965).
3. T. W. SWADDLE and E. L. KING, *Inorg. Chem.*, **4**, 532 (1965).
4. P. MOORE, F. BASOLO and R. G. PEARSON, *Inorg. Chem.*, **5**, 223 (1966).
4a. G. R. BRUBAKER, I. H. LEGG and B. E. DOUGLAS, *J. Am. Chem. Soc.*, **88**, 3446 (1966).
5. J. W. VAUGHN and B. J. KRAINC, *Inorg. Chem.*, **4**, 1077 (1965).
6. W. M. McNEVIN and W. B. CRUMMET, *Anal. Chim. Acta*, **10**, 323 (1954).
7. J. SCHUBERT, *J. Phys. Chem.*, **52**, 340 (1948).
8. J. SCHUBERT and A. LINDENBAUM, *J. Am. Chem. Soc.*, **74**, 3529 (1952).
9. (a) L. G. SILLEN and A. E. MARTELL, *Stability Constants of Metal-Ion Complexes*, Chemical Society, London, 1964. (b) A. RINGBOM, *Complexation in Analytical Chemistry*, Pergamon Press, 1966.
10. S. FRONAEUS, *Acta Chem. Scand.*, **5**, 859 (1951).
11. S. FRONAEUS, *Acta Chem. Scand.*, **6**, 1200 (1952).
12. M. GRIMALDI, A. LIBERTI and M. VICEDOMINI, *J. Chromatog.*, **11**, 101 (1963); M. GRIMALDI and A. LIBERTI, *ibid.*, **15**, 510 (1964).
13. M. LEDERER, *Ann. dell, Inst. Superiore de Sanita*, **2**, 150 (1966).
14. R. M. WALLACE, *J. Phys. Chem.*, **71**, 1271 (1967).
15. R. A. HORNE *et al.*, *J. Phys. Chem.*, **61**, 1651, 1655, 1661 (1957).
16. K. A. KRAUS, F. NELSON, F. B. CLOUGH and R. C. CARLSTON, *J. Am. Chem. Soc.*, **77**, 1391 (1955).
17. U. SCHINDEWOLF, *Z. Elektrochem.*, **62**, 335 (1958).
18. K. A. KRAUS and F. NELSON, Anion exchange studies of metal complexes, in *The Structure of Electrolytic Solutions*, W. J. HAMER (ed.), Wiley, 1959.
19. Y. MARCUS, *J. Phys. Chem.*, **63**, 1000 (1959) and subsequent papers.
20. B. P. NIKOLSKII and V. I. PARAMONAVA, *Proc. 2nd Intern. Conf. on Peaceful Uses of Atomic Energy*, **28**, 75 (1959).
21. S. FRONAEUS, *Acta Chem. Scand.*, **8**, 1174 (1954).
22. M. J. M. WOODS, P. K. GALLAGHER and E. L. KING, *Inorg. Chem.*, **1**, 55 (1962).
23. S. AHRLAND, *Acta Chem. Scand.*, **10**, 723 (1956); see also ref. 15.
24. K. A. KRAUS, F. NELSON, F. B. CLOUGH and R. C. CARLSON, *J. Am. Chem. Soc.*, **77**, 1391 (1955).
25. B. CHU and R. M. DIAMOND, *J. Phys. Chem.*, **63**, 2021 (1959).
26. Y. MARCUS and D. MAYDAN, *J. Phys. Chem.*, **67**, 979, 983 (1963).
27. R. H. STOKES and H. F. WALTON, *J. Am. Chem. Soc.*, **76**, 3327 (1954).
28. K. SHIMOMURA, L. DICKSON and H. F. WALTON, *Anal. Chim. Acta*, **37**, 102 (1967).
29. L. COCKERELL and H. F. WALTON, *J. Phys. Chem.*, **66**, 75 (1962).
30. M. G. SURYARAMAN and H. F. WALTON, *J. Phys. Chem.*, **66**, 78 (1962).

APPENDIX
LIST OF SYMBOLS

A cross-sectional area of a column

a a parameter introduced in the derivation of eqn. (6.28) and later eliminated; also activity; also $q*/c$ (p. 123)

a_w activity of water

Amn   an amine

B a parameter defined as $\bar{D}\Pi^2/r^2$ (p. 55).

C distribution ratio in chromatography, i.e. the amount of a sample constituent in the stationary phase of any plate divided by the amount of the same constituent in the mobile phase of the same plate at equilibrium

C_0 value of C when the eluent is pure water

c concentration of interstitial solution

D diffusion coefficient (cm²/sec); also partition ratio (ml/meq or ml/g)

DVB   divinylbenzene

E selectivity coefficient, i.e. classical equilibrium constant of an ion-exchange reaction

E_A asymmetry potential

E_M membrane potential

EDTA   ethylenediamine tetraacetic acid

F fractional approach to equilibrium; also volume flow rate (ml/min); also faraday

F_M fraction by volume of monomers (i.e. styrene and DVB) in a mixture of styrene, DVB and toluene used to prepare a resin

H height or length of column (cm); also theoretical-plate height (p. 123)

h_1 fraction of the column height traveled by a sample constituent under the influence of the first eluent

J quantity (mmol) of a sample constituent in an elution

K ion-exchange equilibrium coefficient, defined by eqn. (3.5); also $q*/C$

$\mathbf{K}$ thermodynamic ion-exchange equilibrium constant

K_a ratio of the concentration of a solute in the internal solution of a resin to the concentration of the same solute in the external solution

K_b partition ratio, i.e. the amount of solute absorbed per gram of dry resin divided by the amount in 1 ml of the external solution (ml/g)

K_c a parameter defined by eqn. (3.8)

k a constant in salting-out chromatography, the value depending on the electrolyte and the sample constituent

$\mathbf{L}$ generalized chemical symbol for a ligand

L amount of electrolyte absorbed per gram of dry resin (mmol/g)

$L_{n,\,r}$ the fraction of a sample constituent in the interstitial solution of the rth plate after nv ml of eluent has entered the column

L^* maximum value of $L_{n,\,p}$

M   molar

M molarity; also v/aH (p. 126)

M_d molarity of the invading ions in the Donnan invasion

M_r molarity of the fixed ions in the internal phase

M_s molarity of the external solution; also molarity of the salt in salting-out chromatography

N number of plates in a column

N_B mole fraction

n the series of cardinal numbers, 1, 2, 3, etc. (p. 96); also the number of times that v ml of eluent has entered a column

n^* the value of n at the peak of an elution graph

P the number or plates per cm of column; also vapor pressure; also pressure of resin

p the number of plates in a column

275

Q	specific exchange capacity, i.e. the amount of exchangeable ion per gram of dry exchanger
Q_t	extent of an exchange reaction (meq. after t min)
Q_∞	extent of an exchange reaction after infinite time, i.e. at equilibrium
q	amount of a solute in both phases of a unit volume of column
q^*	the value of q when equilibrium between the two phases is established
$\bar{q}$	amount of solute in a unit volume of exchanger
$\bar{q}^*$	value of $\bar{q}$ at equilibrium
q_0	amount of exchanger taken, meq. in the limited-bath method of studying ion-exchange kinetics
R	a monomeric unit of the matrix of an ion-exchange resin, including the fixed ion
R	gas constant
R_f	distance traveled by a spot in paper chromatography divided by the distance traveled by the eluent
R_m	a parameter defined by eqn. (10.9)
r	radius (cm) of resin beads; also the ordinal number of a plate in a column
S	solubility product; also solubility of a nonelectrolyte in a salt solution; also entropy
S_0	solubility of a nonelectrolyte in pure water
$S_{n,\,r}$	fraction of a sample constituent in the stationary phase of the rth plate after nv ml of eluent have entered the column
T	absolute temperature
t	time; also x/σ in the theory of probability; also transport number of a cation
U	volume (ml) of effluent
U^*	peak (or retention) volume in an elution
u	volume of column (p. 129)
V	interstitial volume of a column, ml
V_b	volume of a column (ml)
V'	interstitial volume of a column above the peak of some sample constituent
V_R	volume (ml) in a mixing chamber
V_S	volume (ml) of stationary phase in a column
V_{wr}	volume (ml) of water inside the resin of a column
v	interstitial volume (ml) of one plate; also volume of mobile phase in a unit quantity of paper also volume of liquid that has entered the column
v_{wr}	volume (ml) of water inside the resin of one plate of a column
$\bar{v}$	partial equivalent volume
W	weight (g) of exchanger on a dry basis in a column
W'	weight (g) of exchanger on a dry basis in a column above the peak of a sample constituent
w	amount of water (g) absorbed by 1 g of dry resin; also weight (g) of exchanger on a dry basis in one plate of a column; also the weight (g) of exchanger on a dry basis in a unit quantity of paper
X_B	equivalent fraction
x	distance from top of column to any given plate or segment
z	charge on an ion
α	void fraction of a column
γ	activity coefficient
κ	equilibrium concentration in the resin of the exchanging ion divided by its concentration in the solution in eqn. (4.4)
ν	number of ions derived from one "molecule" of salt
σ	standard deviation
ϕ	volume (ml) of eluent delivered to a column from a mixing chamber; also volume that has passed the peak of a constituent in the column
ϕ^*	value of ϕ at the peak of an elution graph
ω	width (ml) of an elution graph
[]	molarity of enclosed species
()	activity of enclosed species

Bars over symbols denote the exchanger phase, except $\bar{v}$.

Table A.1. U.S. Standard Screens

Sieve number	Opening (mm)
20	0.84
50	0.297
100	0.149
150	0.105
200	0.074
250	0.062
280	0.052
325	0.044
400	0.037

NAME INDEX

SUBJECT INDEX

289

OTHER TITLES IN THE SERIES IN ANALYTICAL CHEMISTRY